Textbook of Developmental Biology

Alok Chandra Bharti • Anna Senrung
Ashok Kumar • Gautam Sethi
Editors

Textbook of Developmental Biology

Laboratory Techniques and Experimental Approaches

Springer

Editors
Alok Chandra Bharti
Molecular Oncology Lab.
Department of Zoology
University of Delhi
New Delhi, Delhi, India

Anna Senrung
Department of Zoology
Daulat Ram College
University of Delhi
New Delhi, Delhi, India

Ashok Kumar
Department of Pharmacological and Pharmaceutical Sciences
College of Pharmacy
University of Houston
Houston, TX, USA

Gautam Sethi
Department of Pharmacology
National University of Singapore
Singapore, Singapore

ISBN 978-981-95-7268-7 ISBN 978-981-95-7269-4 (eBook)
https://doi.org/10.1007/978-981-95-7269-4

This Springer imprint is published by the registered company Springer Nature Singapore Pte Ltd.
The registered company address is: 152 Beach Road, #21-01/04 Gateway East, Singapore 189721, Singapore

Preface

The book *Textbook of Developmental Biology: Laboratory Techniques and Experimental Approaches* is a compilation of the *Developmental Biology* curriculum at the undergraduate and postgraduate levels across universities in India. It arises from the recognized need for a comprehensive handbook that integrates both the *theoretical foundations* and *practical approaches* to developmental biology at undergraduate and postgraduate level. This volume aspires to serve as a definitive academic resource for students and educators, bridging the crucial gap between conceptual understanding and practical application in the study of developmental biology.

The content is organized into two principal sections. The *first section* focuses on the *fundamental aspects* of developmental biology, encompassing the study of gamete types and cleavage patterns in animals, as well as the early *developmental stages of frog and chick embryos* through permanent slides. It also covers methods for *culturing chick embryos*, understanding *indirect development through the developmental stages and the life cycle by culturing* ***Spodoptera litura*** (a lepidopteran insect), and exploring the *life cycles of major model organisms* with deep insights into each developmental stages such as *Drosophila, Danio rerio* (zebrafish), and *Caenorhabditis elegans.*

The *second section* of the book highlights the *application of model organisms*, including *C. elegans*, chick embryos, and mammalian embryonic systems, in *biomedical and health research*, illustrating their relevance in modern experimental biology. It underscores their indispensable role in advancing contemporary experimental biology, emphasizing their significance in understanding developmental mechanisms, disease modelling, and translational research.

The editors express their sincere gratitude to the chapter contributors, *Dr. Sanjeev Mallick, Dr. Meenakshi Thakur, Dr. Madhu, Dr. Sonika Sharma, Dr. Jyotsna Singh, Dr. Sunita Yadav, Dr. Radhika Gupta, and Dr. Narender*, for their valuable academic input. Special thanks are also extended to *Ms. Nidhi Bhardwaj, Ms. Shreya Pandey*, and *Ms. Hangsika Sharma* for their significant assistance in preparing and illustrating the diagrams/videos of chick and frog embryo developmental stages.

The editors welcome thoughtful feedback and constructive suggestions from readers that may help enhance the quality, utility, and pedagogical value of future editions of this work.

New Delhi, Delhi, India — Alok Chandra Bharti
New Delhi, Delhi, India — Anna Senrung
Houston, TX, USA — Ashok Kumar
Singapore, Singapore — Gautam Sethi

Contents

Editors and Contributors

About the Editors

Alok Chandra Bharti a Banaras Hindu University doctorate with 29 years' experience in molecular oncology, conducted pioneering research at MD Anderson Cancer Center, Texas, on NF-κB and STAT3 in carcinogenesis, diagnostics, and therapeutic targeting. Currently serving as professor at the University of Delhi, Dr. Bharti held a senior scientist role at ICMR's National Institute of Cancer Prevention and Research for 12 years. He has published 104 international peer-reviewed papers and 20 book chapters and guided 19 Ph.D. students. His work focuses on developing affordable diagnostics for HPV-associated cancers and advancing research on cervical cancer stem cells, particularly in identifying circulating tumor cells as diagnostic markers. In 2008, he was nominated advisor to the WHO-HPV Laboratory Network. Prof. Bharti's accolades include the INSA Medal for Young Scientist (2009) and Senior Research Fellowships from ICMR and INSA (2012, 2018). He has the expertise in working with mouse and chick embryo models.

Anna Senrung is a faculty in the Dept. of Zoology at Daulat Ram College (DRC), University of Delhi, with over 11 years of teaching and research experience. An active researcher, Dr. Senrung specializes in angiogenesis and the tumor microenvironment of Glioblastoma, utilizing chick embryos as a model system and has authored four publications on chick embryo studies and culture apart from other publications in national and international journals as well as in conference proceedings. Dr. Senrung actively mentors students at DRC in cancer research techniques and *in silico* drug discovery. Dr. Senrung has also contributed as a resource person for chick embryo culture for in-house as well as external workshops. Beyond academics and research, during her tenure as the Nodal officer of the NE Cell at DRC, she organized programs supported by the Ministry of DoNER, ICPR and the P.A. Sangma Foundation, Meghalaya.

Ashok Kumar has been a tenured Else and Philip Hargrove Endowed Professor of Drug Discovery at the Uni. of Houston (UH) since 2019. With over 32 years of research and teaching experience, he is the founding Director of UH's Institute of Muscle Biology and Cachexia. Prior to joining UH, Dr. Kumar served as a professor and distinguished university scholar in Anatomical Sciences and Neurobiology at the University of Louisville School of Medicine. Dr. Kumar's research has made significant contributions to biomedical science, notably identifying TWEAK as a key cytokine involved in muscle wasting. He has elucidated the roles of major signaling molecules, including TRAF6, TAK1, IRE1, PERK, and MyD88, in regulating skeletal muscle mass and cancer cachexia. His research program extensively employs nude, knockout, and transgenic mouse models. Dr. Kumar has authored 120+ peer-reviewed publications, mentored over 60 students, postdocs, and junior faculty, and regularly reviews for 70+ journals and major funding agencies, including the NIH.

Gautam Sethi is an associate professor in the Deparment of Pharmacology at the Yong Loo Lin School of Medicine, National University of Singapore. His research primarily explores the mechanisms by which carcinogens and inflammatory agents activate oncogenic transcription factors, aiming to identify novel inhibitors for cancer prevention and treatment using diverse model systems. His work has led to 500+ scientific publications in high-impact peer-reviewed journals. Additionally, Dr. Sethi contributes to the scientific community as an editorial board member and ad-hoc reviewer for numerous prestigious international journals, supporting advancements in pharmacology and cancer research. He is highly regarded globally, consistently recognized as one of the world's most-cited scientists, and was named among Clarivate's Highly Cited Researchers™ in 2020 and 2021. Additionally, he has been featured in Stanford University World's Top 2% Scientists List for 2020–2025 in oncology and carcinogenesis.

Contributors

Mukesh Behera Wound Healing Research Lab, Department of Zoology, Faculty of Sciences, University of Delhi, New Delhi, India

Alok Chandra Bharti Molecular Oncology Lab., Department of Zoology, University of Delhi, New Delhi, India

Narender K. Dhania Wound Healing Research Lab, Department of Zoology, Faculty of Sciences, University of Delhi, New Delhi, India

Radhika Gupta Department of Biochemistry, Daulat Ram College, Delhi University, New Delhi, India

Suhani Jain Wound Healing Research Lab, Department of Zoology, Faculty of Sciences, University of Delhi, New Delhi, India

Aman Kumar Department of Animal Biotechnology, Laboratory of Molecular Diagnosis, Lala Lajpat Rai University of Veterinary and Animal Sciences, Hisar, Haryana, India

Narendra Kumar Department of Biochemistry, Daulat Ram College, Delhi University, New Delhi, India

Padmshree Mudgal Department of Biochemistry, Daulat Ram College, Delhi University, New Delhi, India

Sanjiv Mullick Department of Zoology, Dyal Singh College, University of Delhi, Delhi, India

Vikas Nishad Department of Biochemistry, Daulat Ram College, Delhi University, New Delhi, India

Srianshu Kumar Panda Wound Healing Research Lab, Department of Zoology, Faculty of Sciences, University of Delhi, New Delhi, India

Anna Senrung Department of Zoology, Daulat Ram College, University of Delhi, Delhi, India

Neuropharmacology, Drug delivery, and Cancer Research Laboratory (NDCRL), Daulat Ram College, University of Delhi, New Delhi, India

Gautam Sethi Department of Pharmacology, National University of Singapore, Singapore, Singapore

Pallavi Sethi Department of Zoology, Daulat Ram College, University of Delhi, Delhi, India

Gaurav Sharma Wound Healing Research Lab, Department of Zoology, Faculty of Sciences, University of Delhi, New Delhi, India

Nikita Sharma Department of Biochemistry, Daulat Ram College, Delhi University, New Delhi, India

Sonika Sharma Drosophila Resource Center, Department of Zoology, Daulat Ram College, University of Delhi, New Delhi, India

Jyotsna Singh Department of Zoology, Daulat Ram College, University of Delhi, Delhi, India

Madhu Singh Drosophila Resource Center, Department of Zoology, Daulat Ram College, University of Delhi, New Delhi, India

Rahul Singh Department of Zoology, Faculty of Sciences, Wound Healing Research Lab, University of Delhi, Delhi, India

Meenakshi Thakur Department of Zoology, Daulat Ram College, University of Delhi, Delhi, India

Tanya Tripathi Molecular Oncology Lab., Department of Zoology, University of Delhi, New Delhi, India

Sunita K. Yadav Department of Zoology, Daulat Ram College, University of Delhi, New Delhi, India

Study of the Different Types of Animal Eggs and Their Corresponding Cleavage Patterns Through Diagrams/Photomicrographs

1

Anna Senrung and Alok Chandra Bharti

Abstract

This practical chapter focuses on the comparative classification of animal eggs and their cleavage patterns, offering foundational insights into early embryonic development. The process of cleavage, rapid mitotic divisions following fertilization, is profoundly influenced by the quantity, distribution, and orientation of yolk within the egg. Eggs are categorized based on yolk concentration (alecithal, microlecithal, mesolecithal, macrolecithal) and distribution (isolecithal, telolecithal, centrolecithal, discoidal), each affecting cleavage symmetry and completeness. Cleavage patterns are broadly divided into holoblastic (complete: equal or unequal) and meroblastic (incomplete: discoidal or superficial), determined by the physical limitations imposed by yolk density. The orientation of cleavage planes, meridional, vertical, equatorial, latitudinal, and oblique, influences embryonic polarity and axis formation. Eggs are further classified based on external coverings as cleidoic (shelled) or non-cleidoic (unshelled), reflecting adaptations to environmental conditions. Students are introduced to classical embryological concepts such as mosaic (determinate) and regulative (indeterminate) development, essential to understanding cell fate and embryonic plasticity. Through examination of diagrams and photomicrographs, students will be able to identify egg types and their associated cleavage patterns. This practical

A. Senrung (✉)
Department of Zoology, Daulat Ram College, University of Delhi, Delhi, India

A. C. Bharti
Molecular Oncology Lab., Department of Zoology, University of Delhi, Delhi, India

A. C. Bharti et al. (eds.), *Textbook of Developmental Biology*,
https://doi.org/10.1007/978-981-95-7269-4_1

chapter/exercise aims to build an understanding of how egg structure governs embryogenesis and contributes to evolutionary diversity in reproductive strategies. Such knowledge is central to developmental biology and underpins further studies in embryology, evolution, and cellular differentiation.

Keywords

Animal eggs · Yolk · Cleavage

Aim To study and compare the different types of animal eggs and their corresponding cleavage patterns using diagrams and photomicrographs, thereby enhancing understanding of early embryonic development across species.

Learning Objectives
By the end of this practical, students will be able to:

- Classify animal eggs based on yolk quantity and distribution
- Know the types of cleavage patterns in fertilized eggs of animals
- Correlate egg types with specific cleavage patterns
- Understand the biological significance of cleavage in embryonic development

1.1 Introduction

The study of animal eggs and their cleavage patterns offers fundamental insights into early embryonic development. Embryogenesis begins with the fertilization of an egg, followed by a sequence of rapid mitotic divisions, the cleavage. The nature, symmetry, and completeness of cleavage divisions are strongly influenced by the type, quantity, and distribution of yolk within the egg, as well as by cytoplasmic factors that regulate the orientation of the mitotic spindle and its formation timing (Gilbert 2000).

Animal eggs exhibit remarkable diversity in yolk content, ranging from yolk-free to heavily yolked eggs, and in how the yolk is spatially distributed. This distribution typically leads to egg polarity, with a yolk-rich vegetal pole and a yolk-poor animal pole, as seen in the frog egg. The zygotic nucleus is usually moved toward the animal pole, where active cleavage occurs.

Cleavage forms the fundamental basis of the development of organisms. Cleavage begins after fertilization with mitosis-promoting factor (MPF) activation. During cleavage, blastomeres undergo a simplified cell cycle comprising only the S (DNA synthesis) and M (mitosis) phases. The MPF activity peaks during mitosis and becomes untraceable during the S phase (Gerhart et al. 1984; Newport and Kirschner 1984).

This practical chapter emphasizes the comparative study of animal eggs based on yolk content, classifying them into isolecithal, mesolecithal, telolecithal, and centrolecithal types, along with their corresponding cleavage patterns, holoblastic (equal or unequal) and meroblastic (discoidal or superficial). Additionally, other variant forms in egg classification and cleavage determination are discussed. By examining diagrams and photomicrographs, students will gain an understanding of how yolk content and distribution govern cleavage patterns and influence early embryonic development.

1.2 Requirements

- Labeled photomicrographs/diagrams or permanent slides of animal eggs, cleavage stages and types
- Observation sheets/notebook, pencils, eraser, sharpener

1.3 Methodology

Carefully examine the diagrams or photomicrographs of animal eggs on the following parameters:

- *Identify the type of egg based on yolk content:* Alecithal, microlecithal (oligolecithal), mesolecithal, macrolecithal (megalolecithal or polylecithal).
- *Identify the type of egg based on yolk distribution*: Isolecithal, telolecithal, centrolecithal, or discoidal eggs.
- *Determine the presence and absence of hard eggshell:* Cleidoic or non-cleidoic egg,
- *Determine the mode of development*: Determinate (mosaic) or indeterminate (regulative) eggs.
- *Observe the cleavage pattern:* Note whether it is holoblastic (equal and unequal), bilateral, spiral, radial, rotational, or meroblastic (discoidal and superficial).
- Record your observations and label important structures such as blastomeres, yolk, animal, and vegetal poles where visible.

1.4 Description of Animal Eggs and Cleavage Patterns

1.4.1 Animal Eggs

Eggs produced by animals serve as the initial developmental stage, containing the necessary cellular components and nutrients to support embryo growth. They vary widely in size, structure, yolk content, and protective coverings, reflecting diverse reproductive strategies across taxa.

1.4.1.1 Kinds of Eggs Centered on the Concentration/Amount of Yolk

The quantity and distribution of egg yolk play a critical role in early embryonic development and cleavage patterns. Yolk serves as the primary nutrient source for the developing embryo, but its dense, viscous nature makes it difficult to cleave, often slowing or restricting cell division in yolk-rich regions. Consequently, yolk distribution is frequently unequal, leading to asymmetrical cleavage and differential cell sizes. Eggs are classified based on yolk concentration into types ranging from those with sparse yolk evenly distributed to those heavily loaded with yolk concentrated at specific poles. This variation reflects evolutionary adaptations aligned with reproductive strategies and environmental conditions.

Alecithal Alecithal eggs (Fig. 1.1a) are characterized by the absence of yolk, or the presence of only a negligible amount. As a result, the developing embryo relies entirely on maternal support, typically through placental nourishment. This egg type is characteristic of eutherian (placental) mammals, where yolk is evolutionarily reduced due to the presence of internal gestation mechanisms.

Microlecithal/oligolecithal Microlecithal eggs (Fig. 1.1b) contain a small quantity of yolk, which is typically distributed uniformly within the cytoplasm. This limited yolk content supports only the earliest stages of development, after which additional nutritional support is required. Such eggs are characteristic of organisms like amphioxus and some primitive chordates.

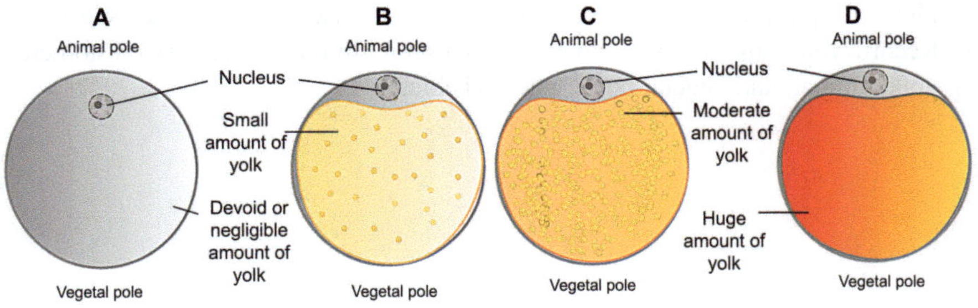

Fig. 1.1 Diagrammatic representation of different egg types based on yolk concentration. (**a**) Alecithal. (**b**) Microlecithal. (**c**) Mesolecithal. (**d**) Macrolecithal. (Created in https://BioRender.com)

Mesolecithal Mesolecithal eggs (Fig. 1.1c) hold a modest to relatively large quantity of yolk sufficient to support extended embryonic development outside the mother's body. Such eggs are typical of amphibians, in which yolk is unevenly distributed, with a greater concentration toward the vegetal pole.

Macrolecithal (Megalolecithal/polylecithal) Macrolecithal eggs (Fig. 1.1d) hold a massive quantity of yolk, which is the primary source of nutrition for the developing embryo. The large yolk volume significantly influences cleavage, typically restricted to blastodisc, a small cytoplasmic disc in the animal pole. This sort of egg is typical of birds, reptiles, and some fishes, where external development occurs with minimal or no maternal support post-laying.

1.4.1.2 Kinds of Eggs Centered on Yolk Distribution/Orientation Pattern

Distribution and orientation of yolk within an egg are fundamental factors influencing early embryonic development and cleavage patterns. The yolk serves as the primary nutrient source for the developing embryo, and its quantity and arrangement influence cell division, embryo size, and developmental timing. Eggs can be classified into different types based on the quantity and spatial distribution of yolk (Fig. 1.2), which in turn reflects adaptations to the species' reproductive strategies and environmental conditions.

Isolecithal eggs These kinds of eggs (Fig. 1.2a) contain a minimal quantity of yolk that is sparsely and evenly distributed throughout the cytoplasm. This yolk pattern is typical of species such as *amphioxus* and many mammals, facilitating holoblastic cleavage.

Telolecithal eggs These kinds of eggs (Fig. 1.2b) hold an unevenly distributed moderate quantity of yolk with a higher concentration localized at the vegetal pole. This asymmetrical yolk distribution results in relatively faster cell divisions at the animal pole than at the vegetal pole. Such eggs are typical of amphibians and many fish species.

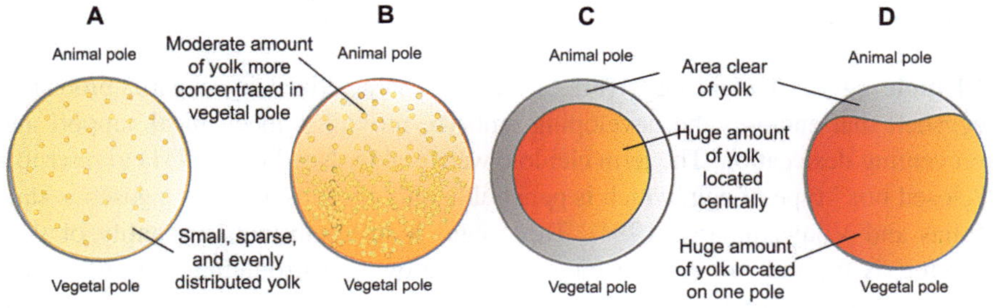

Fig. 1.2 Diagrammatic representation of different types of eggs based on yolk distribution/orientation. (**a**) Isolecithal. (**b**) Telolicithal. (**c**) Centrolicithal. (**d**) Discoidal. (Created in https://BioRender.com)

Fig. 1.3 Image representation for cleidoic and non-cleidoic eggs. (**a**) Cleidoic egg (bird egg). (**b**) Non-cleidoic (frog egg). (Source: Bird egg (Bird Eggs "by David~O is licensed under CC BY 2.0), Frog egg (Common frog egg cells" by talaakso is licensed under CC BY-SA 2.0))

Centrolecithal eggs Centrolecithal eggs (Fig. 1.3c) hold a large quantity of yolk concentrated centrally within the cytoplasm. This central yolk distribution typically results in superficial cleavage. Such eggs are characteristic of many insects.

Discoidal eggs Discoidal eggs (Fig. 1.2d) hold a massive yolk concentration localized to one pole, which occupies the egg's major volume. The embryo develops from blastodisc, a small disc-shaped region of clear cytoplasm, situated at the animal pole atop the yolk mass. This yolk arrangement is typical of birds, reptiles, and some fish species.

1.4.1.3 Kinds of Eggs Centered on Eggshell Covering

The eggs of animals exhibit remarkable diversity in structure, especially in their external coverings, which play a crucial role in protection, gas exchange, and moisture regulation during embryonic development. Based on the presence, absence, and nature of the eggshell or outer protective layers, eggs can be broadly classified into different types. These variations are closely linked to the organism's reproductive strategies and environmental adaptations. From soft, gelatinous coatings in aquatic species to rigid, calcified shells in terrestrial vertebrates, the eggshell structure reflects evolutionary responses to habitat and developmental needs.

Cleidoic eggs Cleidoic eggs (Fig. 1.3a) are characterized by a hard, protective eggshell that encloses the developing embryo, providing mechanical support and preventing desiccation. The term cleidoic was used by Needham (1931) to refer to a "closed box" type of egg, which is penetrable only by substances in a gaseous state (Silas and Vijayakumaran 1984). These eggs hold a substantial quantity of yolk serving as the nutrient reserve for embryo's growth. Surrounding the yolk is the albumin (egg white), which cushions the embryo and supplies additional water and protein. The eggshell is typically calcareous and waterproof, allowing these eggs to develop in terrestrial environments without drying out. Cleidoic eggs are a hallmark of reptiles and birds, representing a major evolutionary adaptation for life on land

(Rawal 2025). Additionally, the eggs of these animals possess specialized extraembryonic structures, the amnion, allantois, chorion, and yolk sac, each fulfilling distinct and essential roles in embryonic development. The fluid-filled amniotic cavity moderates and shields the embryo, while the chorion assists the embryo in gas exchange with the external environment. The yolk sac supplies nutrients derived from the yolk, and the allantois serves in waste storage and also plays a key role in respiration. Collectively, these membranes enable the embryo to develop independently of an aquatic environment, thereby supporting successful terrestrial reproduction.

Non-cleidoic Unlike cleidoic eggs, non-cleidoic eggs (Fig. 1.3b) lack a hard, protective eggshell and are typically laid in aquatic environments. Non-cleidoic eggs, such as amphibian eggs, are encased in a gelatinous jelly layer that provides physical protection, helps retain moisture, and offers some defense against predators and pathogens. The absence of a rigid shell allows for water and mineral absorption (Silas and Vijayakumaran 1984) as well as direct exchange of gases and wastes with the surrounding water but necessitates a moist habitat to prevent desiccation. This type of egg is characteristic of amphibians, whose reproductive cycle is closely tied to freshwater ecosystems.

1.4.1.4 Kinds of Eggs Centered on Mode of Development

Determinate (Mosaic Eggs) In this type of egg, the outcome of every portion of the egg is programmed to become a specific type of cell beforehand or at fertilization (Conklin 1905). If a specific part of the egg is detached, the growing embryo becomes lacking in the specific organ for which the removed portion was determined to develop. This intrinsic differentiation potential of the embryos is due to regionally localized morphogenetic determinants, specific proteins or mRNAs, distributed to daughter cells during cleavage, directing their developmental fate and specifying distinct cell types (Gilbert 2000; Makabe and Nishida 2012). Embryonic development is viewed as the integrated outcome of interactions among individual parts, with each contributing to the formation of the complete organism (Callebaut et al. 2010). Example: Ascidian and mollusk embryos.

Indeterminate (Regulative eggs) In this type of egg, the fate of each region is not predetermined. Development is highly regulative, allowing any major isolated portion to form a normal, symmetrical embryo (Callebaut et al. 2010). The mammalian zygote and early blastomeres (2- to 4-cell stage) are considered totipotent (Sozen et al. 2014); i.e., each blastomere, if separated at the 2- to 4-cell stage, has the ability to develop to a whole embryo (Totipotency) if the right conditions prevail. According to Driesch (1909), regulation is the embryo's ability to restore normal development following disturbance, such as the removal of a part. Later, Sander (1971) expanded this concept, suggesting that regulation also includes instances where embryonic parts alter their fate in response to experimental manipulation (Lawrence and Levine 2006). Example: Vertebrates.

1.4.2 Cleavage

Cleavage is the initial phase of embryonic development, characterized by rapid mitotic cell divisions that begin soon after fertilization. This process transforms the single-celled zygote, which possesses a large cytoplasmic volume, into multiple nucleated cells, the *blastomeres* (Brun-Usan and Salazar-Ciudad 2021). These early divisions occur without an increase in overall embryo size, meaning the cytoplasm is partitioned into progressively smaller cells. Cleavage is a pivotal step in converting the diploid zygote into a multicellular organism, forming the cellular foundation for subsequent embryonic development. In some species that undergo parthenogenesis, cleavage can initiate even in the absence of fertilization. As cleavage progresses, rapid cell divisions reduce the cytoplasmic to nuclear volume ratio and typically end when a new nucleus to cytoplasm balance is reached. The specific cleavage pattern in each species influences the structure and organization of the resulting blastula.

1.4.2.1 Cleavage Planes

Cleavage planes refer to the specific patterns of cell division in animal eggs, progressively dividing the single-celled zygote into multiple smaller cells called *blastomeres*. These divisions follow defined planes, determined by the mitotic spindle fibers' orientation, and play a crucial role in establishing early body axis and symmetry of the growing embryo, determining the overall embryonic organization and establishing the relative positioning of cells and tissues within the developing embryo (Hasley et al. 2016; Kitajima and Hamaguchi 2005). The orientation and nature of cleavage planes differ among species and are influenced by the quantity and yolk distribution in the egg. In many animal species, including starfish and sea urchins, the first cleavage plane passes through the animal-vegetal axis (Kitajima and Hamaguchi 2005). Common types of cleavage planes observed in animals include meridional, vertical, equatorial, latitudinal (or transverse), and oblique.

Meridional Plane In meridional cleavage (Fig. 1.4a, b), the cleavage furrow passes along the animal-vegetal axis, cutting through the median longitudinal plane of the zygote. This results in the division of the egg into two symmetrical halves, each

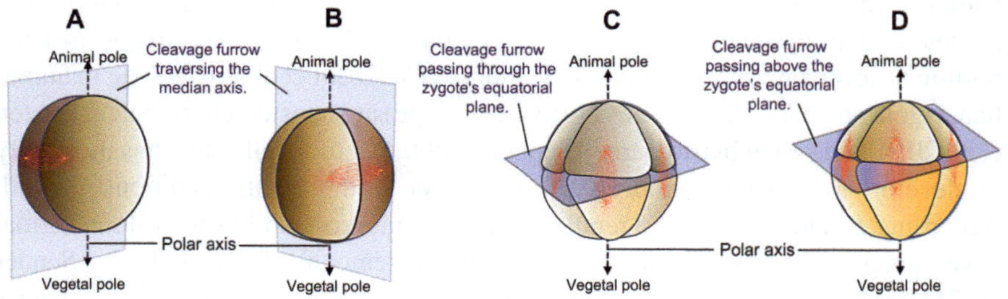

Fig. 1.4 Cleavage planes. (**a**, **b**) Meridional cleavage plane: 2-cell stage (**a**), 4-cell stage (**b**). (**c**) Equatorial cleavage plane: 8-cell stage. (**d**) Transverse cleavage plane: 8-cell stage. (Created in https://BioRender.com)

receiving an equal share of cytoplasm and yolk along the axis. Meridional cleavage is typically the first cleavage in many animals and is crucial for establishing bilateral symmetry. This type of cleavage plane could be observed in the frog (*Xenopus*) embryonic first and second cleavage (Denegre et al. 1998) and the first cleavage of mammalian embryo (Hasley et al. 2016).

Vertical Plane In vertical cleavage, the cleavage furrow extends from the animal pole to the vegetal pole, similar to meridional cleavage, but it does not pass through the median longitudinal axis of the zygote. Instead, the furrow is slightly off-center, resulting in the formation of two unequal or asymmetrical blastomeres. This type of cleavage contributes to asymmetry in early embryonic divisions and helps establish cellular diversity in the developing embryo. This kind of cleavage plane is seen in later divisions in chick eggs post the first and the second cleavage resulting in the formation of the blastoderm (Gilbert 2000).

Equatorial Plane In equatorial cleavage (Fig. 1.4c), the cleavage furrow is oriented perpendicular to the animal-vegetal axis, bisecting the zygote horizontally between the animal and vegetal poles. This plane lies at the equator of the egg and divides it into upper (animal) and lower (vegetal) hemispheres. Equatorial cleavage typically follows one or more meridional divisions and contributes to the further compartmentalization of embryonic cells. This type of cleavage plane is generally observed in the second cleavage of mammalian embryo (Hasley et al. 2016).

Latitudinal or Transverse Plane The latitudinal or transverse cleavage plane (Fig. 1.4d) is oriented vertical to the animal-vegetal axis, similar to the equatorial plane. However, unlike the equatorial plane, which cuts through the midpoint (equator) of the egg, latitudinal cleavage occurs above or below the equator, resulting in asymmetrical division between the animal and vegetal hemispheres. It contributes to differential cytoplasmic distribution and plays a part in axis formation and cell fate specification. This type of cleavage plane could be observed in the frog (*Xenopus*) embryonic third cleavage (Denegre et al. 1998).

Oblique Plane In oblique cleavage, the cleavage furrow is oriented at an oblique angle to the animal-vegetal axis, neither parallel nor strictly perpendicular (Gilbert 2000). This angular orientation results in the formation of asymmetrical or spirally arranged daughter blastomeres, often contributing to the spatial arrangement of cells that is critical for later developmental patterning. This kind of cleavage is typically observed in most mollusks (snails), some annelids, and flatworms.

These cleavage planes collectively define the cleavage pattern of the embryo, such as radial, spiral, bilateral, or rotational cleavage, and are essential for proper embryonic axis formation and body plan organization.

Spindle Fibers and Cleavage Planes

During early embryonic development, cleavage involves a sequence of mitotic divisions where the orientation of cleavage planes is crucial for proper cell

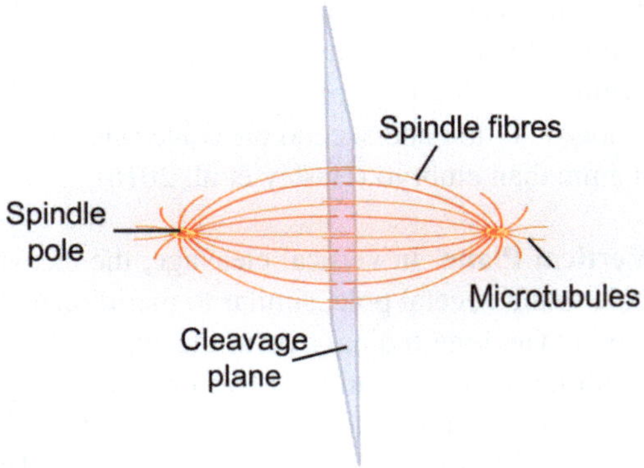

Fig. 1.5 Spindle fiber and cleavage planes. (Created in https://BioRender.com)

partitioning and body axis formation. The spindle fibers, composed of microtubules, play a central role in this process by organizing and segregating chromosomes during mitosis. The position and direction of mitotic spindle apparatus determine the direction of the cleavage plane (Fig. 1.5), guiding where the cleavage furrow forms and ultimately how the zygote divides into blastomeres (Galli and Van Den Heuvel 2008). Cleavage planes can be oriented in various ways relative to the egg animal-vegetal axis, including meridional, equatorial (perpendicular to the axis), vertical, latitudinal (transverse), or oblique angles. The precise alignment of spindle fibers ensures that each daughter cell receives an appropriate set of chromosomes and cytoplasmic material, which is especially important in eggs with uneven yolk distribution. By dictating cleavage patterns, spindle fibers indirectly influence embryonic symmetry, polarity, and subsequent developmental fate. Understanding the interplay between spindle fiber orientation and cleavage planes is fundamental to developmental biology, as it reveals how cellular architecture during early mitosis shapes the embryo's overall body plan and diversity.

1.4.2.2 Cleavage Patterns

The cleavage patterns of animal eggs in different groups follow specific patterns. By convention, the cleavage tends to produce equal daughter cells upon division, and each successive cleavage plane is oriented at a right angle to the preceding one. Secondly, the speed or rate of cleavage of the zygote is inversely proportional to the amount of yolk concentration or distribution (Balfour 1885). The resulting small daughter cells or small blastomeres formed by the divisions of the zygote are called *micromeres*, and the large ones, *macromeres*. By convention, the micromeres are formed or confined to the animal pole, while the macromeres are confined to the vegetal pole. Under the influence of yolk concentration, the cleavage pattern may be broadly divided into *holoblastic* or *meroblastic* cleavage.

1.4.2.2.1 Holoblastic Cleavage

Holoblastic cleavage refers to the complete division of the zygote or egg with the cleavage furrow passing through the egg cytoplasm in its entirety. This kind of cleavage can be observed in eggs with little or moderate yolk, including alecithal,

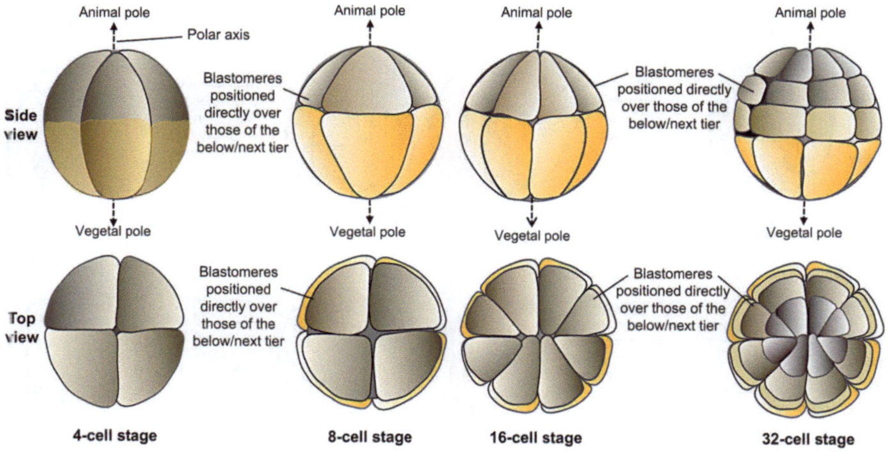

Fig. 1.6 Diagrammatic representation of radial pattern of cleavage. (Created in https://BioRender.com)

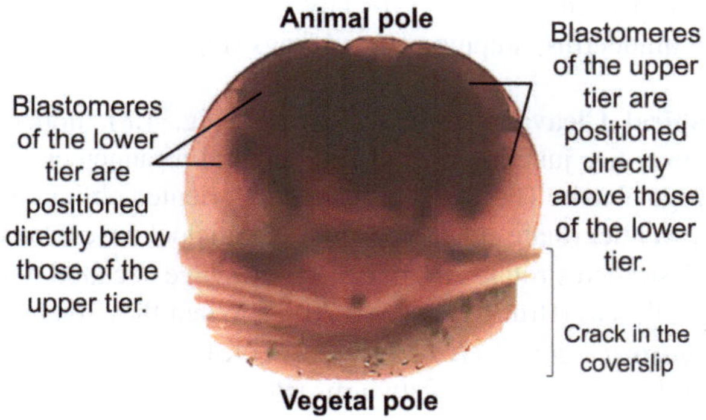

Fig. 1.7 Image of 16-cell stage frog embryo (W.M) depicting radial pattern of cleavage. (Source: dept. of Zoology, Daulat Ram College, Uni. of Delhi, Delhi, INDIA).

isolecithal, microlecithal, and mesolecithal eggs. It is characteristic of species like mammals, amphibians, echinoderms, and some invertebrates. In mammalian eggs with uniformly distributed yolk, early cleavage divisions are equal and complete. In contrast, in amphibians with mesolecithal eggs, the yolk is concentrated more at the vegetal pole, which slows down cleavage in that region, resulting in complete but unequal blastomeres in the animal and vegetal poles. The holoblastic cleavage is further categorized into specific kinds, including radial, bilateral, spiral, or rotational, depending on the organism.

Radial cleavage In this cleavage pattern (Fig. 1.6), the blastomeres of the upper tier are aligned directly above those of the lower tier, resulting in a radially symmetrical arrangement around the embryo's animal-vegetal axis. The cleavage planes are oriented either parallel or perpendicular to the animal-vegetal axis. The daughter cells are stacked precisely on top of each other. This pattern is characteristic of

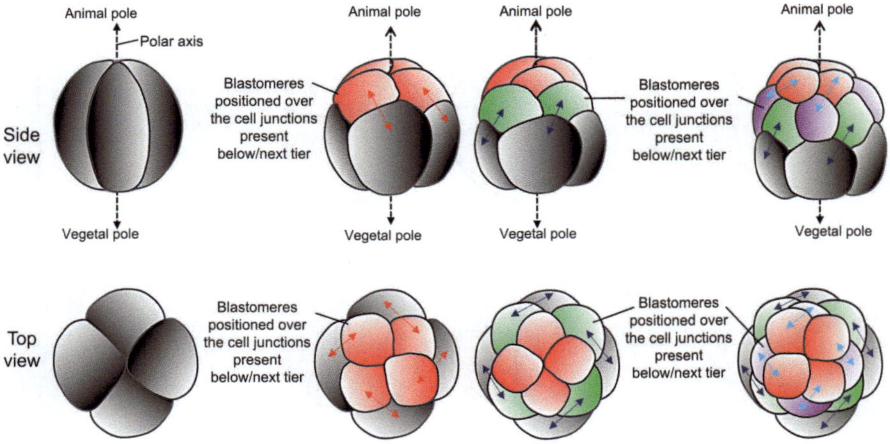

Fig. 1.8 Diagrammatic representation of the spiral pattern of cleavage. Same color cells indicate their simultaneous origin. (Created in https://BioRender.com)

deuterostomes and typically leads to indeterminate cleavage. Example: echinoderms, amphioxus, and frogs (Fig. 1.7).

Spiral Cleavage In this cleavage (Fig. 1.8), upper-tier blastomeres align over lower-tier junctions, producing a spiral arrangement around the embryo's animal-vegetal axis. The cleavage planes are oriented obliquely. The four blastomeres of the lower tier rotate either clockwise or anticlockwise relative to the upper tier. When the blastomeres rotate clockwise, the cleavage is called *dextral spiral cleavage* (e.g., in Mollusca) (Brun-Usan et al. 2017). When they rotate anticlockwise, it is known as *sinistral spiral cleavage*. This cleavage pattern is commonly observed in protostomes such as flatworms (Platyhelminthes) and annelids (Fig. 1.9).

Bilateral Cleavage Bilateral cleavage refers to a pattern of early embryonic cell division in which the arrangement of blastomeres exhibits bilateral symmetry, with each side of the embryo mirroring the other (Fig. 1.10). This cleavage pattern is typically observed in organisms such as tunicates and ascidians, and it plays a crucial role in the establishment of the body's left-right axis at early development.

Rotational Cleavage (Fig. 1.11) A distinctive kind of holoblastic cleavage (complete cleavage). In this pattern, the first cleavage is a typical meridional division, running along animal-vegetal axis, and divides the zygote into two equal blastomeres. However, during the second cleavage, a key feature of rotational cleavage emerges: One blastomere divides meridionally (similar to the first), while the other divides equatorially, i.e., perpendicular to the meridional plane. This results in a 4-cell stage where the orientation of the cleavage planes differs between the daughter cells, a feature unique to rotational cleavage. Unlike species with rapid and synchronous cleavage (e.g., frogs or sea urchins), rotational cleavage is slower and

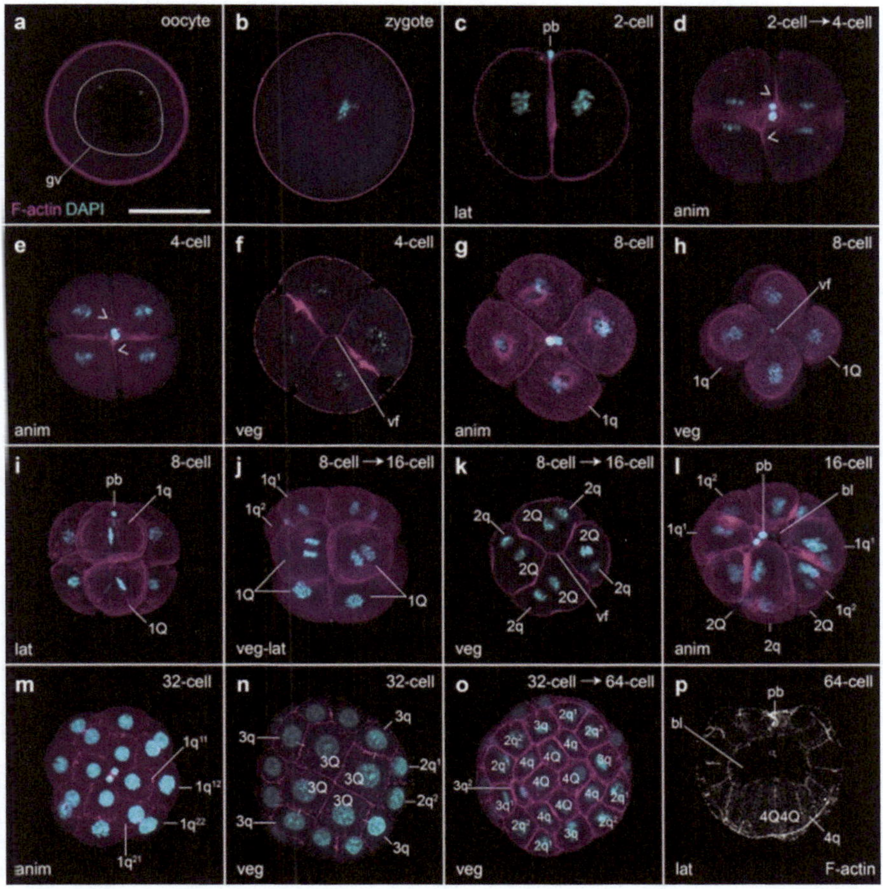

Fig. 1.9 Photomicrographs of early cleavage in *Owenia fusiformis,* an annelid exhibiting equal spiral cleavage. (Adapted from Carrillo-Baltodano et al. (2021). The image is licensed under Creative Commons).

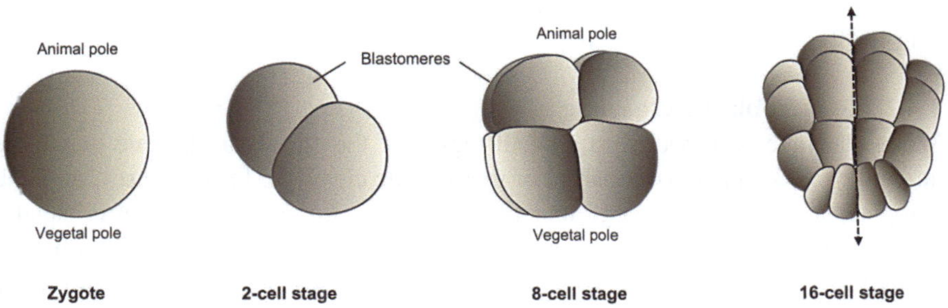

Fig. 1.10 Diagrammatic representation of cleavage showing bilateral cleavage. (Created in https://BioRender.com)

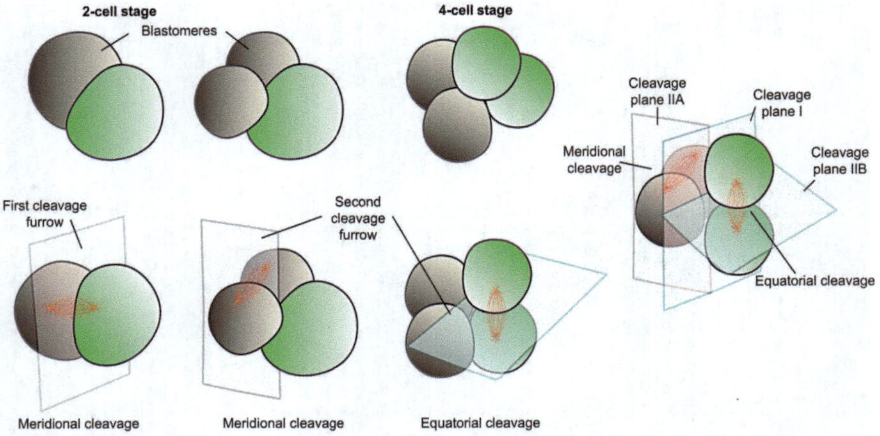

Fig. 1.11 Diagrammatic representation of cleavage showing rotational cleavage. (Created in https://BioRender.com)

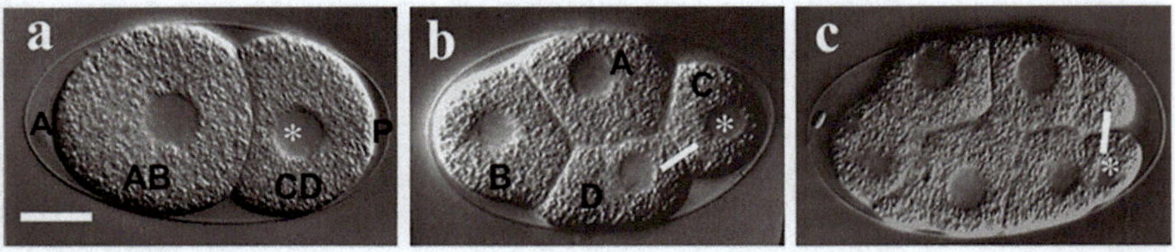

Fig. 1.12 Photomicrographs of early cleavage in *C. elegans* exhibiting rotational division. (**a**–**c**) The AB cell anaphase spindle forms perpendicular to the A–P axis but, constrained by the eggshell during elongation, skews to position one pole anterior to the other (Schonegg et al. 2014). (Image source: "Nematode Variations in early cleavage pattern" by Einhard Schierenberg is licensed under CC BY 2.5).

asynchronous, reflecting the more complex regulation of early mammalian embryogenesis, including the early activation of the zygotic genome. This kind of cleavage is seen in mammals (Humans, mice) and nematodes (*Caenorhabditis elegans*) (Fig. 1.12).

1.4.2.2.2 Meroblastic Cleavage

When cleavage is restricted to a small region of the egg, typically due to a large amount of yolk, it results in partial or meroblastic cleavage. In such cases, only the cytoplasmic portion of the egg undergoes division, forming a blastoderm, usually situated at the animal pole, while the yolk laden region (vegetal pole) remains uncleaved. The large yolk content prevents the cleavage furrows from fully penetrating the egg. This kind of cleavage is characteristic of discoidal and superficial (centrolecithal) cleavage patterns. Common symmetry types observed in incomplete cleavage include: bilateral cleavage (e.g., cephalopods), discoidal cleavage (e.g., birds, reptiles), and superficial cleavage (e.g., arthropods/insects).

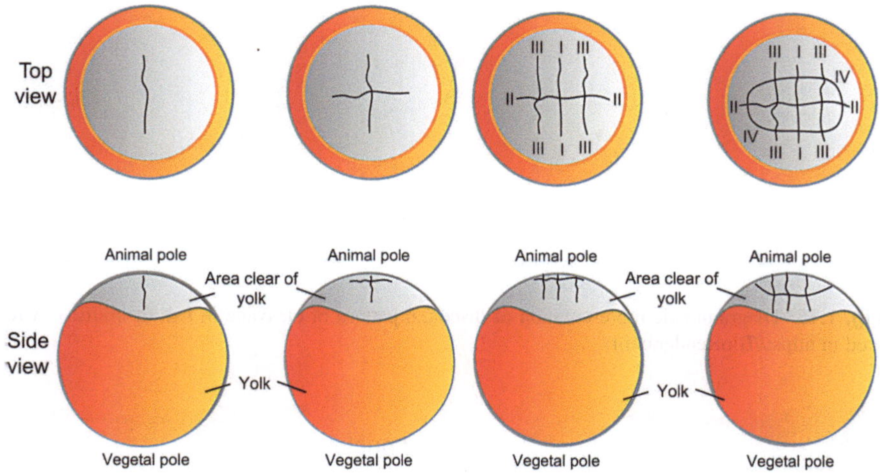

Fig. 1.13 Diagrammatic representation of discoidal pattern of cleavage in birds (chick). (Created in https://BioRender.com)

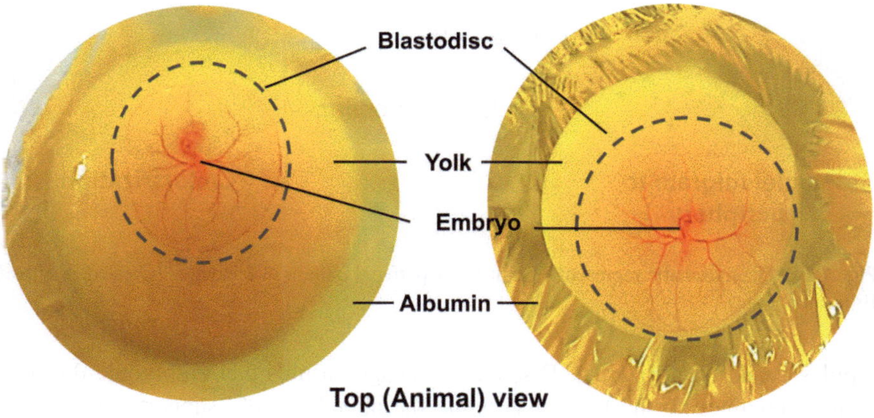

Fig. 1.14 Image of 4 days old chick embryo in an ex ovo culture (boat method)

Discoidal Cleavage It is an incomplete (meroblastic) form of cleavage observed in animal eggs with large yolk concentrated toward the vegetal pole (Fig. 1.13). In such eggs, the yolk is so dense and voluminous that the cleavage furrows are limited to yolk-free small cytoplasmic area at the animal pole, forming a blastodisc, which rests on top of the unsegmented yolk mass. The underlying uncleaved yolk serves as a nourishment source for the embryo. The blastodisc undergoes a sequence of mitotic divisions that eventually lead to the blastoderm establishment, a structure that later contributes to the embryonic body. This cleavage pattern is commonly observed in birds (chicken) (Fig. 1.14), bony fishes (zebrafish) (Gilbert 2000) (Fig. 1.15), and

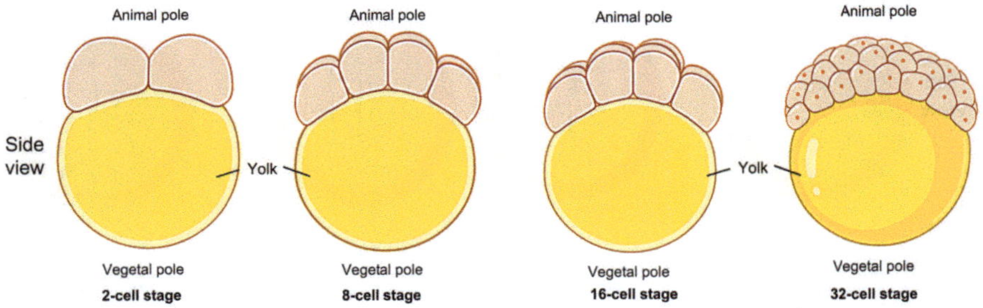

Fig. 1.15 Diagrammatic representation of discoidal pattern of cleavage in fish (zebrafish). (Created in https://BioRender.com)

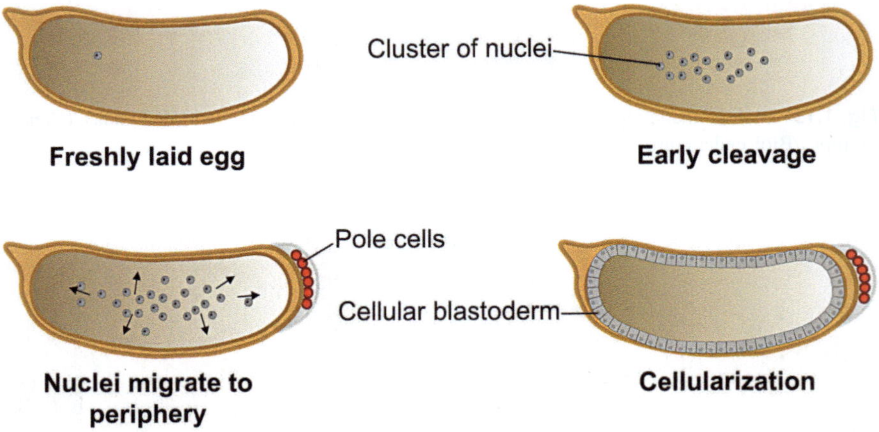

Fig. 1.16 Diagrammatic representation of the superficial pattern of cleavage. (Created in https://BioRender.com)

reptiles (snakes and lizards). Discoidal cleavage ensures that embryonic development is confined to the blastodisc while the abundant yolk supports growth and morphogenesis throughout embryogenesis.

Superficial Cleavage It is a kind of incomplete (meroblastic) cleavage seen in centrolecithal eggs, where the yolk is densely concentrated within the zygote centrally (Fig. 1.16). In this pattern, the cleavage furrows do not pass through the yolk. Instead, only the peripheral cytoplasm undergoes division. During early development, division of the nucleus occurs without cytokinesis, giving rise to a syncytial blastoderm. The nuclei then migrate toward the egg periphery, after which plasma membranes invaginate around them, forming individual cells at the surface and establishing the cellular blastoderm around a central yolk mass. This type of cleavage is characteristic of insects, allowing rapid nuclear proliferation while conserving the yolk for later development.

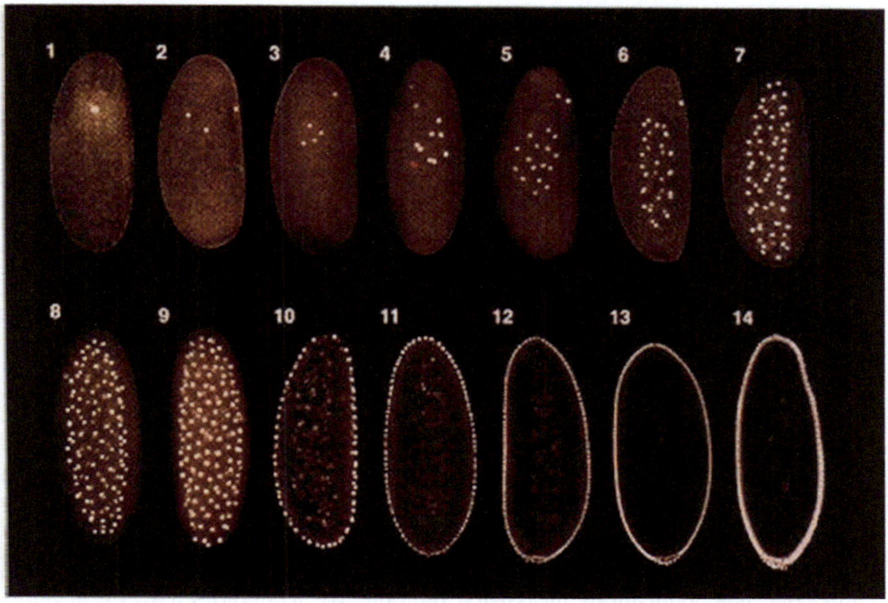

Fig. 1.17 Image of superficial pattern of cleavage in *Drosophila* with the help of trackers to show the formation and movement of the cells. (Source: File: Drosophila Cleavage.jpg—Wikimedia Commons. This file is licensed under the Creative Commons).

Example *Drosophila melanogaster* (fruit fly). In *Drosophila*, cleavage occurs within a yolk-rich egg and is confined to the peripheral cytoplasmic rim surrounding the yolk mass (Fig. 1.17). The first seven nuclear divisions take place centrally within the egg, without cytokinesis, creating a syncytium, an embryo with multiple nuclei sharing a common cytoplasm. These early, synchronous divisions begin 20 minutes post-fertilization and recur every 10 minutes. By cycle 10, nuclei migrate to the periphery, forming the syncytial blastoderm, where molecules diffuse freely. Zygotic genome transcription initiates at this stage. By cell cycle 14 (2.5–3 hours post-fertilization), membranes form around each nucleus, producing the cellular blastoderm. This transition marks the end of syncytial development and the beginning of individual cell formation.

1.5 Significance/Scientific Application

The study of different types of animal eggs and their corresponding cleavage patterns provides fundamental insights into embryonic development. Understanding these early stages is essential, as it forms the basis for comprehending how multicellular organisms arise from a single fertilized cell, provides critical insights into species diversity, highlights evolutionary adaptations to reproductive strategies and ecological niches, and supports phylogenetic and taxonomic studies. Moreover, this

Table 1.1 Representative studies highlighting the significance of studying early developmental and cleavage patterns in animals

S. no.	Type of study/research	Reference
1.	Identification of the role of subcortical maternal complex conserved protein in regulating embryo cleavage and implantation in humans using mouse embryos	Li et al. (2008)
2.	Investigation of the role of the *filia* gene expressed in maintaining euploidy in oocytes during the cleavage stage of mouse embryogenesis to provide insights into the contribution of conserved human homologue defects to recurrent fetal wastage	Zheng and Dean (2009)
3.	The molecular basis of embryonic chromosomal mutation and karyotypic complexity during early mammalian cleavage, often causing developmental arrest or implantation failure analysis using bovine embryos as a model during initial mitotic divisions	Brooks et al. (2022)
4.	Evaluation of the significance of blood serum supplementation in culture media for human in vitro fertilization, early embryo development, and embryo transfer, focusing on its influence on developmental outcomes	Menezo et al. (1984)

knowledge has applications in assisted reproductive technology (ART), developmental genetics, and teratology. Table 1.1 presents a few representative studies highlighting the benefits of understanding early development and cleavage patterns across different organisms for human research and health.

1.6 Precautions

- Handle photomicrographs, diagrams, or slides carefully.
- Ensure proper labelling of the diagrams while recording observations.

1.7 Quick Review Questions

1.7.1 Multiple Choice Questions

1. Based on yolk content and distribution, a bird's egg is classified as:
   (a) Microlecithal
   (b) Mesolecithal
   (c) Macrolecithal
   (d) Alecithal
   Answer: c) Macrolecithal
2. Which kind of cleavage occurs in amphibian (frog) eggs?
   (a) Meroblastic cleavage
   (b) Superficial cleavage

(c) Holoblastic unequal cleavage
(d) Discoidal cleavage
Answer: c) Holoblastic unequal cleavage

3. Insects exhibit which type of cleavage?
(a) Holoblastic cleavage
(b) Radial cleavage
(c) Superficial cleavage
(d) Meroblastic discoidal cleavage
Answer: c) Superficial cleavage
4. Alecithal eggs are generally found in:
(a) Birds
(b) Mammals
(c) Amphibians
(d) Reptiles
Answer: b) Mammals
5. Meroblastic cleavage is a characteristic feature of:
(a) Eggs with little yolk
(b) Eggs with moderate yolk
(c) Eggs with a huge amount of yolk
(d) Alecithal eggs
Answer: c) Eggs with a huge amount of yolk
6. In zebrafish, the cleavage pattern is:
(a) Discoidal meroblastic
(b) Holoblastic radial
(c) Superficial
(d) Spiral cleavage
Answer: a) Discoidal meroblastic
7. Which of the following is a distinguishing feature of discoidal cleavage?
(a) Entire egg divides equally
(b) Cleavage occurs in yolk-free cytoplasm at the animal pole
(c) Cleavage furrows pass wholly through the egg
(d) Cleavage nuclei migrate to the periphery
Answer: b) Cleavage occurs in yolk-free cytoplasm at the animal pole
8. Which type of cleavage is seen in frog eggs?
(a) Spiral holoblastic
(b) Radial holoblastic
(c) Superficial
(d) Discoidal
Answer: b) Radial holoblastic
9. Cleidoic egg refers to:
(a) An egg without any yolk
(b) An egg enclosed within a protective shell, adapted for terrestrial life
(c) An egg with evenly distributed yolk
(d) An egg undergoing superficial cleavage
Answer: b) An egg enclosed within a protective shell, adapted for terrestrial life

10. Indeterminate (regulative) eggs are those in which:
    (a) The fate of each blastomere is fixed early
    (b) Each blastomere can regulate and form a complete embryo
    (c) Cleavage is always superficial
    (d) Yolk is concentrated at one pole
    Answer: b) Each blastomere can regulate and form a complete embryo

1.7.2 Short Answer Questions

1. Give one example each of organisms possessing microlecithal, mesolecithal, and macrolecithal eggs.
2. What are centrolecithal eggs? Name one organism in which they are found.
3. Distinguish between equal and unequal cleavage patterns.
4. Differentiate between *holoblastic* and *meroblastic* cleavage?
5. Define superficial cleavage. Mention one animal group where it occurs.
6. Differentiate between determinate (mosaic) and indeterminate (regulative) eggs.
7. Explain the correlation between egg type and the mode of development.
8. What is a cleidoic egg? Give two examples of organisms possessing this type of egg.
9. Explain why yolk distribution influences the pattern of cleavage.
10. Distinguish between radial and spiral cleavage patterns.

Glossary

Albumin (egg white) The clear protein-rich viscous fluid around the yolk that protects and nourishes the embryo.

Allantois An extraembryonic membrane that stores waste and facilitates gas and nutrient exchange for the developing amniote embryos.

Amnion A thin, protective extraembryonic membrane that surrounds the embryo in amniotes, enclosing the fluid-filled amniotic cavity, cushioning the embryo, preventing desiccation, and allowing free movement of the embryo during development.

Amniote Vertebrate with embryos developing within an amnion, e.g., reptiles, birds, mammals.

Amphibians Cold-blooded vertebrates adapted to a dual mode of life, with aquatic larvae and terrestrial adults, e.g., frogs.

Amphioxus Small marine invertebrate (Cephalochordata) with chordate features.

Animal pole The region of the ovum or egg with the least yolk, the site of active cleavage, forming the embryo.

Animal-vegetal axis An imaginary line extending from the animal to the vegetal pole, establishing egg polarity. It represents a primary body axis determined during oogenesis.

Arthropods Invertebrates with segmented bodies, jointed appendages, and a chitinous exoskeleton, e.g., insects and spiders.

Birds Warm-blooded, hollow-boned, feathered vertebrates with wings and beaks, laying cleidoic eggs.

Blastoderm Cell layer formed on the yolk surface in a fertilized egg, which gives rise to the embryo.

Blastodisc A disc-shaped cytoplasmic region on a yolky egg where embryonic development takes place.

Blastomeres Cells produced by zygote cleavage (division) during early embryogenesis.

Body Axis Imaginary line defining the spatial arrangement of body parts (e.g., head-to-tail, back-to-belly).

Calcareous A substance or structure that is composed of or contains calcium carbonate ($CaCO_3$), often giving it a hard, stony nature.

Cellularization Formation of individual cells from a multinucleate cytoplasm (syncytium).

Chorion Outermost extraembryonic membrane in amniotes (e.g., birds, reptiles, mammals) facilitating embryo's gas exchange with the environment.

Chromosomes Thread-like structures of DNA and proteins (mainly histones) located in the nucleus carrying genetic information.

Cleavage The series of rapid mitotic divisions of a zygote forming blastomeres without increasing overall embryo size.

Cleavage furrow Surface indentation in a dividing zygote that separates daughter blastomeres.

Cytokinesis Division of the cytoplasm to form two daughter cells, usually following nuclear division.

Eggshell Protective outer layer of an egg made of calcium carbonate, allowing gas exchange and preventing dehydration.

Equatorial cleavage Division along the egg's equator, forming roughly equal daughter cells.

Eutherian (placental) mammals Mammals with embryos nourished via a placenta inside the uterus, allowing longer gestation and more developed offspring at birth.

Extraembryonic membranes Membranes outside the embryo proper that support embryonic development (amnion, chorion, yolk sac, allantois).

Fertilization Fusion of sperm and egg to form a zygote.

Flatworms (Platyhelminthes) Soft-bodied, unsegmented, flattened invertebrates; acoelomate, free-living (planarians) or parasitic (like tapeworms and flukes).

Gestation Period from fertilization to birth when the embryo/fetus develops in the uterus.

Human homologue A gene, protein, or structure in humans that is evolutionarily related to a corresponding gene, protein, or structure found in another species.

Meridional cleavage Embryonic cell division in which the cleavage plane passes along the egg's animal-vegetal axis.

Microtubules Hollow, cylindrical filaments composed of tubulin proteins forming part of the cytoskeleton, aiding structure, transport, and cell division, especially in creating the mitotic spindle.

Mitotic cell divisions Cell division producing two identical daughter cells with the same chromosome number.

Mitotic spindle A structure composed of microtubules and associated proteins that forms during mitosis to segregate chromosomes into the two daughter cells. It ensures accurate chromosome alignment and distribution during cell division.

Mollusks Invertebrates with soft bodies, usually with a shell (e.g., snails, clams, sepia).

Parthenogenesis Asexual reproduction where an egg develops without fertilization.

Polar axis An imaginary line connecting the animal pole and the vegetal pole.

Polar cells Specialized cells located at the poles (usually the vegetal or animal poles) of an egg or early embryo, often playing a role in signaling, axis formation, or guiding development rather than forming the main body of the embryo.

Predators Organisms that hunt and eat other organisms for survival.

Protostomes Animals whose mouth forms from the blastopore.

Reptiles Cold-blooded, scaly vertebrates that lay eggs on land.

Spindle fibers Microtubules that separate chromosomes during cell division.

Spindle pole The region at each end of the mitotic or meiotic spindle from which spindle fibers radiate, helping to organize and separate chromosomes during cell division.

Symmetry Balanced arrangement of body parts around an axis or plane.

Syncytium A single cell with many nuclei.

Terrestrial Living or growing on land.

Vegetal pole Yolk-rich pole of the egg opposite the animal pole.

Yolk Nutrient-rich yellow material within an egg that provides food reserves, mainly proteins, lipids, and vitamins, to support the growth and development of the embryo.

Yolk sac An extraembryonic membrane that surrounds and absorbs nutrients from the yolk, supplying them to the developing embryo during early growth.

Zygote The first diploid cell formed after fertilization.

References

Balfour FM (1885) The works of Francis Maitland Balfour, vol 1. Macmillan and Company

Brooks KE, Daughtry BL, Davis B, Yan MY, Fei SS, Shepherd S, Carbone L, Chavez SL (2022) Molecular contribution to embryonic aneuploidy and karyotypic complexity in initial cleavage divisions of mammalian development. Development 149(7). https://doi.org/10.1242/dev.198341

Brun-Usan M, Marin-Riera M, Grande C, Truchado-Garcia M, Salazar-Ciudad I (2017) A set of simple cell processes is sufficient to model spiral cleavage. Development 144(1):54–62. https://doi.org/10.1242/dev.140285

Brun-Usan M, Salazar-Ciudad I (2021) The evolution of cleavage in metazoans. In: Evolutionary developmental biology: a reference guide. Springer, pp 529–543

Callebaut M, Van Nueten E, Harrisson F, Hubens G (2010) Mosaic and regulation phenomena during the early formation of the chick blastoderm. Int J Zool 2010(1):572502

Carrillo-Baltodano AM, Seudre O, Guynes K, Martin-Duran JM (2021) Early embryogenesis and organogenesis in the annelid Owenia fusiformis. EvoDevo 12(1):5. https://doi.org/10.1186/s13227-021-00176-z

Conklin EG (1905) Mosaic development in ascidian eggs. J Exp Zool 2(2):145–223

Denegre JM, Valles JM Jr, Lin K, Jordan WB, Mowry KL (1998) Cleavage planes in frog eggs are altered by strong magnetic fields. Proc Natl Acad Sci USA 95(25):14729–14732. https://doi.org/10.1073/pnas.95.25.14729

Galli M, Van Den Heuvel S (2008) Determination of the cleavage plane in early C. elegans embryos. Annu Rev Genet 42(1):389–411

Gerhart J, Wu M, Kirschner M (1984) Cell cycle dynamics of an M-phase-specific cytoplasmic factor in Xenopus laevis oocytes and eggs. J Cell Biol 98(4):1247–1255. https://doi.org/10.1083/jcb.98.4.1247

Gilbert SF (2000) An introduction to early developmental processes. In: Developmental biology, 6th edn. Sinauer Associates

Hasley A, Chavez S, Danilchik M, Wühr M, Pelegri F (2016) Vertebrate embryonic cleavage pattern determination. In: Vertebrate development: maternal to zygotic control, pp 117–171

Kitajima A, Hamaguchi Y (2005) Determination of first cleavage plane: the relationships between the orientation of the mitotic apparatus for first cleavage and the position of meiotic division-related structures in starfish eggs. Dev Biol 280(1):48–58

Lawrence PA, Levine M (2006) Mosaic and regulative development: two faces of one coin. Curr Biol 16(7):R236–R239

Li L, Baibakov B, Dean J (2008) A subcortical maternal complex essential for preimplantation mouse embryogenesis. Dev Cell 15(3):416–425. https://doi.org/10.1016/j.devcel.2008.07.010

Makabe KW, Nishida H (2012) Cytoplasmic localization and reorganization in ascidian eggs: role of postplasmic/PEM RNAs in axis formation and fate determination. Wiley Interdiscip Rev Dev Biol 1(4):501–518

Menezo Y, Testart J, Perrone D (1984) Serum is not necessary in human in vitro fertilization, early embryo culture, and transfer. Fertil Steril 42(5):750–755. https://doi.org/10.1016/s0015-0282(16)48202-6

Newport JW, Kirschner MW (1984) Regulation of the cell cycle during early Xenopus development. Cell 37(3):731–742. https://doi.org/10.1016/0092-8674(84)90409-4

Rawal DD (2025) Cleidoic eggs: a key to water to land transition. Available at SSRN 5161634.

Schonegg S, Hyman AA, Wood WB (2014) Timing and mechanism of the initial cue establishing handed left-right asymmetry in Caenorhabditis elegans embryos. Genesis 52(6):572–580. https://doi.org/10.1002/dvg.22749

Silas E, Vijayakumaran M (1984) Are turtle eggs cleidoic or non-cleidoic. CMFRI Bull 35:34–40

Sozen B, Can A, Demir N (2014) Cell fate regulation during preimplantation development: a view of adhesion-linked molecular interactions. Dev Biol 395(1):73–83

Zheng P, Dean J (2009) Role of Filia, a maternal effect gene, in maintaining euploidy during cleavage-stage mouse embryogenesis. Proc Natl Acad Sci USA 106(18):7473–7478. https://doi.org/10.1073/pnas.0900519106

Study of Whole Mounts and Sections of Developmental Stages of the Frog

2

Anna Senrung and Alok Chandra Bharti

Abstract

Frogs, as amphibians, exhibit a dual mode of life, initially aquatic during embryonic and larval stages, and later terrestrial as adults. Their externally developing embryos and transparent early stages make them ideal models for studying vertebrate embryology. The life cycle of the frog begins with external fertilization, followed by a series of well-defined developmental processes, cleavage, blastulation, gastrulation, neurulation, and organogenesis, leading to distinct embryonic stages, including the morula, blastula, gastrula, neurula, tailbud, and the free-swimming herbivorous tadpole stages with external and internal gills. Subsequent metamorphosis transforms the tadpole into a carnivorous, lung-breathing adult frog, involving significant anatomical and physiological changes such as tail resorption, limb development, and lung formation. Practical study through whole mounts and histological sections allows detailed observation of structural and tissue-level differentiation, offering insights into conserved developmental processes across vertebrates. Frog embryology has broad applications, from understanding early axis formation and neural tube development to gene function, cell signaling, and modeling congenital disorders. Additionally, frog embryos are useful in toxicological studies and regenerative biology, contributing to drug testing and stem cell research. The tadpole stage, with its fish-like traits, provides evidence of evolutionary ancestry, highlighting the transition from aquatic to terrestrial life. This chapter deals with the study of frog embryogenesis, emphasizing its relevance in developmental biology. Through observation, dia-

A. Senrung (✉)
Department of Zoology, Daulat Ram College, University of Delhi, Delhi, India

A. C. Bharti (✉)
Molecular Oncology Lab., Department of Zoology, University of Delhi, Delhi, India

A. C. Bharti et al. (eds.), *Textbook of Developmental Biology*,
https://doi.org/10.1007/978-981-95-7269-4_2

grammatic analysis, and microscopy, learners will gain an experience comparable to hands-on practice in tracing the stages and mechanisms of vertebrate development, using the frog as a representative model.

Keywords

Frog embryo · Gosner's staging · Cleavage · External gill tadpole · Internal gill tadpole

Aim To identify and describe the different developmental stages in frog development from a fertilized egg to the internal gill stage from permanent slides by observing under the microscope.

Learning Objectives

By the end of this experiment, students should be able to:

- Identify and distinguish the various developmental stages of a frog, including cleavage patterns and stages, blastula, gastrula, neurula, tailbud stage, as well as the external and internal gill stages according to Gosner's staging.
- Identify and interpret key anatomical structures in whole mounts and histological sections of frog embryos.
- Correlate the developmental stages of the frog embryo with the functional differentiation of its tissues and organs.
- Explain the significance and application of the frog embryo in understanding the concepts of vertebrate development, evolutionary biology, and biomedical research.

2.1 Introduction

Frogs are amphibians that exhibit a dual mode of life, beginning in water and later transitioning to land. They are tailless members of the order *Anura*, commonly belonging to the family *Ranidae*, and are characterized by smooth, moist skin, protruding eyes, and webbed hind limbs. Frogs have long been a valuable model organism in the study of vertebrate development, physiology, and genetics. Their ease of availability, external fertilization, large and transparent embryos, and relatively rapid embryonic development make them ideal for educational and research purposes.

Frog embryology is especially useful for understanding development in species with mesolecithal and moderately telolecithal eggs, where the yolk is moderately concentrated at the vegetal pole (V.P.). Despite some species-specific differences, the embryological development of anurans follows a conserved sequence. It begins with external fertilization, followed by cleavage, blastula, gastrula, and neurula stages. These early phases lead to the formation of a free-living larva, the tadpole, which is aquatic, gill-breathing, herbivorous, and tailed.

Maintaining live frog cultures for daily classroom observation is often impractical. Therefore, the study of whole mounts and sections of frog embryos through permanent slides at different developmental stages provides an effective alternative. It enables students to visualize key morphological changes and patterns of tissue differentiation with clarity. This experience, comparable to a hands-on practice approach, reinforces theoretical knowledge while enhancing observational and analytical skills crucial for understanding developmental biology.

2.2 Requirements

- Prepared whole mount slides of frog developmental stages (cleavage, blastula, gastrula, neurula, tail-bud, external, and internal gill stages)
- Prepared sectional (histological) slides of the corresponding stages of whole mounts
- Compound and stereo zoom microscope
- Photomicrographs and labelled diagrams for comparison
- Observation record sheet, pencil, and eraser

2.3 Methodology

Note the changes through successive stages. Compare whole mount observations with histological sections to correlate surface morphology with internal structure. Sketch the observed structures in your notebook and label significant parts of specific developmental stages.

2.3.1 Observation/Study of Whole Mounts

- Place the permanent slides of the whole mounts of frog embryos of different developmental stages under a microscope of appropriate magnification to identify overall morphology.
- Identify key features in each stage.
- *Early cleavage stages up to morula*: Symmetry and number of blastomeres.
- *Blastula stages*: Blastocoel, blastoderm arrangement, micromeres, and macromeres.
- *Gastrula stages*: Blastopore on the posterior side of the embryo, circumference of the blastopore, and the size of the yolk plug.

- *Neurula stages*: Identify the neural plate, neural fold, neural groove, and neural tube.
- *Tail-bud stages*: Tail bud, body form (arched/non-arched), protuberance or cement gland, gill arches and buds, yolky abdomen, tail length, optic, otic, gill, and pronephric protuberances.
- *External gill stage*: External gill filaments, elongated tail, transparent tail fin.
- *Internal gill stage*: Absence of external gill filaments, presence of operculum, spiracle, elongated tail, transparent tail fin.

2.3.2 Observation/Study of Histological Sections

- Examine stained prepared sections under the microscope for cellular and tissue-level organization.
- Identify the following:
  - *Early cleavage stages up to morula*: Absence of blastocoel, number, and position of blastomeres.
  - *Blastula stages*: Position of the blastocoel, size, number, and position of blastomeres.
  - *Gastrula stages*: Invagination forming the archenteron, dorsal and ventral lip of the blastopore, archenteron and blastocoel size, formation of germ layers, and yolk plug.
  - *Neurula stages*: Neural plate, neural fold, neural groove, neural tube, notochord, presence of archenteron, closure of the blastopore, mesoderm or somites.
  - *Tail-bud stages*: Formation of somites, unsegmented tailbud tissue or mesoderm at the posterior end of the embryo, and elongation of the tail.
  - *External gill stage*: Gill filaments protruding freely on either side of the head margin, absence of opercular chamber, presence of myotomes, notochord, neural tube, and other fundamental organ parts such as liver, optic, otic, gastrointestinal tract, and heart.
  - *Internal gill stage*: Internal gill filaments within the operculum chamber, presence of myotomes, notochord, neural tube, and other fundamental organ parts such as liver, optic, otic, gastrointestinal tract, and heart.

2.4 Description of Frog Development (Gosner's Staging)

The developmental stages of frogs are commonly described using Gosner's staging system (1960), which provides a standardized and simplified framework widely accepted in anuran research based on key morphological landmarks. This system, introduced in 1960 by Kenneth Gosner, outlines 46 stages, from fertilization through metamorphosis (Gosner 1960). Stages 1 to 25 represent the embryonic phases, including cleavage, blastula, gastrula, and neurula stages, leading to the formation of a free-swimming, feeding tadpole. Stages 26 to 46 encompass metamorphic changes such as limb development, tail resorption, and transformation of mouthparts and skin, culminating in the adult form by stage 46.

Here, the study focuses on stages 1 to 25, which are particularly valuable for understanding early vertebrate development. These early stages are broadly conserved across anuran species; however, some inter- and intraspecific variations exist, particularly in egg characteristics such as size, color, and the structure of jelly envelopes, as well as in the rate and timing of embryonic development.

2.4.1 Ovum of a Frog

Before the ovum (primary oocyte) is released from the ovary (Fig. 2.1), its nucleus is located in the pigmented region near the periphery, within a clear zone at the animal pole (A.P.). By the time the egg enters the oviduct, the first polar body has already formed and remains inside the vitelline membrane, close to the A.P. The second maturation division is initiated but only completes upon fertilization. The entry of a spermatozoon into the unfertilized egg triggers the completion of the second meiotic (maturation) division. As the egg travels through the oviduct, it becomes surrounded by a thick coat of albumin secreted by the albuminous glands. This albumin layer swells upon contact with water, causing the eggs to adhere to one another in clusters called *spawn*. Albumin serves protective functions, including insulation of the eggs, retaining heat, and acting as a barrier against harmful microorganisms.

The frog's egg measures approximately 1–10 mm in diameter. It exhibits distinct polarity: the animal hemisphere (upper half) is deeply pigmented, while the vegetal hemisphere (lower half) is unpigmented and rich in yolk granules, making it less metabolically active. The dark pigment in the animal hemisphere is thought to aid in heat absorption.

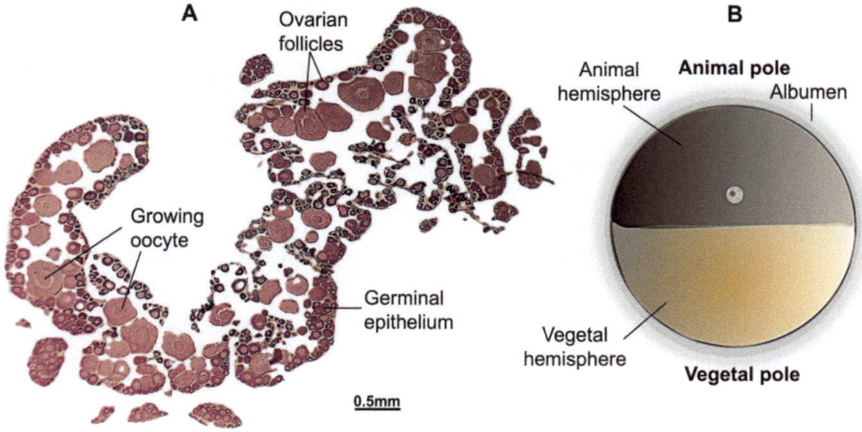

Fig. 2.1 Frog ovary and ovum. (**a**) Cross-section of frog ovary. (**b**) Diagrammatic representation of unfertilized frog ovum. (Source: Department of Zoology, Daulat Ram College, University of Delhi (T.S. of Frog ovary); Unfertilized egg diagram was created in https://BioRender.com)

2.4.2 Fertilization and Gray Crescent Formation

Gosner Stage 1 (Fertilization)

Gosner stage 1 (GS1) corresponds to the process of fertilization (Fig. 2.2a). During the breeding season, the male frog clasps the female in amplexus, stimulating her to release eggs. This, in turn, prompts the male to release sperm over the eggs, leading to external fertilization, typically near the A.P. Upon sperm entry, a block to polyspermy is triggered to prevent additional sperm from entering. The entire sperm enters the egg, and fluid accumulates between the egg and the vitelline membrane, enabling the egg to rotate freely. The sperm nucleus then migrates toward the egg nucleus, and their fusion forms the diploid zygote nucleus.

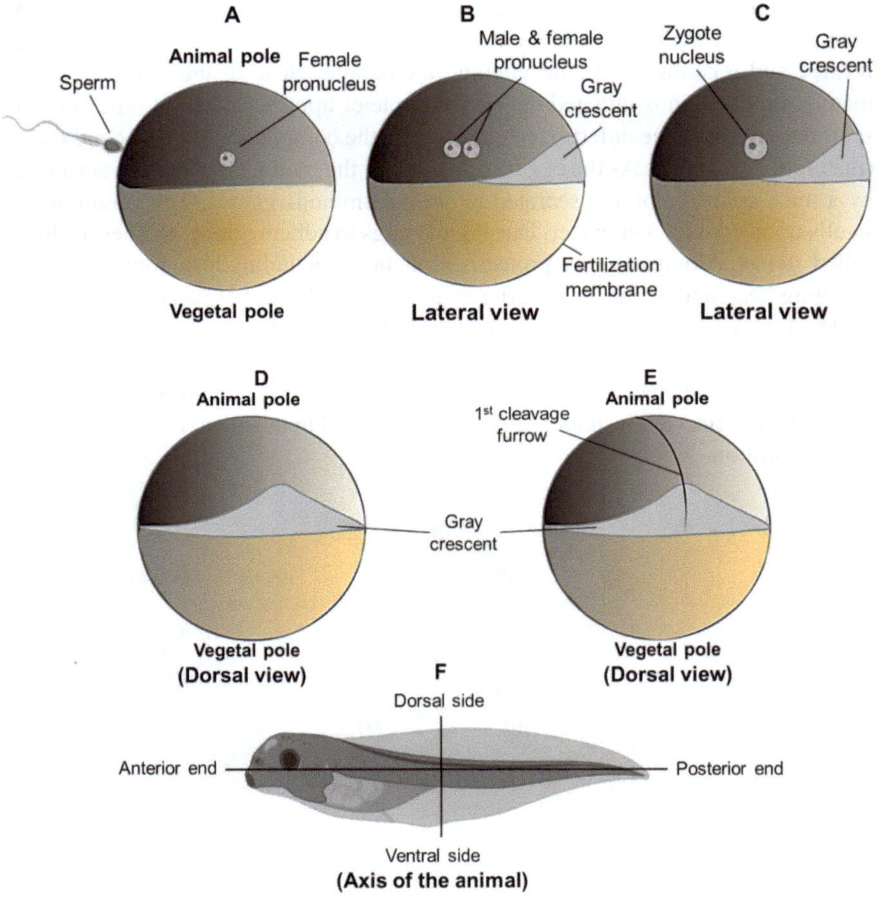

Fig. 2.2 Schematic representation of the fertilization and gray crescent formation in frog. (**a**) Fertilization of frog egg by sperm. (**b**) Fertilized egg of frog (lateral view). (**c**) Zygote of frog (lateral view). (**d**) Fertilized egg (dorsal view). (**e**) Fertilized egg showing first cleavage (Dorsal view). (**f**) Axis of the animal. (The diagrams have been prepared based on information gathered from Balinsky (1960), Gilbert (2000), Gosner (1960), Holmes (1927), Rugh (1951). Created in https://BioRender.com)

Gosner Stage 2 (Gray Crescent Formation) (Fig. 2.2b–f)
Upon sperm entry into the animal hemisphere, the egg's vegetal cortex undergoes a cortical rotation of approximately 30 degrees toward the A.P. relative to the underlying cytoplasm. This rotation leads to the formation of a lighter, grayish region caused by the redistribution of diffuse pigment granules in the underlying cytoplasm, known as the *gray crescent*. In species such as *Xenopus*, the gray crescent is not visibly distinct and requires dye labeling for detection. While the unfertilized frog egg is radially symmetrical, the appearance of the gray crescent marks the onset of bilateral symmetry, establishing the embryo's future dorsal side and allowing early identification of the left-right and anterior-posterior axis. The formation of the gray crescent also signals the activation of the Nieuwkoop Centre, a critical signaling region that initiates dorsal development, including the induction of the Spemann-Mangold organizer. In *Hoplobatrachus rugulosus*, a flattening of the upper region of the fertilized egg is observed at this stage (Traijitt et al. 2021).

2.4.3 Early Cleavage Stages of the Frog

Following fertilization, the zygote of the frog undergoes a series of rapid mitotic divisions known as *cleavage* (Fig. 2.3). These early divisions are holoblastic and unequal, influenced by the yolk content of the egg. Cleavage transforms the single-celled zygote into a multicellular structure composed of smaller cells called *blastomeres*, laying the foundation for subsequent embryonic development.

Gosner stage 3 (2-cell stage) The frog egg is mesolecithal, containing a moderate amount of yolk that is predominantly concentrated in the vegetal hemisphere (telolecithal). The first cleavage is meridional (vertical), aligned with the animal-vegetal axis and perpendicular to the spindle formed by the centrioles of the entering sperm. Cleavage begins at the A.P., where the cytoplasm is less yolky, and progresses toward the yolk-rich V.P. This division results in the formation of two equal-sized blastomeres (Fig. 2.3c), representing the left and right halves of the developing embryo.

Gosner stage 4 (4-cell stage) The second cleavage is also meridional (vertical) but oriented at a right angle to the first cleavage plane. It divides each of the two blastomeres into halves, passing from the less yolky animal pole to the yolky V.P. Together, the first and second cleavages produce four equal blastomeres (Fig. 2.3d–f).

Gosner stage 5 (8-cell stage) The third cleavage is horizontal (equatorial) but is shifted toward the animal pole due to the unequal distribution of yolk. It divides each of the four blastomeres into dorsal and ventral halves, with the Nieuwkoop Centre located on the dorsal side. This results in an 8-cell stage comprising two distinct cell types: four pigmented, smaller, yolk-poor cells at the animal pole called *micromeres*, and four lighter, larger, yolk-rich cells at the V.P. called *macromeres* (Fig. 2.3g–i).

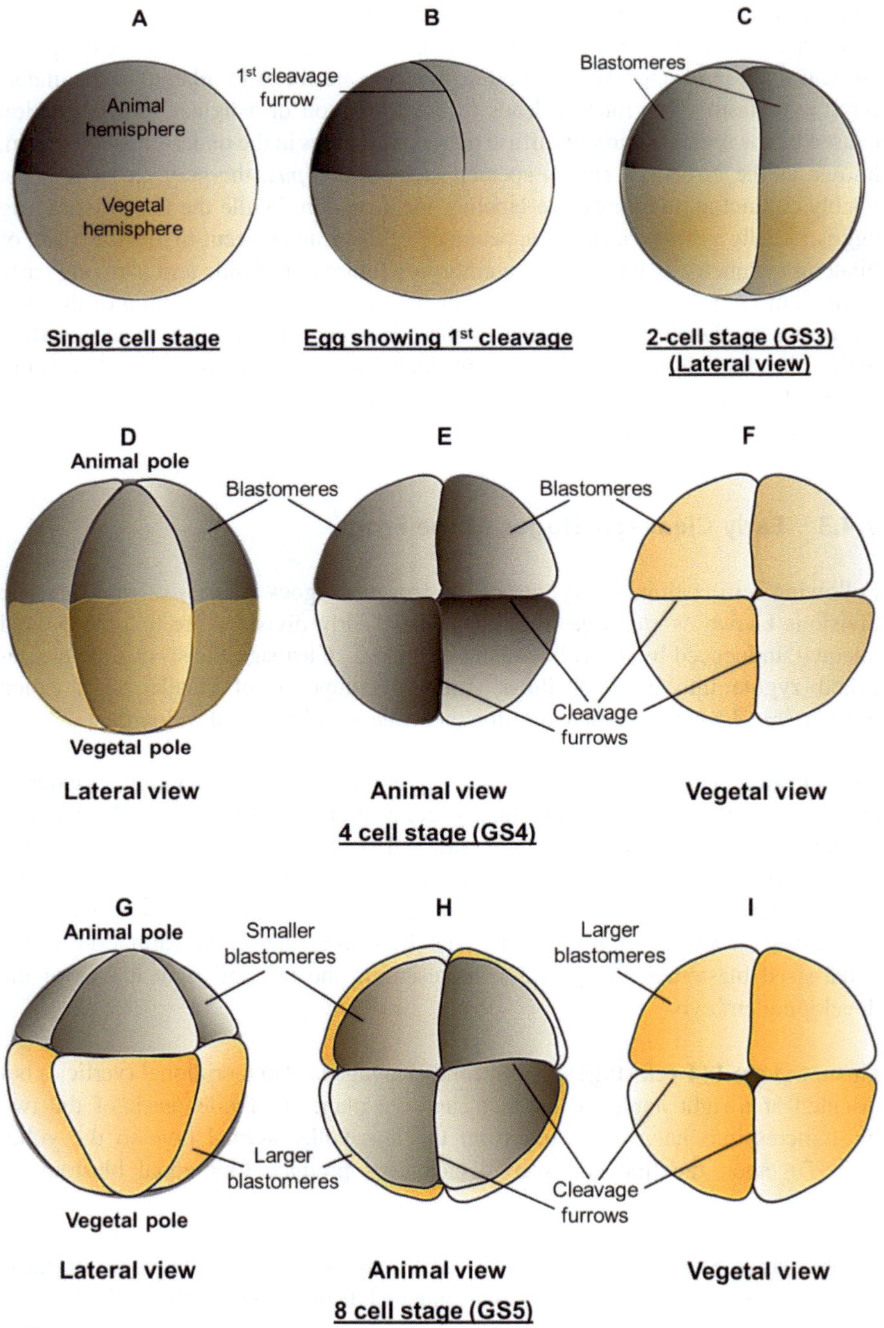

Fig. 2.3 Illustration of initial cleavage stages following fertilization up to the morula or 64-cell stage. GS–Gosner stage. (The diagrams have been prepared based on information gathered from Balinsky (1960), Gilbert (2000), Gosner (1960), Holmes (1927), Rugh (1951), Zahn et al. (2022). Created in https://BioRender.com)

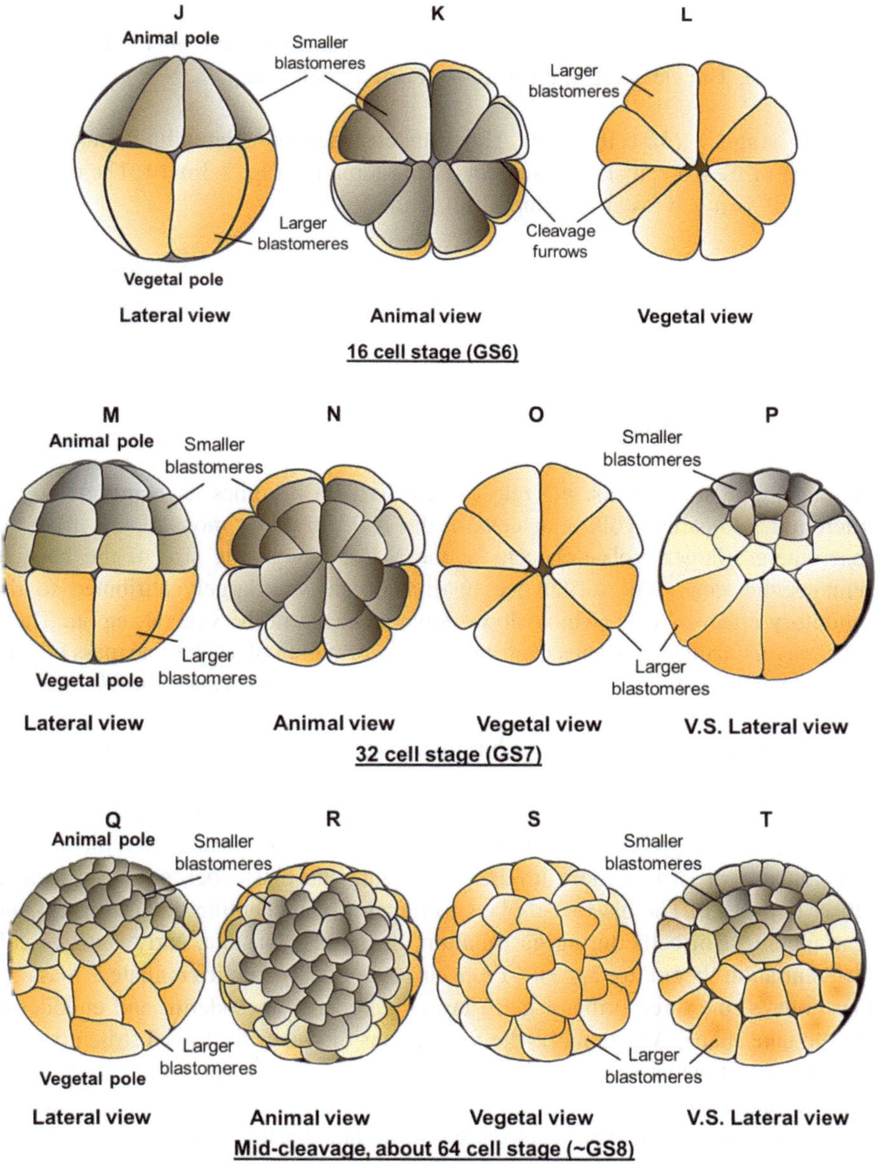

Fig. 2.3 (continued)

This asymmetry reflects the influence of yolk on cleavage dynamics and marks the beginning of size differentiation among blastomeres.

Gosner stage 6 (16-cell stage) The fourth cleavage in frog embryonic development is vertical, holoblastic (complete), and unequal due to the influence of yolk distribution. This cleavage further divides both the micromeres at the animal pole and the macromeres at the V.P., resulting in the formation of 16 blastomeres (Fig. 2.3j–l). The cleavage at the V.P. is generally delayed.

Gosner stage 7 (32-cell stage) The fifth cleavage or a latitudinal (horizontal) cleavage of the 16-cell stage results in 32 blastomeres (Fig. 2.3m–p). The cleavage at the A.P. is much faster in comparison to the V.P.

Gosner Stage 8 (mid-cleavage) This stage represents the cleavage stage, consisting of 64 cells (Fig. 2.3g–t). This stage is commonly referred to as the *morula* stage; however, embryos consisting of 16–64 cells are likewise termed the *morula.*

2.4.4 Late Cleavage/Blastula

As development proceeds, the rate of cell division becomes asynchronous, with micromeres (smaller, yolk-poor cells at the A.P.) dividing more rapidly than the macromeres (larger, yolk-rich cells at the V.P.), though the size of the blastula remains the same. This difference in division rates is largely attributed to the inhibitory effect of yolk, which slows down cleavage in the vegetal region. These successive, unequal cleavages contribute to the formation of a spherical mass of cells with apparent visibility of the fluid-filled cavity (blastocoel) called the *blastula.*

Gosner stage 9 (blastula) The blastula (Fig. 2.4a–h) is a key stage in early embryonic development, following the cleavage of the fertilized egg or zygote. It marks the transition from a solid mass of cells (*Morula*) to a hollowed, structured form without change in size. It comprises 128 or more cells. Structurally, the blastula is composed of two or more layers surrounding the blastocoel, which is located closer to the A.P. The blastocoel plays a critical role in enabling the movement and rearrangement of cells during gastrulation. In the late blastula, early signs of cell differentiation begin to appear. From the previously undifferentiated mass of blastomeres, the three primary germ layers, ectoderm, mesoderm, and endoderm, start to take shape. Additionally, the foundations of two critical axial structures, the notochord and the nerve cord, begin to form. These regions can be identified using specific staining techniques and are essential for understanding the early patterning of the vertebrate body plan. Each region of the late blastula is fated to give rise to specific tissues and organs in the embryo. These regions are known as *presumptive organ-forming areas* or *prospective organ-forming areas*, and they can be mapped using *fate maps* (Fig. 2.4i–j). For example, the micromeres in the animal hemisphere give rise to the ectoderm, while the region of the gray crescent contributes to the mesoderm. The endoderm originates from the deeper yolk-rich regions.

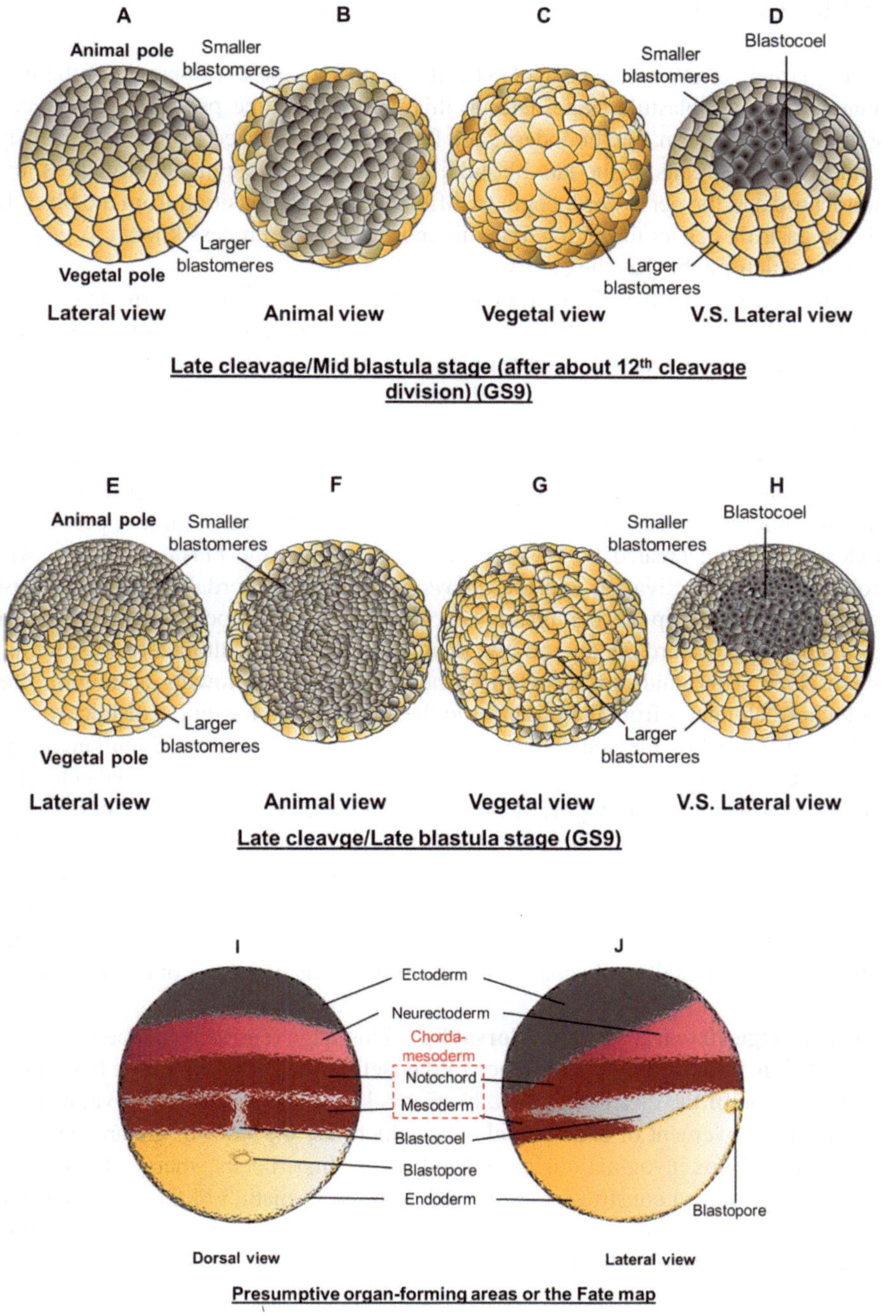

Fig. 2.4 Schematic representation of blastula stage and fate map. (The diagrams have been prepared based on information gathered from Balinsky (1960), Freeman and Bracegirdle (1967), Gilbert (2000), Gosner (1960), Holmes (1927), Rugh (1951), Zahn et al. (2022). Created in https://BioRender.com)

2.4.5 Gastrula

The gastrula comprises the GS10-GS12. It is a stage in the development of the frog that follows the blastula stage. During this phase, the three primary germ layers, ectoderm, mesoderm, and endoderm, are formed and organized from the previously undifferentiated cells of the late blastula. In addition, two key axial structures, the notochord and the nerve cord, begin to differentiate. The transition from the blastula to the gastrula occurs through a dynamic morphogenetic process known as *gastrulation*, which establishes the embryo's basic body plan.

Gastrulation involves the rearrangement of presumptive organ-forming regions, enabling them to move into their specific positions within the developing embryo. Such rearrangement is achieved through a coordinated series of large-scale cell movements. It begins with the inward movement/folding *(invagination/emboly)* of cells (endodermal cells), resulting in the formation of a small transverse cleft located midway between the equator and the V.P. As gastrulation progresses, the cleft enlarges and extends downward, eventually forming a crescent-shaped structure. Within this crescent, a small cavity, the *archenteron* (primitive gut), appears. The external opening created by this invagination is called the *blastopore*. This was followed by the activation and the *convergence* of notochordal cells located just above the dorsal lip of the blastopore, toward the blastopore and involute or gradually roll inward in an orderly manner *(involution)*, ultimately positioning themselves in the mid-dorsal region of the archenteron. Following involution, the notochord elongates from the posterior to the anterior end, forming a distinct row of cells situated below the neurectoderm and above the archenteron. Simultaneously, the presumptive mesodermal cells located on either side of the notochordal cells also converge toward the blastopore. Upon reaching the sides of the blastopore, these mesodermal cells involute spread away *(divergence)* from the blastopore to occupy their respective positions within the developing embryo. Toward the end of gastrulation, the pigmented cells (presumptive epidermal cells) of the animal hemisphere rapidly multiply and spread to grow over the larger, yolk-rich vegetal cells and the underlying neurectodermal cells *(Epiboly)*, forming the embryo's outer covering.

Gosner stage 10 (early gastrula/dorsal lip) This stage represents the beginning of gastrulation, indicated by the presence of the cleft or the dorsal lip of the blastopore (upper region of the cleft) (Fig. 2.5a, b). Gastrulation begins with the invagination, the inward movement of endodermal cells, which line the sides of the archenteron. At the site of this invagination, specialized bottle-shaped cells emerge. These cells arise through apical constriction, which leads to the formation of the initial slit-like blastopore. They play a pivotal role in initiating and directing the invagination process, contributing to the formation and deepening of the blastopore.

Gosner stage 11 (mid-gastrula) This stage represents the inward migration of notochordal cells, their arrangement along the mid-dorsal line of the archenteron, and the subsequent involution of the mesodermal cells and their positioning symmetrically on either side of the notochord (Fig. 2.5c–f). The archenteron steadily

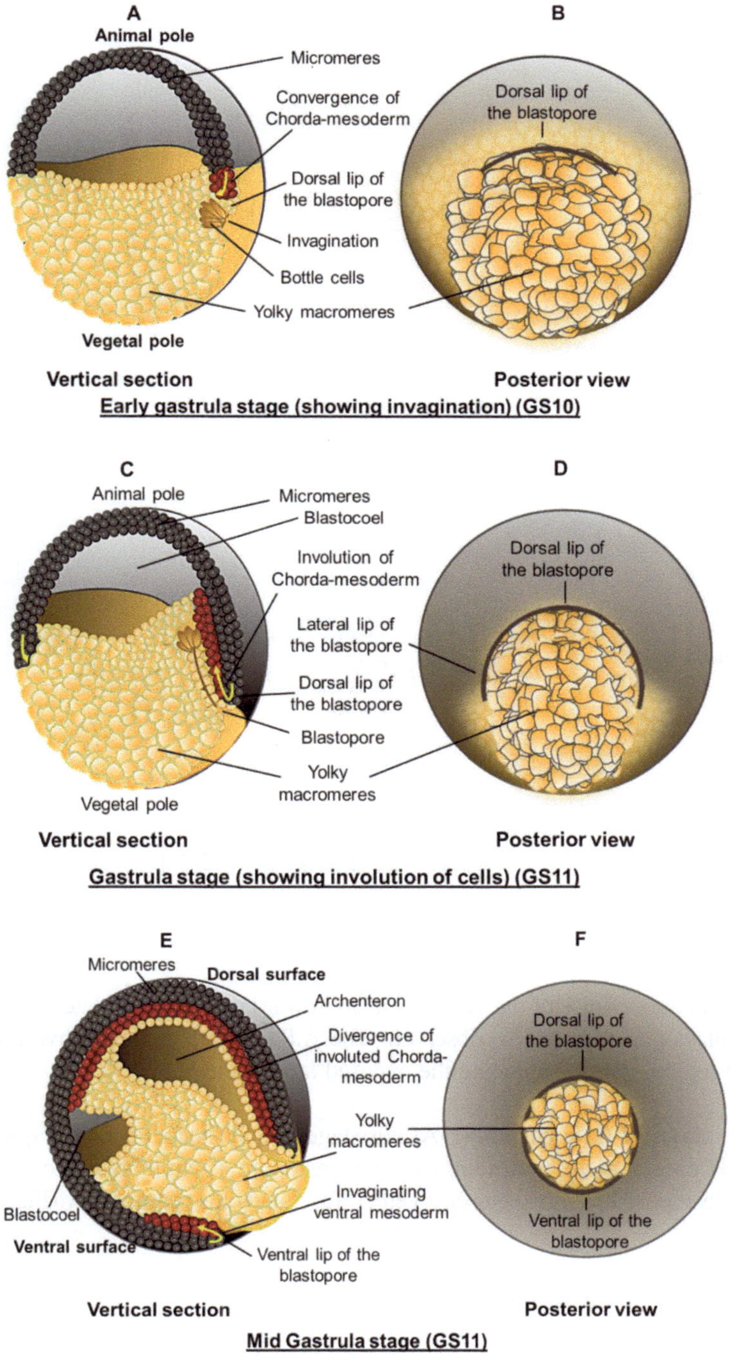

Fig. 2.5 Diagrammatic representation of the stages of gastrulation. (The diagrams were prepared, modified, and adapted based on information gathered from Balinsky (1960), Freeman and Bracegirdle (1967), Gilbert (2000), Gosner (1960), Holmes (1927), Rugh (1951), Zahn et al. (2022). Created in https://BioRender.com)

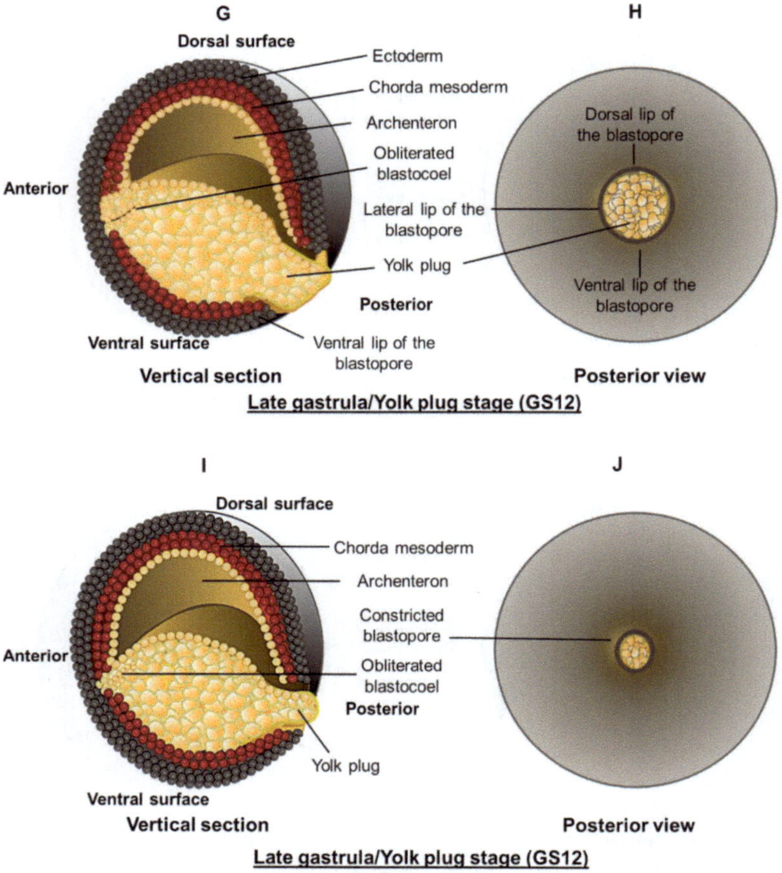

Fig. 2.5 (continued)

expands at the expense of the blastocoel. Downward movement or lateral movement of the blastopore lips is also observed. The A.P. surface gradually overgrows the vegetal region, thereby reducing the exposed area of the V.P.

Gosner stage 12 (late gastrula) At this stage, the blastocoel becomes obliterated. The yolk plug, initially protruding into the blastopore, is withdrawn into the interior as the blastopore lips move closer together, fuse, and close completely (Fig. 2.5g–j).

2.4.6 Neurula Stage of the Frog

Neurula is the stage in embryonic development in which the neural tube is formed by a process referred to as *neurulation*. It comprises broadly of subsequent phases: neural plate, neural fold, neural groove, and neural tube formation. It begins with the thickening of the ectoderm to form the neural plate. Neural plate cells elongate, become columnar, and align tightly. Simultaneously, the lengthening along the anterior-posterior axis and narrowing of the embryo occur to facilitate neural tube formation through bending. During this process, neuroepithelial cells at the edges of the neural plate undergo apical constriction, changing from a columnar to a wedge-shaped, pyramidal form. This shape change, caused by a reduction in apical surface area relative to the basal surface, facilitates the rolling and closure of the neural plate into the tube. Once the neural tube is formed, it detaches from the overlying surface ectoderm, which reestablishes a continuous outer epithelial layer.

Gosner stage 13 (neural plate) This stage is represented by the formation of the neural plate (Fig. 2.6a–d). It involves the thickening of the ectoderm along the mid-dorsal line and flattening of the thickened ectoderm to form a plate-like structure, with the margins of the plate being visible at the advanced neural plate stage. The neural plate is induced by signals from both the overlying ectoderm and underlying mesoderm during gastrulation. In *Xenopus*, the organizer region and notochord play essential roles in neural induction and establishing the embryo's anterior-posterior axis.

Gosner stage 14 (neural fold) At this stage, the lateral edges of the neural plate thicken and elevate above the surrounding surface, giving rise to *neural folds* (Fig. 2.6e–h).

Gosner stage 15 (neurular rotation or neural groove) As the neural folds continue to rise and curve inward, they form a *neural groove* along the midline of the embryo (Fig. 2.6e–h). The neural folds closed gradually along the dorsal midline. This stage is typically characterized by neurular rotation or active ciliary rotation of the embryo within its capsule according to Gosner staging; however, studies have reported the absence of this particular embryonal movement in certain frog species such as *Hyla arborea* (Sayim and Kaya 2008). Species of frog reported to have exhibited the ciliary rotation include *Lithobates pipiens* (Shumway 1940), *Polypedates teraiensis* (Tamuly and Dey 2014), *Hydrophylax leptoglossa* (Saha and Gupta 2011), *Hoplobatrachus rugulosus* (Traijitt et al. 2021), and others. The elongation of the posterior end of the embryo is observed at this stage.

Gosner stage 16 (neural tube) The neural folds converge and fuse at the dorsal midline, transforming the flat neural plate into a closed neural tube (Fig. 2.6i–l). The cavity of the neural tube is called the *neural canal* or *neurocoel*. The newly formed neural tube becomes regionalized, its anterior (cephalic) end broadens and shows distinct swellings and constrictions that will give rise to different regions of the brain, whereas the caudal (posterior) end remains narrow and simple, extending into the developing tail. This narrow end will form the spinal cord, which remains filled with cerebrospinal fluid in adults.

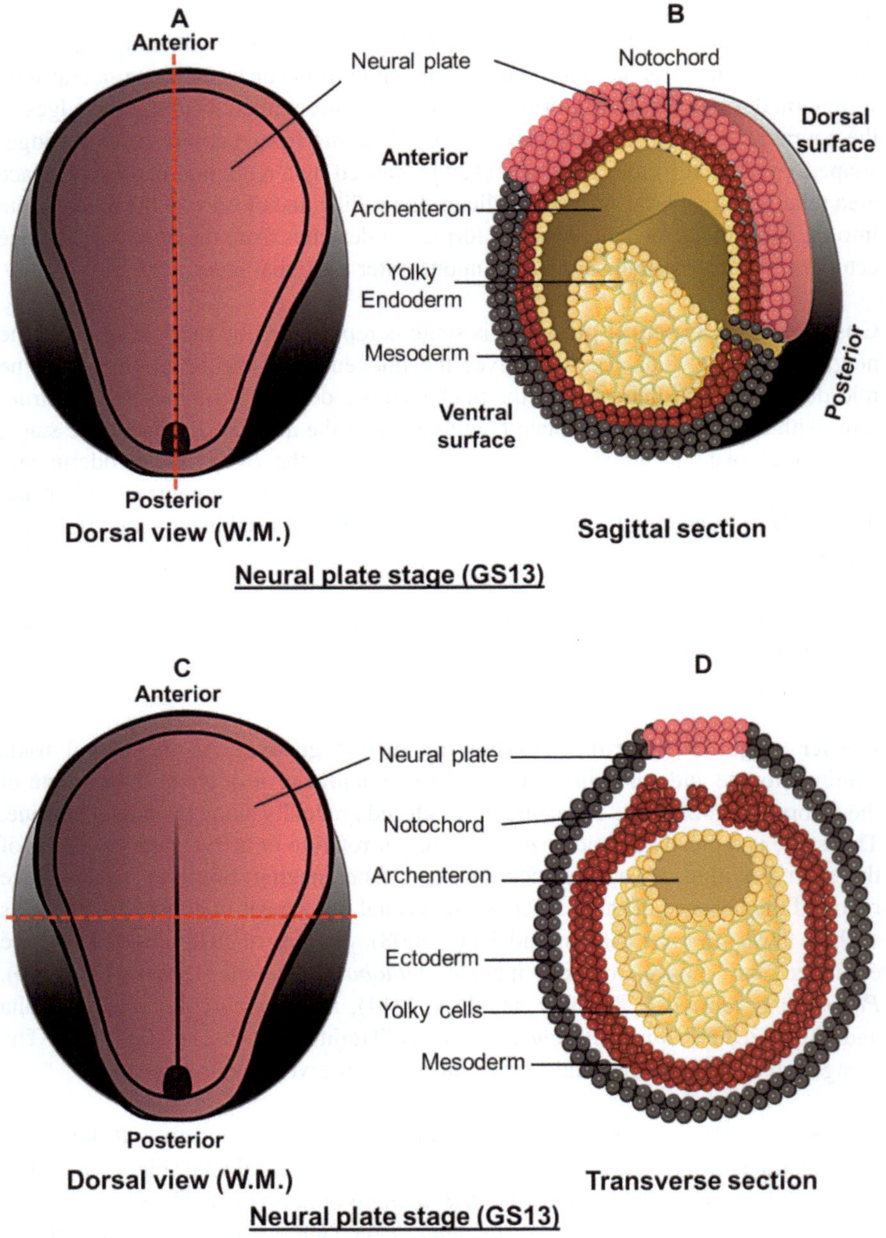

Fig. 2.6 Diagrammatic representation of the stages of Neurula. GS - Gosner stage. (The diagrams were prepared, modified, and adapted based on information gathered from Balinsky (1960), Freeman and Bracegirdle (1967), Gilbert (2000), Gosner (1960), Holmes (1927), Rugh (1951), Zahn et al. (2022). Created in https://BioRender.com)

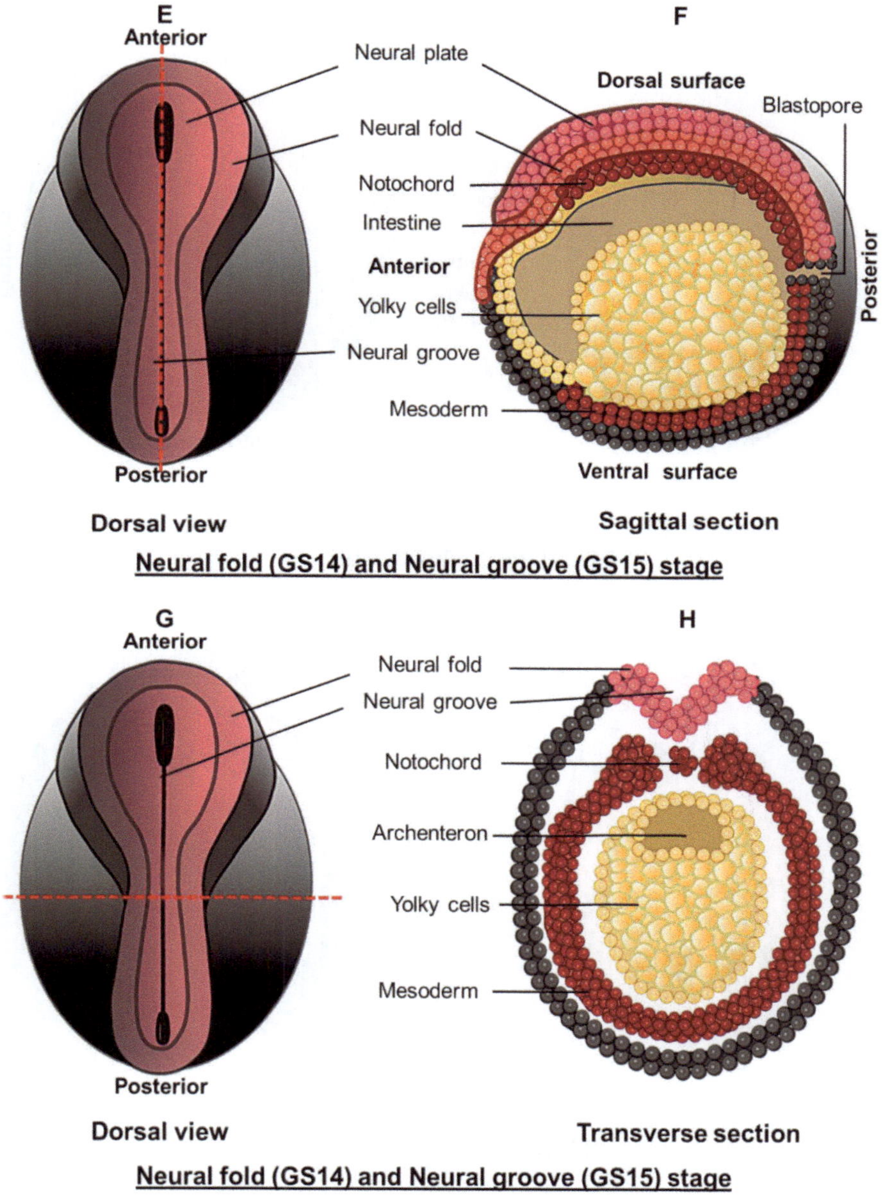

Fig. 2.6 (continued)

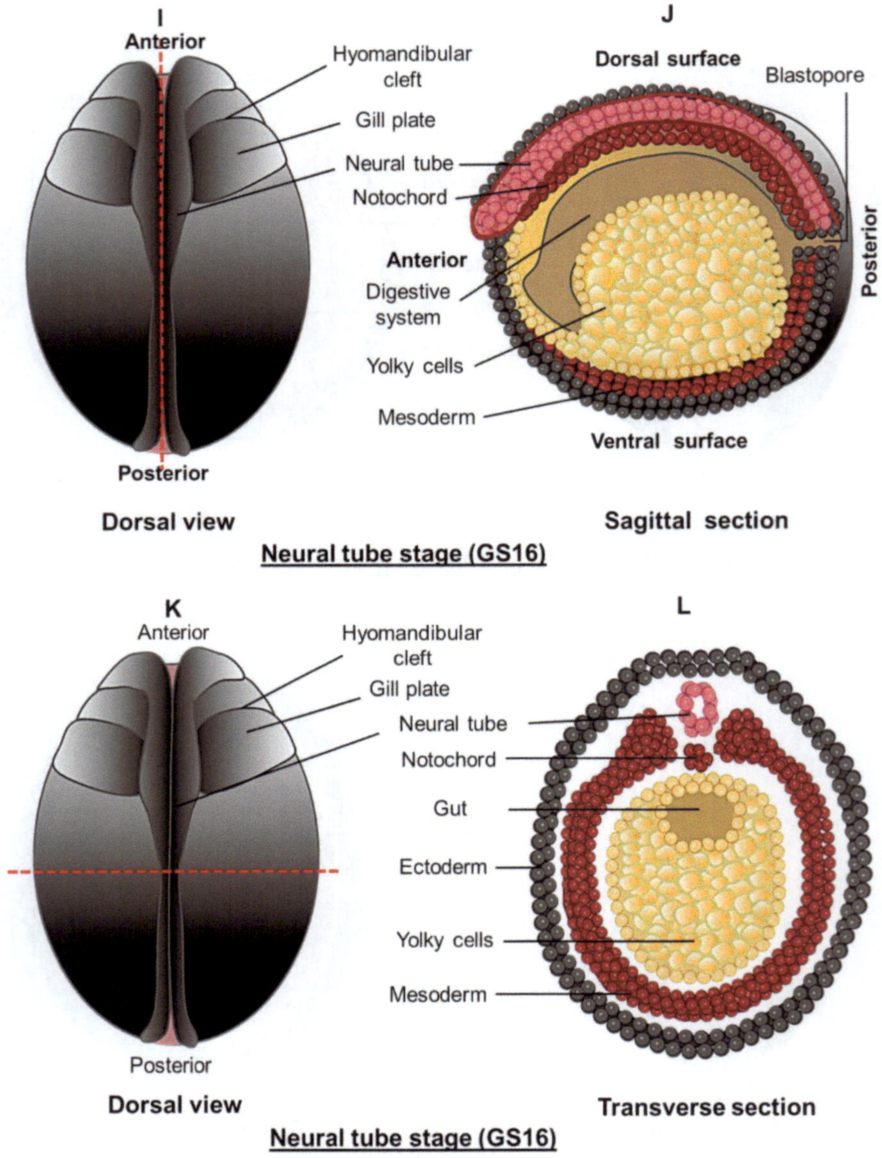

Fig. 2.6 (continued)

2.4.7 Tailbud Stages

The tailbud stage broadly represents the stage or period following the neurula stage until the transformation to a tadpole. This stage is characterized by the differentiation of the cell mass at the posterior end of the embryo into tail structures of the tadpole. Additionally, other sensory structures or responses develop at this stage. Arbitrarily it comprises GS17 to GS19.

Gosner stage 17 (tailbud stage) This stage is the typical tailbud stage (Fig. 2.7a), characterized by the development or presence of a mass of mesenchymal cells at the posterior end of the embryo. The shape or form of the embryo might differ in several species. For example, in the case of congeneric *Hyla* species, *H. arborea* and *H. japonica (*Iwasawa and Futagami 1992; Sayim and Kaya 2008*)*, *Microhyla ornata* (Narzary and Bordoloi 2013), the embryo at the tail bud is arched; however, in the case of *Rana pipiens* (Shumway 1940), *Lithobates pipiens* (Shumway 1940), *Mantidactylus betsileanus* (Scheld et al. 2013), *Polypedates teraiensis* (Tamuly and Dey 2014), *Hydrophylax leptoglossa* (Saha and Gupta 2011), *Hoplobatrachus rugulosus* (Traijitt et al. 2021), it is not arched. At this stage, the adhesive organs also begin to develop.

Development of adhesive organs (oral suckers) is generally observed between Gosner stages 17 and 21, showing variability both at the species and individual level. The oral sucker initially might be seen as a united crescent-shaped ridge, which might become bifid at full development. In some cases, such as *Scaphiopus*, the sucker remains U or Y-shaped. Post GS21, the oral suckers rapidly disappear, in certain cases, remain as scars, past stage 26, such as in *Scaphiopus*.

Gosner stage 18 This stage (Fig. 2.7b) comprises the beginning of simple flexures, such as spasmodic muscular responses or wriggling of the embryo. Other changes include the segregation and differentiation of the gill plate into ridges or the visceral arches, the appearance and presence of olfactory pits, the stomodeum, suckers, a rounded tail, and the thinning of the vitelline membrane.

Gosner stage 19 (gill buds) At this stage (Fig. 2.7c, d), there is the appearance of the gill buds, and the heartbeat pulsation about the gills is apparent.

2.4.8 Gill Stage Tadpole of the Frog

Gill stage in the frog development refers to the larval forms or tadpole stage, where breathing/respiration is achieved via the gills (external and internal gills). Some frog species, such as *Raorchestes longchuanensis*, have no external gill tadpole stage (Yan et al. 2021).

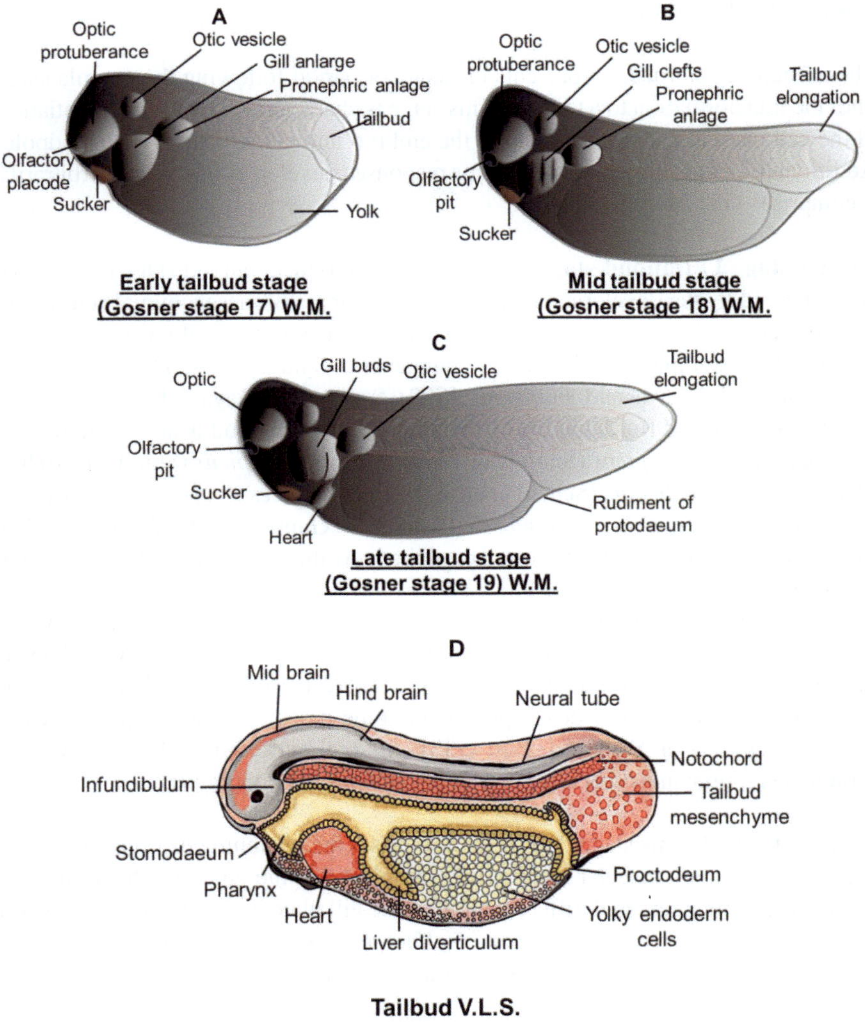

Fig. 2.7 Schematic representation of the tailbud stage of the frog. (The diagrams were prepared, modified, and adapted based on information gathered from Balinsky (1960), Freeman and Bracegirdle (1967), Gilbert (2000), Gosner (1960), Holmes (1927), Rugh (1951), Zahn et al. (2022). Created in https://BioRender.com and (Fig. 2.7d) further refined using Infinite Painter (Android application).)

2.4.8.1 External Gill Stages

The external gill stage period broadly comprises GS20 to GS22. The embryos are characterized by the presence of a head, trunk, tail, and pairs of external gills. The gills are crucial for respiration. Progressive development of external gills is generally seen through stages 19 to 20, though variability has been observed both individually and across species. Complete development generally occurs between stages 21 and 23. The external gills function for a short time and are replaced by internal gills. Four vertical slits (gill slits or branchial clefts or gill clefts) appear on either side of the head below the external gills. They put the pharynx in communication with the exterior.

A few days after hatching, the stomodaeum develops a connection with the pharynx, and the mouth is formed. The surrounding parts of the mouth form the jaws and lips. Horny papillae or teeth develop on the jaws and lips. The tadpole scrapes the weeds and starts feeding on algae and other aquatic vegetation. In some frog species such as *Limnonectes limborgi*, the larvae forms are non-feeding, retaining their yolk nutrient until stage 37 (Rowley and Altig 2012). The alimentary canal is long and coiled, which can be seen through translucent skin and muscles of the abdomen. The tail becomes longer and develops a tail fin. By undulating movements of its tail, the tadpole swims like a fish. Tadpole changes into the internal gill stage.

Gosner stage 20 This stage (Fig. 2.8a) is categorically characterized by the formation of the external gill filaments and initiation of the gill circulation. Additionally, the appearance of the optic vesicles, the elongation of the tail, and the commencement of hatching could be observed at this stage. The embryo, by rigorous wiggling, ruptures the surrounding gelatinous membrane and emerges from the membrane or hatches. Hatching of embryos in the majority of anuran species takes place between GS17 and GS20, showing variability in both intra- and interspecific terms. In the case of the congeneric *Hyla* species *H. arborea, H. japonica,* and *H. avivoca,* hatching occurs between GS20 and GS21 (Iwasawa and Futagami 1992; Sayim and Kaya 2008; Volpe et al. 1961). Similarly, in the case of *Hoplobatrachus rugulosus,* hatching takes place during stages 20–21 (Traijitt et al. 2021). In the case of *R. pipiens,* the hatching occurs at GS20 (Shumway 1940).

Gosner stage 21 At this stage (Fig. 2.8b), the external gills continue to grow, the eyes become distinguishable with a transparent cornea, the mouth begins to form, the tail fins remain opaque, and the embryo exhibits activity with spontaneous swimming movements. From this stage through GS25, the larva gradually transition to a free-swimming and actively feeding phase.

Gosner stage 22 At this stage (Fig. 2.8c), the external gills continue to grow, the tail fins become transparent with the onset of circulation, the mouth attains a rounded shape, and initial pigmentation appears along the dorsal side.

For the internal characterization of the frog embryo of the external gill stage, a longitudinal section of the frog embryo body and the transverse section (T.S.) through the otic region at the rhombencephalon level are represented in Fig. 2.8d, e.

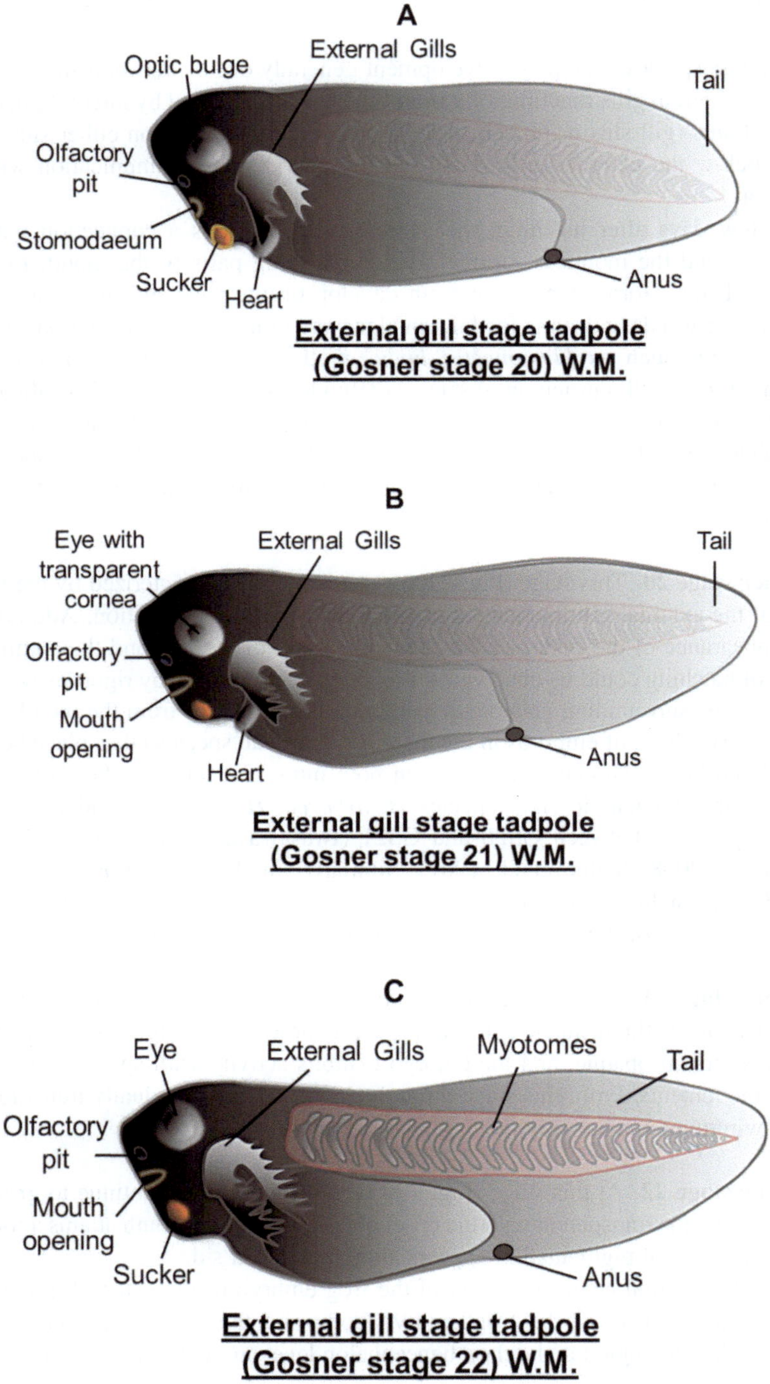

Fig. 2.8 Schematic representation of the external gill stage tadpole of the Frog. (The diagrams were prepared, modified, and adapted based on information gathered from Balinsky (1960), Freeman and Bracegirdle (1967), Gilbert (2000), Gosner (1960), Holmes (1927), Rugh (1951), Zahn et al. (2022). Created in https://BioRender.com and further refined (Fig. 2.8d, e,) using Infinite Painter (Android application).)

D

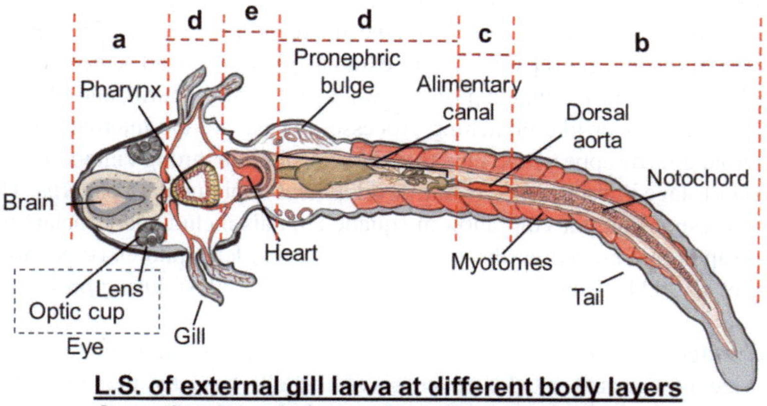

L.S. of external gill larva at different body layers from the dorsal view. a-brain/neural tube layer, **b**-notochord layer, **c**-dorsal aorta layer, **d**-intestine layer

E

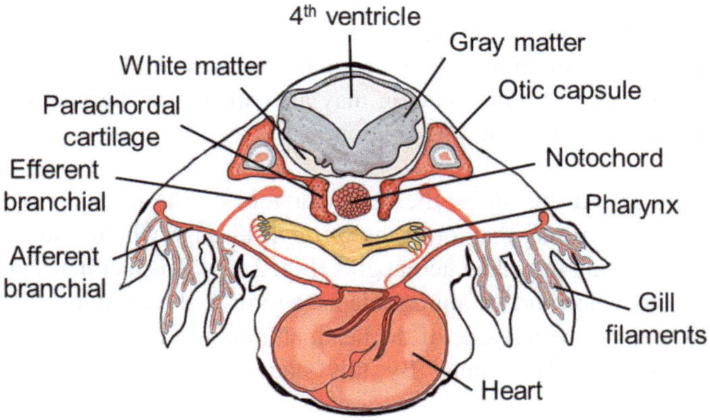

External gill larva through the otic region (T.S.) at the Rhombencephalon level

Fig. 2.8 (continued)

2.4.8.2 Internal Gill Stage

This stage is typically characterized by the disappearance of the external gills and the development of the operculum. This stage broadly comprises GS23 to GS25. The development of the operculum begins soon after the formation of branchial clefts, when a fold of skin starts growing from below the second visceral arch (the hyoid arch). The fold of the skin grows and covers the ventral and lateral surfaces of the head, external gills, and branchial clefts of both sides. The operculum finally fuses with the skin of the posterior part of the head except at one point. This opening is called the spiracle, seen at GS25. The external gills now start shrinking. As the external gills start shrinking, the posterior walls of the third, fourth, and fifth visceral arches produce finger-like branching processes. These are the internal gills. Soon, the external gills disappear, and the larva starts respiration through internal gills. The internal gill stage is also marked by the development of pigmentary patterns. The key pigments responsible for coloration in aquatic animals include dark melanophores, yellow xanthophores, red erythrophores, iridophores, leucophores or guanophores (Zhang et al. 2024). However, these colors are generally lost during preservation.

Gosner stage 23 The growth of external gills reaches its maximum at this stage. This stage is signified by the coverage of the gill base by the opercular fold (Fig. 2.9a). At this stage, the labia and teeth differentiate, the adhesive gland or sucker degenerates, and the gut develops, changes that are more evident on the ventral side. Certain pigmentation patterns are observed on the dorsal surface of the larvae.

Gosner stage 24 At this stage (Fig. 2.9b), the external gills begin to atrophy, and the opercular fold on the right side of the embryo closes, while the external gills on the left remain exposed. Pigmentation may also appear on the ventral side of the larvae, particularly around the belly region.

Gosner stage 25 At this stage, the opercular fold on the left side of the embryo closes, leaving an opening that forms the *spiracle*, through which water exits the gill chamber after passing over the internal gills (Fig. 2.9c). This facilitates gas exchange until the lungs develop and the animal adapts to terrestrial respiration. The spiracle is typically located on the left side of the head, although in microhylids it appears on the ventral side near the anus. By this stage, the mouthparts are also well distinguished.

For the internal characterization of the frog embryo of the internal gill stage, a transverse section (T.S.) through the gill region at the rhombencephalon level is represented in Fig. 2.9d.

For greater accuracy and realistic representation of frog embryonic development, photographs of *Lepidobatrachus laevis* embryos at various developmental stages are provided in Figs. 2.10, 2.11, and 2.12.

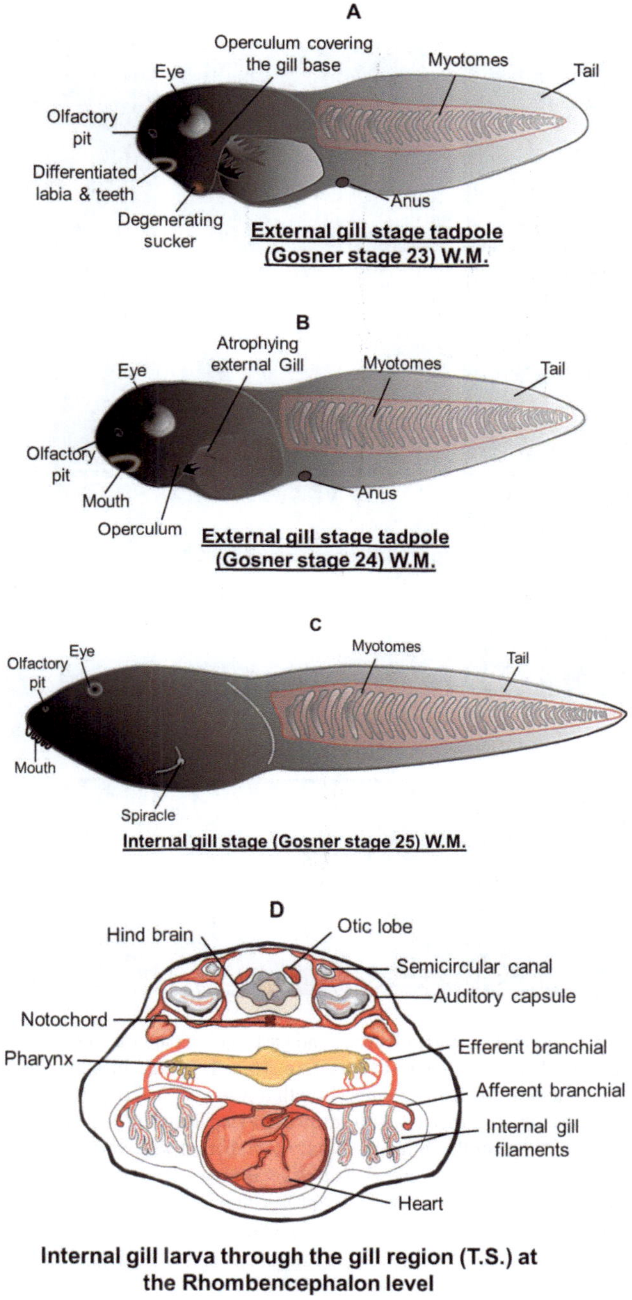

Fig. 2.9 Schematic representation of the internal gill stage tadpole of the Frog. (The diagrams were prepared, modified, and adapted based on information gathered from Balinsky (1960), Freeman and Bracegirdle (1967), Gilbert (2000), Gosner (1960), Holmes (1927), Rugh (1951), Zahn et al. (2022). Created in https://BioRender.com)

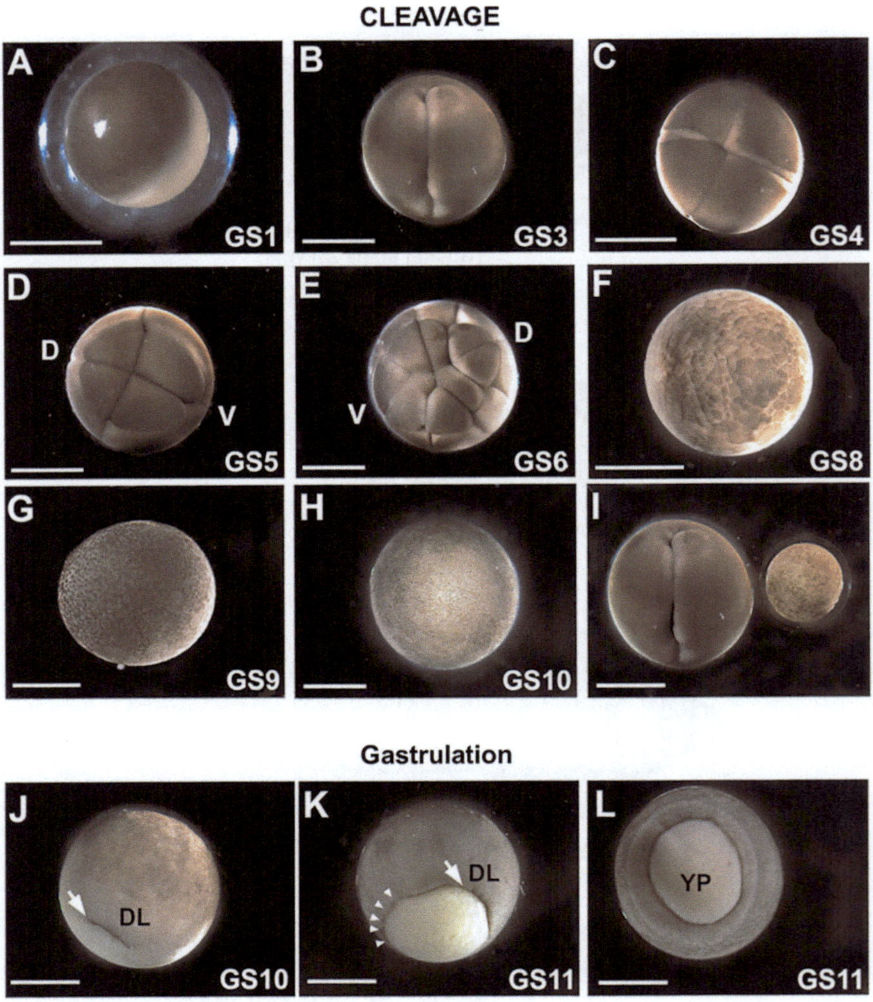

Fig. 2.10 Photomicrographs of early cleavage and gastrulation patterns in *Lepidobatrachus laevis*. (**a**) Fertilized egg; (**b**) 2-cell; (**c**) 4-cell; (**d**) 8-cell; (**e**) 16-cell; (**f**) mid-blastula; (**g**) late blastula; (**h**) early gastrula; (**i**) *L. laevis* 2-cell stage embryo with a *Xenopus laevis* gastrula; (**j**) Early gastrula with dorsal lip of blastopore indicated with arrow. (**k**) Expansion of the dorsal lip (arrowheads) circumferentially; (**l**) Mid gastrula with still slightly protuberant yolk plug. (**a**–**i**) are animal views; (**j**) and (**k**) are latero-vegetal views; and (**l**) is a vegetal view. Scale bar - 1 mm. D - prospective dorsal side of the embryo; GS - Gosner stage, V - prospective ventral side of the embryo; DL - dorsal lip. (Figure adapted from (N. M. Amin et al. 2015) with permission)

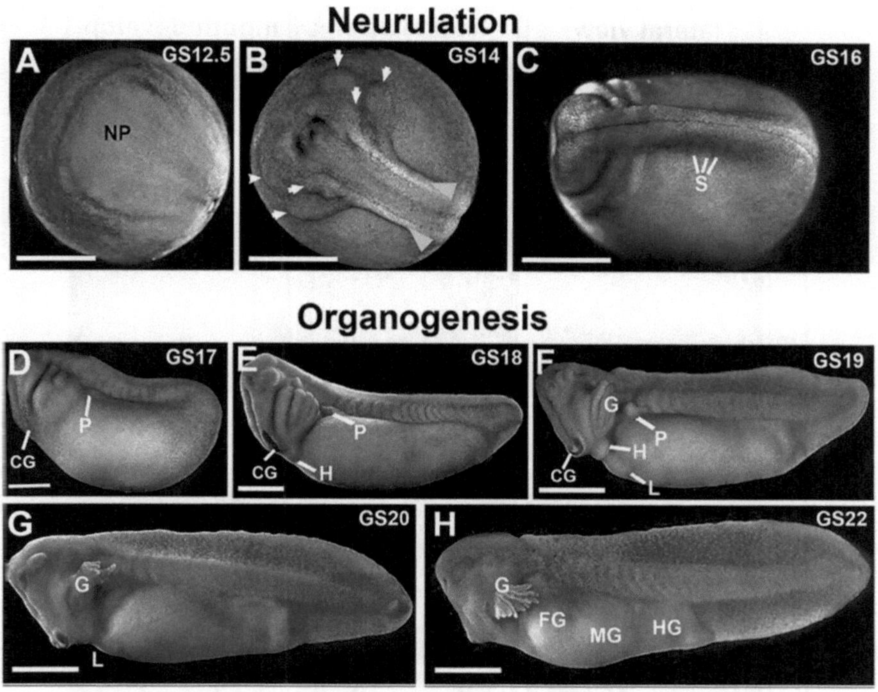

Fig. 2.11 Photomicrographs showing the neurulation and organogenesis processes in *Lepidobatrachus laevis*. (**a**) Neural plate; (**b**) neural folds (yellow arrowheads) and migrating neural crest (arrows); (**c**) Neural tube closure and appearance of somites; (**d**–**f**) Earliest tailbud stages; (**g**, **h**) Later tailbud stages. NP - neural plate; H - heart; L - liver; G - gills; FG - foregut; MG - midgut; HG - hindgut, CG - cement gland, P - pronephros, and S - somites. Scale bar - 1 mm. (Figure adapted from N. M. Amin et al. (2015) with permission)

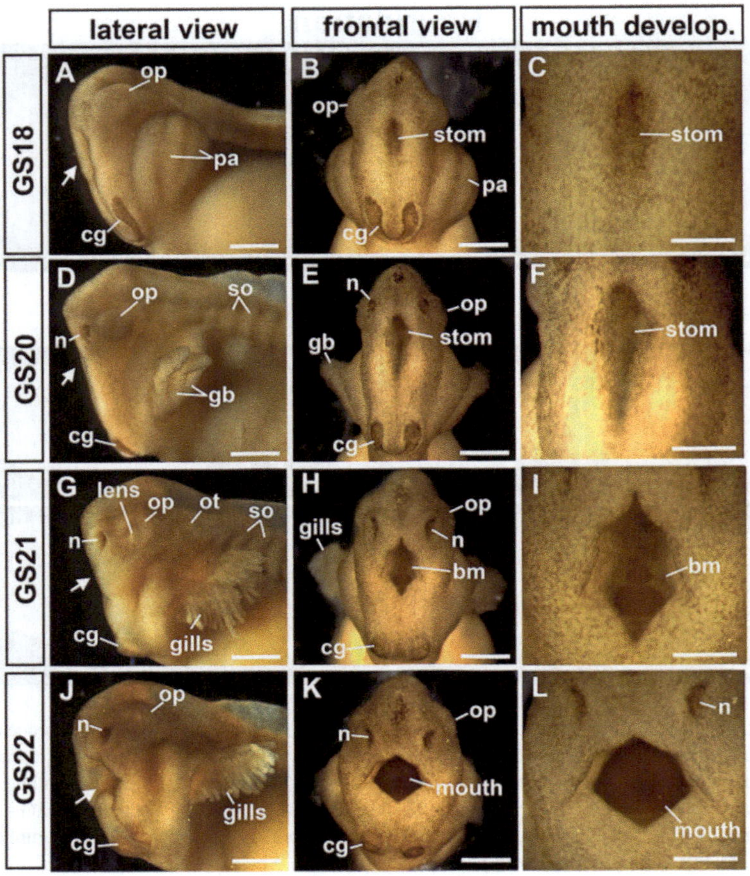

Fig. 2.12 Photomicrographs showing craniofacial morphogenesis of *Lepidobatrachus laevis*. (**a**, **d**, **g**, **j**) Embryonic head in lateral view (arrows indicate stomodaeum/mouth), Scale bar - 500 μm; (**b**, **e**, **h**, **k**) frontal views, Scale bar - 500 μm; (**c**, **f**, **i**, **l**) Development of mouth including buccopharyngeal membrane perforation, Scale bar - 250 μm. GS - Gosner stage, gb - gill buds, n - nasal pits, op - optic cup, ot - otic vesicle, pa - pharyngeal arches, so - somites. (Figure adapted from N. M. Amin et al. (2015) with permission)

2.4.9 Metamorphosis of the Frog Tadpole

The transformation of a tadpole into an adult frog occurs through GS26–GS46. Around stage 26 (Fig. 2.13), the hind limbs begin to emerge and are fully formed by stage 40. Forelimb buds appear at stage 42, coinciding with the tadpole reaching its maximum length. Subsequently, tail resorption begins, and metamorphosis is completed by stage 46. This phase represents a remarkable transition from a distinct aquatic, gill-breathing, herbivorous larva (tadpole) to a terrestrial, lung-breathing, carnivorous adult frog.

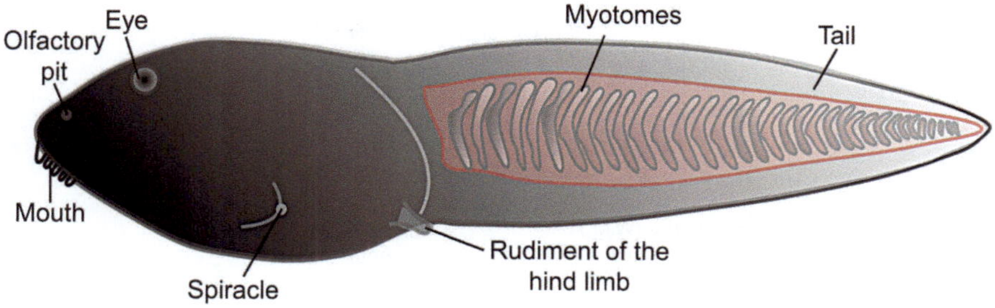

Fig. 2.13 Schematic representation of the GS26 tadpole of the Frog. (The diagrams were prepared, modified, and adapted based on information gathered from Balinsky (1960), Gilbert (2000), Gosner (1960), Holmes (1927), Rugh (1951), Zahn et al. (2022). Created in https://BioRender.com)

2.4.10 Embryonic Development Period

The embryonic development period, beginning from fertilization to GS25, varies significantly, and the physical factors such as temperature appear to play a significant role. The congeneric *Hyla* species, *H. arborea*, *H. japonica*, and *H. avivoca* have a developmental period of 8.8 days at 20 ± 1 °C, 4.4 days at 24 ± 1 °C, and 4.7 days at 28.5–31.5 °C, respectively (Iwasawa and Futagami 1992; Sayim and Kaya 2008; Volpe et al. 1961). In the case of congeneric *Rana* species *R. leptoglossa* and *R. pipiens*, the developmental period lasted for about 26–27 days and 11.8 days at ~25–30 °C and 18 °C, respectively (Shumway 1940; Singh 2015). In the case of *Bufo valliceps* raised at 25 °C, the incubation period lasted approximately 115 hours (~4.8 days) (McDiarmid and Altig 1999).

2.5 Permanent Slides: Whole Mounts and Sections of Frog Embryology

Permanent slides of frog embryos (Fig. 2.14a–z, aa–ax), prepared as whole mounts (W.M.) and sections, serve as long-term teaching aids for laboratory studies and are invaluable for comprehending the dynamic process of development. Whole mounts allow observation of the entire embryo without sectioning, typically at early developmental stages when the embryo is small and sufficiently transparent for light to pass through, thereby highlighting external morphological changes. As development progresses, the embryo becomes opaque and structurally complex, necessitating the use of histological sections to study internal tissue organization, differentiation, and organ formation. Together, whole mounts and sections provide a comprehensive visual representation of amphibian embryogenesis.

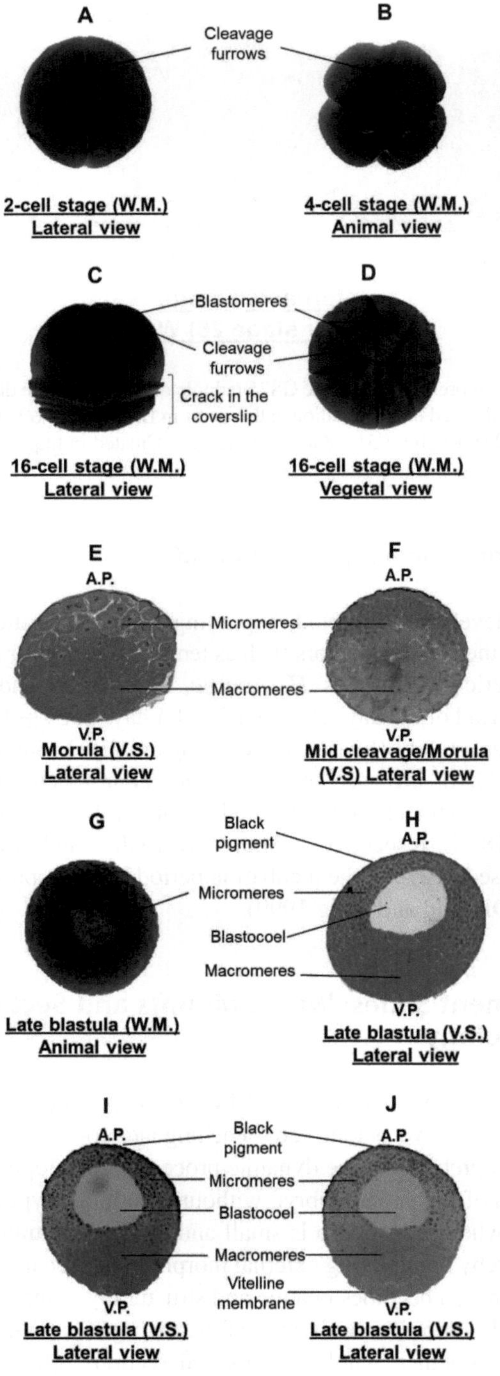

Fig. 2.14 Images of the permanent slides of the whole mount and histological sections of frog developmental stages. A.P. - Animal pole, V.P. - Vegetal pole, W.M. - Whole mount, V. S. - Vertical section, T.S. - Transverse section. (Source: Department of Zoology, Daulat Ram College, University of Delhi)

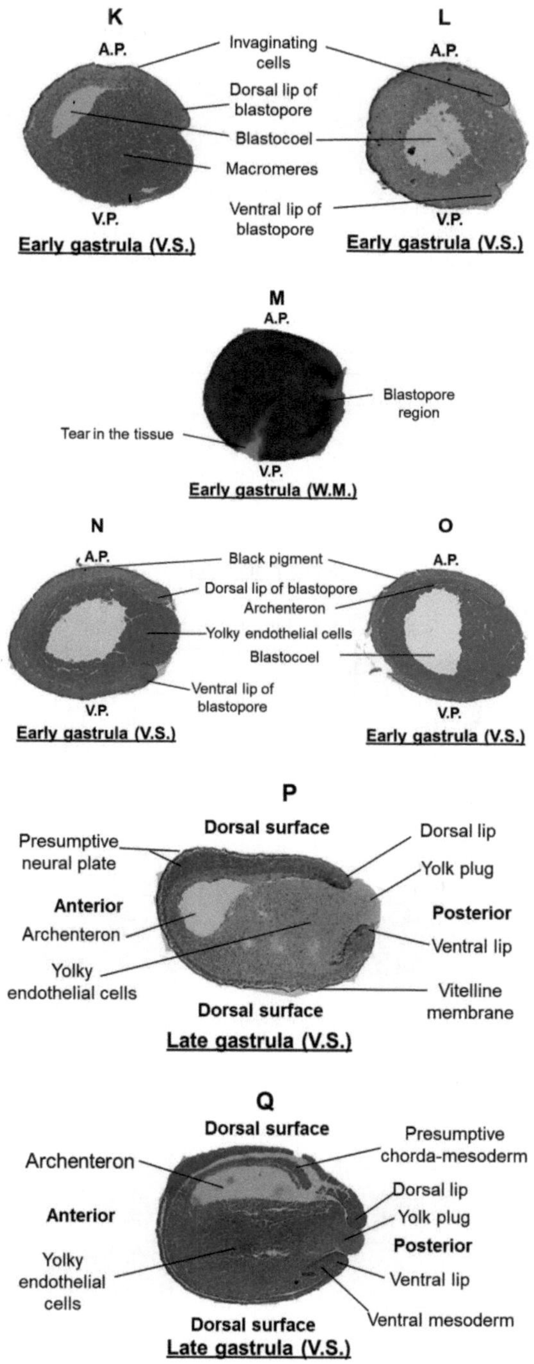

Fig. 2.14 (continued)

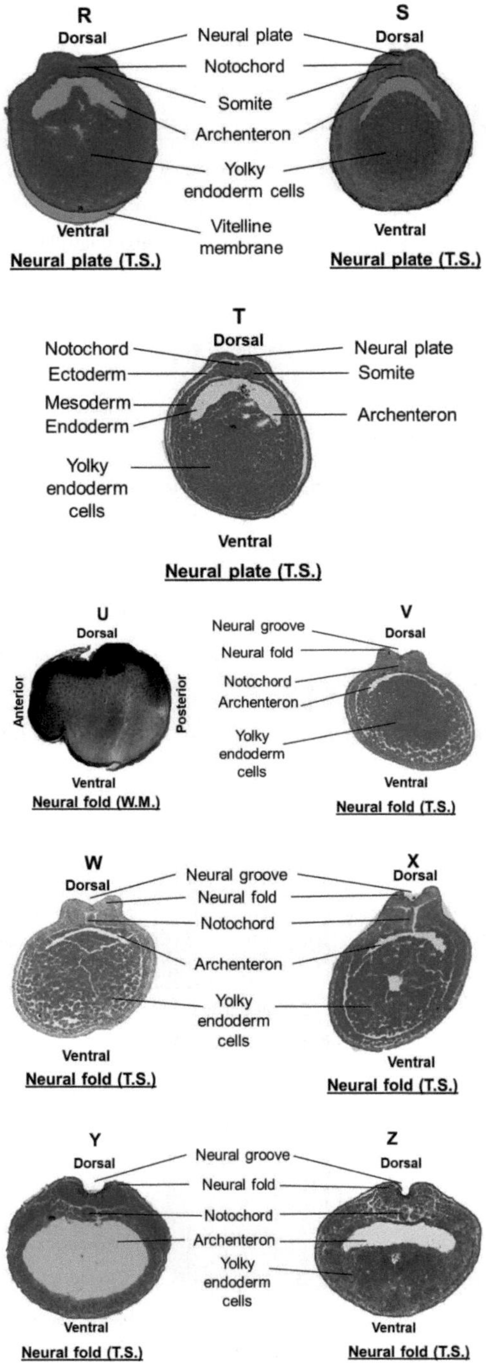

Fig. 2.14 (continued)

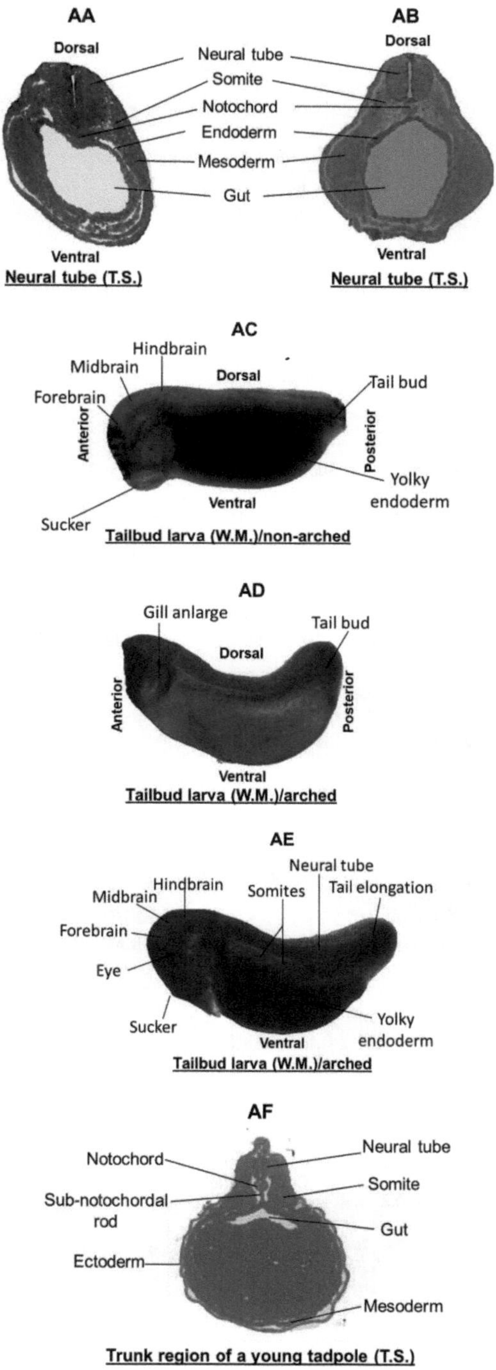

Fig. 2.14 (continued)

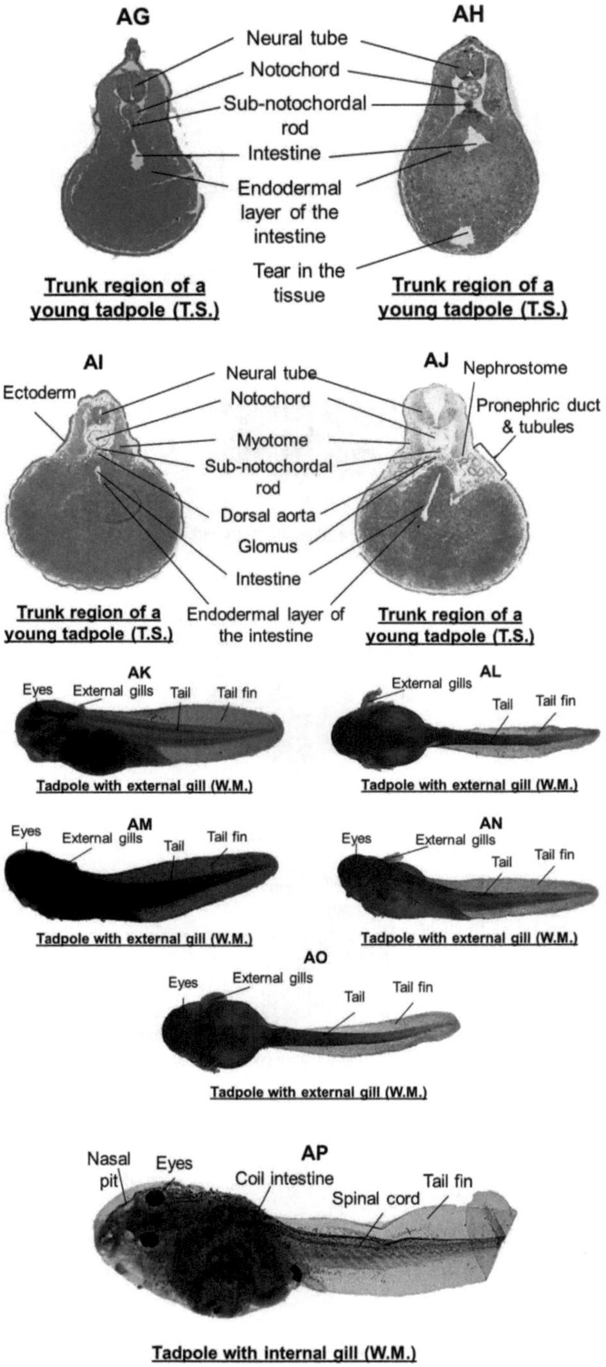

Fig. 2.14 (continued)

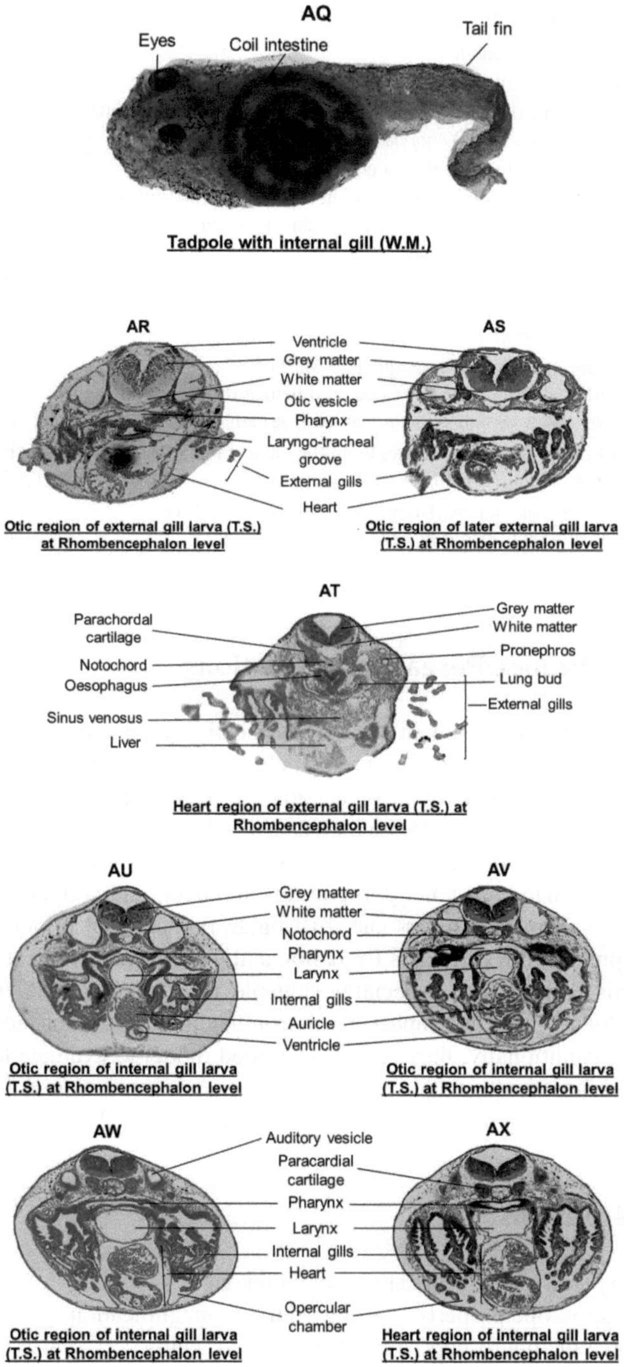

Fig. 2.14 (continued)

Table 2.1 Representative studies highlighting the significance of frog embryos

S. no.	Type of study/research	Reference
1.	Investigation of the impact of ubiquitous chemicals, at concentrations detected in human amniotic fluid, on embryonic brain development in *Xenopus* embryo	Fini et al. (2017)
2.	Evaluation of F0 CRISPR as a cost-effective, efficient gene-editing tool for modeling human congenital malformations or disease phenocopies, and elucidating gene developmental roles in *Xenopus tropicalis*	Bhattacharya et al. (2015)
3.	Exploring scar-free healing or tissue regeneration mechanisms to develop novel therapies enhancing healing and regenerative potential in human patients in *Xenopus* tadpole	Chen et al. (2014)
4.	Assessing the potential teratogenic hazard of 2-acetylaminofluorene, rifampicin, and benzo[a]pyrene using *Xenopus laevis* (South African clawed frog) embryos	Fort et al. (1989)
5.	Impact of acetamiprid insecticide on the environment or other non-target aquatic organisms in *Xenopus laevis*	Kerner et al. (2023)
6.	*Lepidobatrachus laevis* (Budgett's frog) as a model for morphogenesis research or developmental biology	Nirav M Amin et al. (2015)

2.6 Significance/Research Applications

The frog embryonic development holds immense biological and research value. The most significant features of frog embryos include their large size, external development, rapid development, and transparent embryos (which enable direct observation and manipulation). Their developmental stages and core processes of development closely parallel those of higher vertebrates, making this study foundational for understanding human embryology, organ development, and congenital anomalies. Through this system, researchers can investigate fundamental principles of vertebrate development from early axis formation and germ layer specification to organogenesis. This makes them especially valuable for studying gene function, cell signaling pathways, and the impact of environmental or chemical factors on early development. Additionally, they served as a good model to evaluate the impact of pesticides and insecticides on the environment or other organisms. A few representative studies involving frog embryology are listed in Table 2.1.

2.7 Precautions

- Handle prepared slides with care to avoid damage.
- Use the microscope properly, starting with low magnification.
- Do not force the microscope focus knob, adjust gently.
- Record all observations neatly and clearly.

2.8 Quick Review Questions

2.8.1 Multiple Choice Questions

1. What type of cleavage is observed in frog embryos?
   (a) Meroblastic and superficial
   (b) Holoblastic and equal up to the morula stage
   (c) Holoblastic and unequal from the second cleavage onward
   (d) Discoidal and meroblastic
   Answer: c) Holoblastic and unequal from the second cleavage onward
1. What is the fate of the blastopore in frog development?
   (a) It develops into the mouth
   (b) It disappears completely
   (c) It develops into the anus
   (d) It forms the notochord
   Answer: c) It develops into the anus
2. Define the blastocoel.
   (a) The opening of the archenteron
   (b) A solid mass of cells formed after cleavage
   (c) A fluid-filled cavity within the blastula
   (d) The space between the yolk and the embryo
   Answer: c) A fluid-filled cavity within the blastula
3. Significance of the gray crescent in symmetry formation.
   (a) It determines the future anterior end of the embryo
   (b) It marks the site of sperm entry
   (c) It establishes radial symmetry in the embryo
   (d) It indicates the future dorsal side and helps establish bilateral symmetry
   Answer: d) It indicates the future dorsal side and helps establish bilateral symmetry
4. The process of formation of the three germ layers during frog embryonic development is called:
   (a) Cleavage
   (b) Gastrulation
   (c) Organogenesis
   (d) Neurulation
   Answer: b) Gastrulation
5. The most prominent feature of a frog embryo at the neurula stage is:
   (a) Formation of gill arches
   (b) Appearance of the neural tube
   (c) Closure of the blastopore
   (d) Formation of limb buds
   Answer: b) Appearance of the neural tube
6. In which hemisphere is the yolk in a frog egg predominantly concentrated?
   (a) Animal hemisphere
   (b) Vegetal hemisphere

   (c) Both equally
   (d) None of the above
   Answer: b) Vegetal hemisphere
7. In the late gastrula stage of the frog, which structure indicates the future digestive tract?
   (a) Notochord
   (b) Neural groove
   (c) Archenteron
   (d) Blastocoel
   Answer: c) Archenteron
8. In the early stages of practical embryology, whole mounts are preferred because:
   (a) Sections are not feasible in early embryos
   (b) Early embryos are small and transparent
   (c) Whole mounts give a 3D view of development
   (d) All of the above
   Answer: d) All of the above
9. The purpose of studying sections of frog embryonic development stages is to:
   (a) Observe pigmentation only
   (b) Examine internal organ formation
   (c) Identify the blastopore position
   (d) Avoid using whole mounts
   Answer: b) Examine internal organ formation

2.8.2 Short Answer Questions

1. Distinguish between a fertilized and an unfertilized frog egg.
2. How does the yolk distribution affect the cleavage pattern and blastomere size?
3. Define the blastula stage and name its main cavity.
4. Describe the difference between micromeres and macromeres.
5. Which germ layers are formed during the gastrula stage?
6. Why are frog embryos preferred for studying vertebrate development?
7. Differentiate between the animal and vegetal poles of a frog egg.
8. Describe the neurula stage and its key visible features in a whole mount.
9. Explain the role of the notochord in embryonic development.
10. List the key external features of a frog tadpole visible in a whole mount.

Glossary

Afferent branchial The blood vessels that carry deoxygenated blood from the heart to the gills for oxygenation.

Albumen Refers to the gelatinous, protein-rich layer around the egg that protects against damage, prevents desiccation, cushions from stress, and aids species-specific sperm attraction and activation during aquatic fertilization.

Amplexus Reproductive behavior in which the male clasps tightly around the back or behind the forelimbs of the female with forelimbs, to facilitate external fertilization.

Animal hemisphere It is the pigmented, yolk-poor upper half of the egg where cell division is faster during cleavage and gives rise mainly to the ectoderm and parts of the mesoderm.

Archenteron The primitive gut cavity formed during gastrulation, lined by endoderm, opening externally via the blastopore, and later develops into the digestive tract.

Auditory capsule A cartilaginous structure that surrounds and protects the inner ear.

Black pigment Black pigment in frog eggs consists of melanin granules concentrated in the animal hemisphere, providing UV protection and aiding thermal regulation for proper embryonic development.

Blastocoel A fluid-filled cavity formed within the blastula during early embryonic development.

Blastomeres The cells resulting from the cleavage division.

Blastopore The gastrulation opening where cells invaginate to form the archenteron; it later develops into the anus.

Blastulation A process of blastula formation.

Bilateral symmetry Body symmetry in which an organism can be divided into two equal and mirror-image halves along a single plane, typically the sagittal plane, dividing the body into left and right sides.

Chorda-mesoderm A group of mesodermal cells that migrate during gastrulation to form the notochord, a rod-like, supportive structure along the embryo's dorsal midline.

Cornea In a frog embryo cornea is the transparent, protective outer layer covering the developing eye, essential for light transmission.

Dorsal aorta A paired vessel along the dorsal side that distributes oxygen and nutrients during development.

Dorsal lip of the blastopore The upper, curved edge of the blastopore that forms during early gastrulation.

Ectoderm The outermost germ layer formed during gastrulation, giving rise to the skin, nervous system, and sensory organs.

Efferent branchial The blood vessels that carry oxygenated blood away from the gills (branchial arches) after gas exchange and deliver it to the systemic circulation.

Endoderm The innermost germ layer in embryos that forms the digestive tract lining, respiratory, and other internal organs.

External gill Feathery, vascularized, temporary respiratory projections located outside the body that facilitate aquatic respiration in early embryo development.

Germinal epithelium The surface layer of the ovary that contributes to egg cell (oogonia) formation and follicle development.

Gill anlage The early embryonic primordia or rudiments of the gills.

Gill clefts The external grooves between branchial arches that connect the pharyngeal pouches internally to the external environment, enabling water flow over the developing gills for respiration.

Gill filaments Gill filaments are vascularized, thin, thread-like projections arising from the branchial arches that maximize respiratory gas exchange in frog larvae.

Gill plate Flattened, sheet-like supportive structures that hold and protect gill filaments in frog larvae on each branchial arch.

Glomus A developing cluster of capillaries within the pronephros (early kidney) that functions as a filtration unit in the developing frog kidney.

Gray matter Regions of the developing nervous system (brain and spinal cord) composed mainly of neuronal cell bodies, dendrites, and unmyelinated fibers.

Gray crescent Light-colored, crescent-shaped region of a fertilized frog egg, opposite the site of sperm entry, which marks the future dorsal side of the embryo and plays a key role in body axis formation.

Hyomandibular cleft A groove or slit located between the hyomandibular and mandibular branchial arches. It forms part of the pharyngeal clefts (gill clefts).

Infundibulum Funnel-like structure linking the brain and pituitary gland in the developing embryo.

Internal gill Respiratory structures located within the branchial (pharyngeal) chambers, enabling aquatic respiration in frog larvae before lung development.

Labia Paired folds of tissue that form the edges of the developing oral cavity, helping shape the mouth opening.

Laryngo-tracheal groove A longitudinal groove that appears on the ventral side of the foregut during development, marking the beginning of the larynx and trachea formation.

Lateral lip of the blastopore The side edges of the blastopore opening during gastrulation.

Lung bud An early outgrowth from the ventral wall of the foregut, marking the initial stage of lung development.

Mesoderm The middle germ layer formed during gastrulation. It gives rise to muscles, bones, the circulatory system, kidneys, and reproductive organs.

Mesolecithal egg Egg with a moderate amount of yolk concentration.

Myotome Segment of paraxial mesoderm in the embryo that gives rise to the skeletal muscles of the body.

Nasal pit A small depression on the developing head that marks the beginning of the nasal cavity, formed from the invagination of the ectoderm in the frontonasal region.

Nephrostome A ciliated funnel-like opening connecting the coelomic cavity to the pronephric tubules (early kidney structures).

Neural fold A raised ridge of ectodermal tissue on either side of the neural groove during early neurulation. The neural fold eventually fuses to form the neural tube, which develops into the central nervous system (brain and spinal cord).

Neural groove A longitudinal depression along the midline of the dorsal ectoderm during early neurulation forms between the raised neural folds.

Neural plate A thickened, flat region of ectoderm on the dorsal side of the embryo during early neurulation. The neural plate will fold to create the neural groove and neural folds, eventually forming the neural tube.

Neural tube A hollow, tubular structure formed by the folding and fusion of the neural folds during neurulation.

Neurectoderm The specialized region of ectoderm that gives rise to the nervous system. It forms the neural plate during early development.

Notochord A flexible, rod-shaped structure from the chorda-mesoderm that provides early structural support and signals neural development. The vertebral column later replaces it.

Olfactory pit Same as the nasal pit.

Olfactory placode A thickened area of ectoderm located on the head region, the precursor to the olfactory pit and nasal cavity.

Operculum A thin, flap-like structure that develops over the gill region. It covers and protects the internal gills or branchial arches.

Oral sucker A specialized, glandular structure located on the ventral surface of the head near the mouth that secretes a sticky substance that helps the early tadpole temporarily attach or anchor to surfaces in the aquatic environment.

Optic protuberance A bulging structure on the developing head that will develop into the optic cup and eventually the retina and other eye components.

Otic placode A localized ectodermal thickening forming the otic vesicle.

Otic vesicle A fluid-filled sac formed by the invagination of the otic placode near the hindbrain that will form the inner ear.

Ovarian follicles Ovarian follicles in frogs are spherical structures in multiple stages within the ovary containing a developing oocyte surrounded by one or more layers of follicular cells. These follicles provide nourishment, mechanical protection, and hormonal support for the growth and maturation of the oocyte.

Parachordal cartilage A pair of cartilaginous rods located on either side of the notochord at the base of the developing brain. Provide structural support to the brain and notochord region.

Polyspermy A condition in which an egg is fertilized by more than one sperm.

Pronephric anlage The early developmental mass of cells that gives rise to the pronephros, the embryo's first functional kidney.

Pronephric duct and tubules Components of the early excretory system (pronephros).

Pronephros The first, simplest form of the kidney that develops during early embryogenesis.

Pronucleus The haploid nucleus of the sperm or egg before they fuse during fertilization.

Radially symmetrical Type of body symmetry in which body parts are arranged around a central axis, such that any longitudinal plane passing through the central axis divides the organism into roughly equal halves.

Rhombencephalon The hindbrain region of the developing brain.

Rudiment of proctodeum The early embryonic structure that gives rise to the mouth opening and the anterior part of the digestive tract.

Semicircular canal Looped, tubular structures that develop from the otic vesicle, form part of the inner ear, aiding in maintaining balance.

Sinus venosus A thin-walled, posterior heart chamber that collects deoxygenated blood from veins.

Somite A paired block of mesodermal tissue that forms along both sides of the neural tube during segmentation, which forms the vertebral column and associated muscles.

Spemann-Mangold organizer Found on the dorsal lip of the blastopore is a critical region in the early amphibian embryo that controls the formation of the body axis and organizes the development of surrounding embryonic tissues.

Spinal cord The dorsal tubular structure formed from the neural tube.

Spiracle A small opening located on the first branchial (gill) arch region. It serves as a water exit pathway from the gill chamber.

Sub-notochordal rod A rod-like structure of mesodermal cells located just beneath the notochord supports embryonic tissue organization.

Tail fin A thin, membranous extension of the body that surrounds and supports the tail and helps frog tadpoles swim effectively.

Tailbud The posterior swelling during late neurulation, containing proliferating cells that will form tail structures like muscles, connective tissues, and other structures in the developing embryo.

Tailbud mesenchyme A loosely organized group of highly proliferative and migratory mesodermal cells located within the tailbud region. Play a key role in tail elongation and morphogenesis.

Telolecithal egg Egg in which the yolk is more concentrated towards the V.P.

Vegetal hemisphere The lower half of the egg or embryo, opposite the animal hemisphere, consisting of larger, yolk-rich cells or nutrient-providing region.

Ventral lip of the blastopore The lower edge of the blastopore opening during gastrulation.

Ventricle A fluid-filled cavity located in the hindbrain (rhombencephalon) region forms from the neural tube lumen.

Vitelline membrane A thin, transparent, non-cellular membrane that closely surrounds the egg cell and yolk. Regulates sperm entry and safeguards the developing embryo.

White matter The portion of the central nervous system consisting primarily of myelinated axons. It facilitates fast communication between different brain and spinal cord regions as neural connections mature.

Yolk plug A mass of yolk-rich endodermal cells visible at the blastopore during late gastrulation, gradually internalized as gastrulation completes.

Zygote nucleus Diploid nucleus formed after the fusion of the male and female pronuclei during fertilization, marking the beginning of embryonic development.

References

Amin NM, Womble M, Ledon-Rettig C, Hull M, Dickinson A, Nascone-Yoder N (2015) Budgett's frog (Lepidobatrachus laevis): a new amphibian embryo for developmental biology. Dev Biol 405(2):291–303. https://doi.org/10.1016/j.ydbio.2015.06.007

Balinsky BI (1960) An introduction to embryology. W.B. Saunders Company, Philadelphia, London. https://ia801907.us.archive.org/7/items/in.ernet.dli.2015.145984/2015.145984.An-Introduction-To-Embryology_text.pdf

Bhattacharya D, Marfo CA, Li D, Lane M, Khokha MK (2015) CRISPR/Cas9: an inexpensive, efficient loss of function tool to screen human disease genes in Xenopus. Dev Biol 408(2): 196–204. https://doi.org/10.1016/j.ydbio.2015.11.003

Chen Y, Love NR, Amaya E (2014) Tadpole tail regeneration in Xenopus. Biochem Soc Trans 42(3):617–623. https://doi.org/10.1042/BST20140061

Fini JB, Mughal BB, Le Mevel S, Leemans M, Lettmann M, Spirhanzlova P, Affaticati P, Jenett A, Demeneix BA (2017) Human amniotic fluid contaminants alter thyroid hormone signalling and early brain development in Xenopus embryos. Sci Rep 7:43786. https://doi.org/10.1038/srep43786

Fort DJ, James BL, Bantle JA (1989) Evaluation of the developmental toxicity of five compounds with the frog embryo teratogenesis assay: Xenopus (FETAX) and a metabolic activation system. J Appl Toxicol 9(6):377–388. https://doi.org/10.1002/jat.2550090603

Freeman WH, Bracegirdle B (1967) An atlas of embryology, 3rd edn. Heinemann Educational Book Ltd, London

Gilbert SF (2000) Developmental biology, 6th edn. Sinauer Associates, Sunderland

Gosner KL (1960) A simplified table for staging anuran embryos and larvae with notes on identification. Herpetologica 16(3):183–190

Holmes SJ (1927) The biology of the frog. Macmillan. https://archive.org/details/biologyoffrog00holm_0/page/108/mode/1up?view=theater

Iwasawa H, Futagami J (1992) Normal stages of development of a tree frog, Hyla japonica Günther. Jpn J Herpetol 14:129–142

Kerner M, Flach H, Dietmann P, Kuhl M, Kuhl SJ (2023) The impact of the insecticide acetamiprid on the embryogenesis of the aquatic model organism Xenopus laevis. Environ Toxicol Pharmacol 103:104278. https://doi.org/10.1016/j.etap.2023.104278

McDiarmid RW, Altig R (1999) Tadpoles: the biology of anuran larvae. University of Chicago Press

Narzary J, Bordoloi S (2013) Study of normal development and external morphology of tadpoles of Microhyla ornata and Uperodon globulosus of the family Microhylidae (Amphibia: Anura) from North East India. Int J Adv Biol Res 3(1):61–73

Rowley JJ, Altig R (2012) Nidicolous development in Limnonectes limborgi (Anura, Dicroglossidae). Amphibia-Reptilia 33(1):145–149

Rugh R (1951) The frog; its reproduction and development. The Blakiston Company, Philadelphia. https://archive.org/details/frogitsreproduct00rugh/page/n7/mode/2up

Saha B, Gupta B (2011) The development and metamorphosis of an endangered frog, Rana leptoglossa (Cope, 1868). Int J Adv Res 1:67–76

Sayim F, Kaya U (2008) Embryonic development of the tree frog, Hyla arborea. *Biol* 63(4): 588–593

Scheld S, Bina Perl R, Rauhaus A, Karbe D, Van der Straeten K, Hauswaldt S, Randrianiania R, Gawor A, Vences M, Ziegler T (2013) Larval morphology and development of the Malagasy frog Mantidactylus betsileanus. Salamandra 49(9):186–200

Shumway W (1940) Stages in the normal development of Rana pipiens I. External form. Anat Rec 78(2):139–147

Singh CC (2015) Survey on the morphology of Rana Leptoglosssa cope's severly affected frog

Tamuly D, Dey M (2014) Larval morphology and development of tree frog Polypedates teraiensis (Dubois, 1987). Curr World Environ 9(1):182

Traijitt T, Kitana N, Khonsue W, Kitana J (2021) Chronological changes in the somatic development of Hoplobatrachus rugulosus (Wiegmann, 1834)(Anura: Dicroglossidae). Trop Nat Hist 21(1):184–199

Volpe EP, Wilkens MA, Dobie JL (1961) Embryonic and larval development of Hyla avivoca. Copeia 1961(3):340–349

Yan, F., Liu, X., Zhang, Y., & Yuan, Z. (2021). Direct development of the bush frog Raorchestes longchuanensis (Yang and Li 1978) under laborary conditions in Southern China. J Nat Hist, 55(1–2), 125–132

Zahn N, James-Zorn C, Ponferrada VG, Adams DS, Grzymkowski J, Buchholz DR, Nascone-Yoder NM, Horb M, Moody SA, Vize PD, Zorn AM (2022) Normal Table of Xenopus development: a new graphical resource. Development 149(14). https://doi.org/10.1242/dev.200356

Zhang K, Li N, Wang Z, Feng D, Liu X, Zhou D, Li D (2024) Recent advances in the color of aquatic products: evaluation methods, discoloration mechanism, and protection technologies. Food Chem 434:137495

Study of Whole Mounts of Early Developmental Stages of Chick from 13 Hours to 96 Hours (Hamburger and Hamilton Stages)

3

Anna Senrung, Meenakshi Thakur, and Alok Chandra Bharti

Abstract

The chick embryo has long been established as a classical vertebrate model in developmental biology, offering accessibility, ease of manipulation, and close parallels to mammalian embryogenesis. Developmental progression of the chick embryo is best described by the Hamburger–Hamilton (HH) staging system (1951), which defines 46 stages based on morphological features. This practical study focuses on chick embryos incubated for 13–96 hours (HH3–HH24). During these stages, students can examine whole-mount prepared slides to identify characteristic features such as the primitive streak, Hensen's node, somite formation, neural tube closure, brain vesicle differentiation, blood island formation, visceral arches, limb buds, and body flexures. Early embryos are small, soft, and transparent, enabling complete visualization of spatial relationships between developing structures. Whole mounts, prepared through fixation, staining, dehydration, and mounting, provide permanent teaching and research specimens. The exercise familiarizes students with staging criteria and enhances their ability to correlate morphological features with embryonic processes. Beyond teaching, the chick embryo remains an invaluable model for experimental embryology, developmental genetics, and toxicological research. Its large size, external development, year-round availability, and low cost make it especially suited for laboratory investigations. Applications range from the study of neural, limb, and cardiovascular development to assessing teratogenic and toxic effects of compounds. Thus, the chick embryo continues to bridge classical embryology with modern developmental biology.

A. Senrung (✉) · M. Thakur
Department of Zoology, Daulat Ram College, University of Delhi, Delhi, India

A. C. Bharti
Molecular Oncology Lab., Department of Zoology, University of Delhi, Delhi, India

A. C. Bharti et al. (eds.), *Textbook of Developmental Biology*,
https://doi.org/10.1007/978-981-95-7269-4_3

Keywords

Chick embryo · Primitive streak · Allantois · Hamburger–Hamilton staging

Aim To identify and describe chick embryo developmental stages from 13 hours (HH13) to 96 hours (HH24) using permanent slides examined under a microscope.

Learning Outcomes
By the end of this practical exercise, students will be able to:

- Identify and describe the key morphological features of chick embryos from 13 to 96 hours.
- Categorize chick embryos of 13–96 hours according to the Hamburger-Hamilton staging system.
- Understand the sequence of organogenesis, including the formation and development of major systems such as the neural tube, somites, heart, and limb buds.
- Distinguish between developmental stages of chick embryos using whole mount preparations and correlate them with corresponding anatomical and structural changes.
- Apply knowledge of chick embryology to broader concepts in vertebrate development, evolutionary biology, and biomedical research.

3.1 Introduction

The chick embryo serves as a classical and extensively studied model in developmental biology due to its accessibility, ease of manipulation, and striking similarity to early vertebrate development, including that of mammals. The staged progression of chick embryogenesis has been meticulously documented by Hamburger and Hamilton (1951) (Hamburger and Hamilton 1992), who established a standardized system of 46 stages based on morphological criteria rather than time alone.

Chick embryonic development begins with the formation of the zygote following fertilization, which occurs in the upper part of the oviduct (infundibulum) shortly after the ovum is released from the single functional ovary on the left side of the hen. The mature ovum, enclosed by a thin vitelline membrane, is drawn into the oviduct via the ostium. Being a macrolecithal egg, the ovum is large due to the massive accumulation of yolk, which serves as the nutrient source for external development.

Initially, the nucleus of the egg occupies the central region of the ovum, surrounded by a small amount of white yolk called the *latebra*. As development continues, concentric layers of yellow yolk are deposited around the white yolk, displacing the nucleus toward the periphery. This migration creates a funnel-shaped structure known as the *neck of the latebra*. The peripheral location of the nucleus

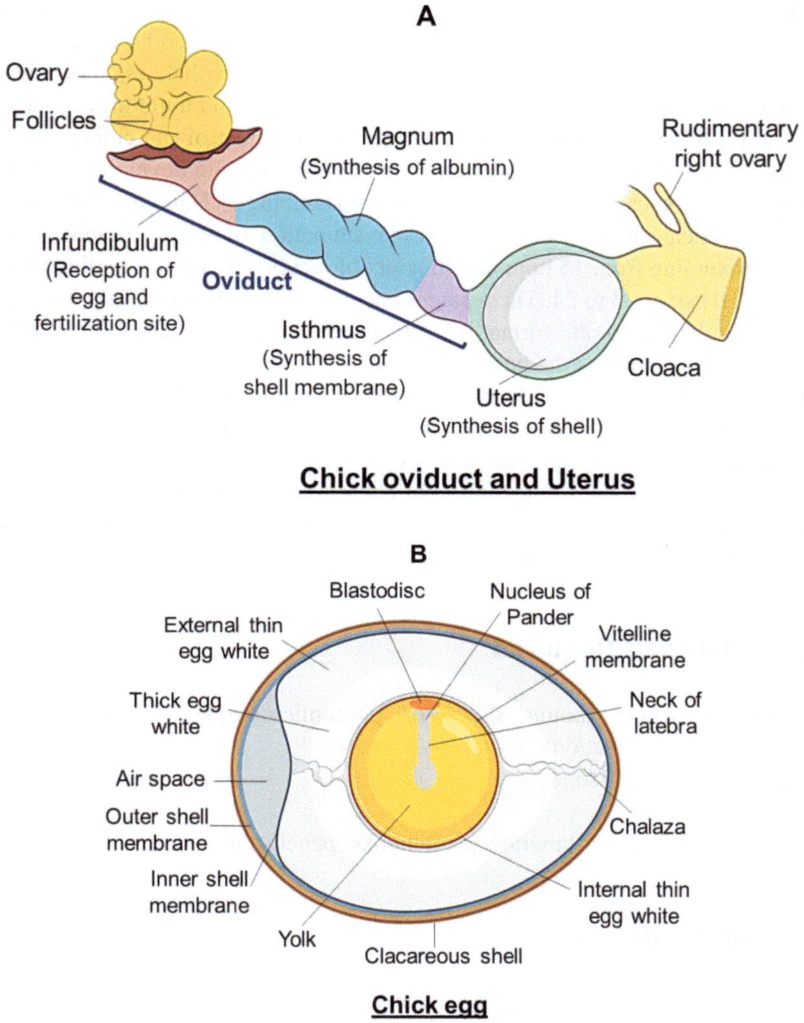

Fig. 3.1 Chick embryo oviduct and egg. (Created in https://BioRender.com)

forms a zone of clear cytoplasm called the *blastodisc* or *blastoderm*, where cleavage and embryogenesis occur post-fertilization. This region is also referred to as the *nucleus of Pander*.

As the egg travels down the oviduct (Fig. 3.1), it acquires several protective layers. In the magnum region, layers of albumen are secreted. The first layer, chalaziferous albumen, forms around the vitelline membrane and contains spiral chalazae that help suspend the yolk. Subsequent layers include dense and watery albumen. In the isthmus, two shell membranes (inner and outer) are formed. These are closely apposed, except at the broad end of the egg, where they separate to form an air chamber. Finally, in the uterus, a calcareous shell is secreted, which is initially soft but hardens upon contact with air after laying. These layers provide mechanical protection and aid in gas exchange, essential for embryonic development.

By the time the egg is laid, at about 24–25 hours after fertilization, they are at the blastula stage. Upon incubating at optimum temperature and humidity, generally at 37°C and a relative humidity ≥60%, further development of the embryo resumes after about 4 to 5 hours of incubation. The chick embryonic developmental stages during incubation can be best described by the Hamburger–Hamilton (HH) staging.

This practical chapter focuses on the examination of whole mounts of chick embryos spanning from 13 hours to 96 hours of incubation, corresponding approximately to HH stages 3 to 24. These stages encompass the formation of primary germ layers, neurulation, somite formation, organogenesis, and the initial differentiation of major systems such as the heart and brain.

Through this exercise, students will learn to identify and describe key morphological features during early development and to relate these features to the corresponding HH stages. The hands-on observation of permanent slides or photomicrographs of the whole mounts of chick embryo will enhance understanding of dynamic embryonic processes, fostering skills in microscopy, anatomical identification, and developmental staging.

3.2 Materials Required

- Permanent whole mount slides or photomicrographs of chick embryos (13–96 hours incubation).
- Stereo zoom microscope.
- HH staging chart.
- Observation sheets or laboratory notebooks, pencils, and eraser.

3.3 Methodology

- Prepare the microscope for observation with proper illumination and focus adjustments.
- Observe each slide beginning from the earliest stage (HH stage 3, ~13 hours) to later stages up to 96 hours (HH stage 24).
- Draw and label the embryos in your notebook, indicating key morphological features observed.
- Identify key structures for each stage:
- *HH stages 3 to 6:* Identify the presence and development of the primitive streak, including the shape, length, development of the primitive groove and fold, Henson's node, and pit. Other features to identify include the shape of the area pellucida, head and neural fold, and spreading of the mesoderm laterally from the primitive streak.
- *HH stages 7 to 14:* Identifying the number of somites. Additionally, other features include the primitive streak, differentiation of the brain and optic vesicles,

proamnion, head and neural fold, and wall of the foregut, margin of the anterior intestinal portal, blood islands, development of the heart and blood vessels, body flexures, visceral arches and clefts, and early stage of branching of vitelline vein (Omphalomesenteric vein) and artery (Omphalomesenteric artery).

- *HH stages 15 to 24:* For these stages, apart from the presence of the somites, look out for other morphological features such as limb buds, visceral arches, and additional distinct structures such as body flexures or curvatures (cranial, cervical, trunk, and lumbar), rotation of the body, tail bud, allantois, eye, auditory vesicle, and fifth, seventh and eighth nerve ganglia.

3.4 Description of Developmental Stages Based on Hamburger–Hamilton (HH) Staging (13–96 Hours)

The HH staging system is a standardized sequence that describes the embryonic development of the white Leghorn chick (*Gallus gallus*) for the entire incubation period, beginning from fertilization to hatching. It comprises 46 stages, beginning with stage 1 (Pre-streak) and ending at stage 46 (pre-hatching), and provides a precise comparative framework for describing chick embryogenesis, independent of incubation temperature or environmental conditions. Introduced in 1951 by Viktor Hamburger and Howard L. Hamilton, this system classifies stages based on distinct morphological features. In the early stages (up to HH6) (Fig. 3.2), the primitive streak, including Hensen's pit and node, head, or notochordal process, serves as the key identifying feature. Between HH7 and HH14 (Fig. 3.3), the number of somite pairs provides reliable staging. Beyond HH14 (Fig. 3.4), somite counts become less dependable due to mesoderm diffusion in the anterior and tail regions, and staging instead relies on additional morphological traits such as limb buds, visceral arches, eye development, and organogenesis.

In this chapter, chick embryo development is documented according to the HH staging system, covering embryos from 13 to 96 hours of incubation, corresponding to HH stages 3 to 24.

HH stage 3 (12–13 hours) (Fig. 3.2c) This stage is characterized by the presence or appearance of a relatively broad primitive streak which extends to about the mid region of the blastoderm or the area pellucida from the posterior inner border of the area opaca. The primitive streak formation begins with the thickening of the blastoderm at the posterior edge of area pellucida, which then gradually extends anteriorly *(HH1–2)* (Fig. 3.2a, b). As the primitive streak forms, numerous epiblast cells migrate inward to initiate germ layer differentiation. The primitive streak lies along the embryo's longitudinal axis, extending from the posterior (caudal) end toward the anterior (cephalic) region. The streak flares, touching the inner margins of the area opaca. The streak lacks the primitive groove at this stage.

HH stage 4 (18–19 hours) At this stage, the primitive streak is well-defined, with a central linear depression, the primitive groove, flanked by ridge-like thickenings or primitive folds (Figure 3.2D). The streak at this stage reaches its maximum average length of ~1.88 mm. Together, the groove and folds constitute the primitive streak. At the anterior end of the groove lies the primitive pit, a deep, slightly expanded depression. Just beyond it, in the cephalic region, is *Hensen's node*, a small, rounded elevation formed by the merging of the primitive folds from either side of the groove. Named after Viktor Hensen, who first documented it in 1875, the node, though significant, does not give rise to any specific structure. The area pellucida appears pear-shaped, and the streak now extends over two-thirds to three-fourths of its length, marking active gastrulation.

HH stage 5 (19–22 hours) (Fig. 3.2e) The primitive streak is shifted toward the caudal region due to rapid growth and differentiation in the cephalic area. The notochord appears as a rod-shaped condensation of mesoderm, arising from the rapidly dividing, undifferentiated cells at the cephalic end of the primitive streak beginning from Hensen's node and extending anteriorly. The notochord is not differentiated from adjacent loose mesoderm cells. Together, the notochord and primitive streak mark the embryo's midline, establishing its definitive longitudinal axis. The primitive groove, primitive pit, and Hensen's node are distinct at this stage.

HH stage 6 (23–25 hours) (Fig. 3.2f) A distinct crescentic head fold appears, marking the embryo's anterior boundary. This first prominent epithelial fold initiates the formation of the anterior gut, heart, eye, and brain. At this stage, mesoderm arises from the sides of the primitive streak and spreads laterally, leaving an anterior mesoderm-free area called the *proamnion* (which is not a primordium of the amnion). The proamnion's posterior margin is bordered by the thickened anterior region of the embryo and the anterior edges of the mesoderm. This stage is considered transitional, as the head fold develops concurrently with the formation of the first somite pair.

HH stage 7 (23–26 hours) (Fig. 3.3a)

At this stage, 1–2 somite pairs are present, with the second somite typically better defined than the first; the stage is generally characterized as having one somite. The mesodermal somites are the first segmented structures to appear in the chick, which is formed by division of the dorsal mesoderm. The consistent addition of somites as the embryo develops makes somite number the most reliable indicator of developmental stage. Besides the anterior head fold, notable features include paired neural folds flanking the neural groove, visible as dark bands. The neural folds extend up to the cephalic half of the embryo, while caudally they diverge and become less distinct. The proximal zone at the boundary between the area opaca and area pellucida, where mesoderm has grown in, is called *area opaca vasculosa*, as this mesoderm gives rise to the yolk-sac blood vessels.

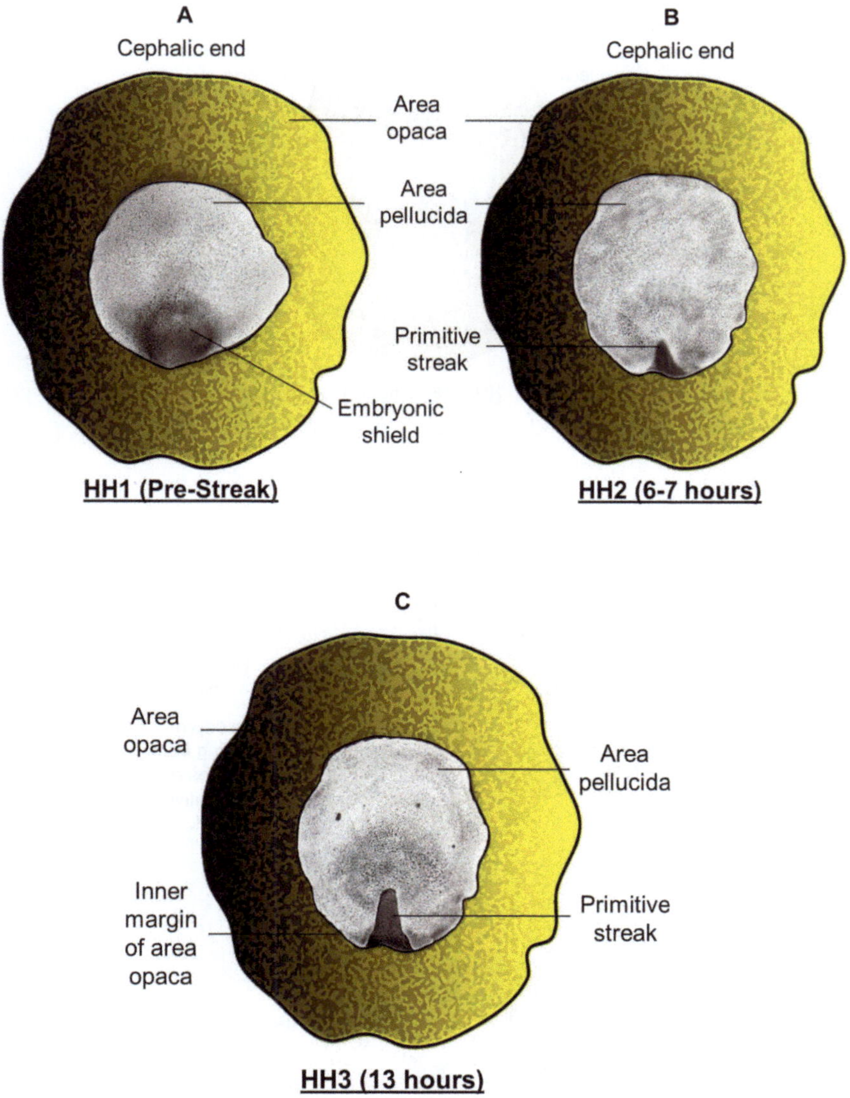

Fig. 3.2 Diagrammatic representation of the whole mount of chick embryos from approximately 13 to 25 hours of incubation. During these stages (HH 3 to HH 6), the primitive streak plays a key role in defining embryonic development. (The diagrams have been prepared based on information gathered from Bellairs and Osmond (2005), Gilbert (2000), Hamburger and Hamilton (1992), Lillie (1919), Patten (1920). Created in https://BioRender.com and further refined using Infinite Painter (Android application).)

D

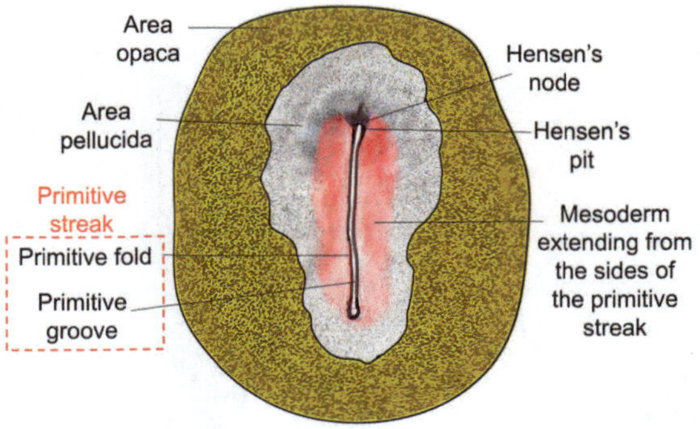

HH4 (18-19 hours)

E

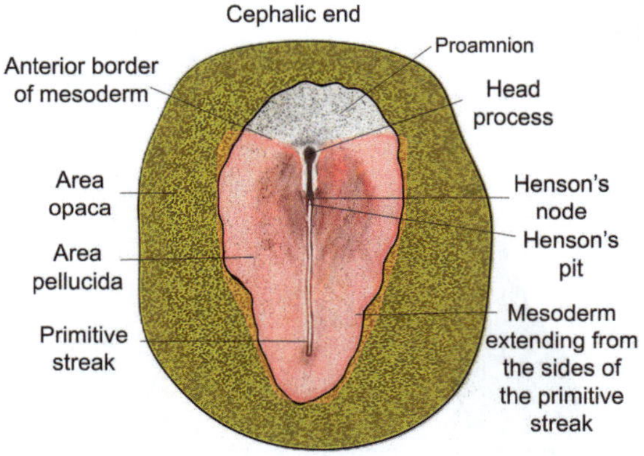

HH5 (19-22 hours)

F

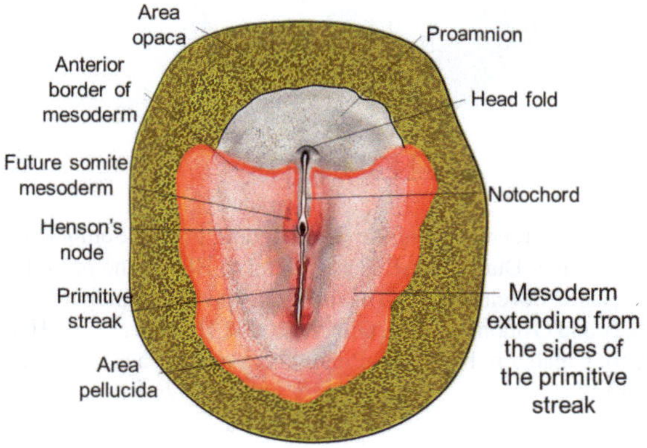

HH6 (23-25 hours)

Fig. 3.2 (continued)

HH stage 8 (26–29 hours) (Fig. 3.3b) This stage is characterized by the presence of approximately four somite pairs. In addition to the proamnion, neural folds, Hensen's node, and primitive streak, blood islands are visible mainly in the posterior region of the embryo; these blood islands will eventually merge to form blood vessels. Dorsally, the neural folds are nearly in contact. In the developing head, the endoderm forms a pocket (the foregut) within the ectoderm. This endodermal pocket is the first region of the digestive tract. The portion of the gut posterior to the foregut, where yolk remains the sole floor, is termed the midgut. The opening connecting the midgut to the foregut is known as the *anterior intestinal portal.*

HH stage 9 (29–33 hours) (Fig. 3.3c) This stage is marked by the presence of about seven somite pairs. Significant cephalic developments include thickening walls, elongation, and expansion of the anterior portion of the neural tube, which will form the brain, as well as the separation of the head region from the blastoderm. The remaining posterior undilated portion of the neural tube will develop into the spinal cord. Although the neural folds begin converging mid-dorsally, the neural groove remains open along its entire length. The optic vesicle primordia emerge, fusion of the paired heart primordia begins, and the primitive streak is markedly reduced compared to earlier stages. With continued growth, the primitive streak becomes progressively less distinct and eventually disappears as the caudal end of the body becomes fully defined.

HH stage 10 (33–38 hours) (Fig. 3.3d) This stage is defined by the presence of about ten somites, with the first somite beginning to degenerate. The anterior region of the neural canal differentiates into three primary brain vesicles: the forebrain (prosencephalon), which is markedly dilated; the midbrain (mesencephalon), separated from the forebrain by a constriction; and the hindbrain (rhombencephalon), demarcated from the midbrain by a faint constriction. The rhombencephalon transitions posteriorly into the spinal cord region of the neural tube without a distinct boundary. The optic vesicles remain broadly connected with the forebrain, without clear separation. The first cranial flexure becomes apparent, the anterior neuropore narrows considerably, and the developing tube-like heart shows a slight rightward bend. The heart, if viewed from the ventral side, is visible; however, on the dorsal side, it is covered by the neural tube.

HH stage 11 (40–45 hours) (Fig. 3.3e) This stage is marked by approximately 13 somites, with the cranial flexure still visible. The three primary brain vesicles begin subdividing into the five regions seen in the adult. At this stage, the division of the rhombencephalon into the metencephalon and myelencephalon is apparent. Closure of the anterior neuropore occurs, while the neural groove posterior to the somites remains open, with the neural folds diverging on either side of Hensen's node. The optic vesicles are constricted at their base, and the heart is bent toward the right.

HH stage 12 (45–49 hours) (Fig. 3.3f) At this stage, approximately 16 somites are present. The head is slightly bent to the embryo's right, and the anterior neuropore is closed, leaving a temporary scar-like fissure that will gradually disappear. The auditory placode appears and invaginates to form a depression. As the pit's opening constricts, its epithelium separates from the head ectoderm and positions near the lateral wall of the myelencephalon. The optic vesicle with its stalk is well developed, and a deep auditory pit with a wide opening is evident. The heart exhibits a slight "S" shape, and the amnion completely covers the forebrain region.

HH stage 13 (48–52 hours) (Fig. 3.3g) Approximately 19 somites are present at this stage. The head is bent to the left, exhibiting a range of flexion from partial to complete. Broad curves are evident in both the cranial and cervical flexures. Enlargement of the telencephalon becomes apparent, accompanied by a slight narrowing of the auditory pit opening. The atrioventricular canal is visible, and the amnion head fold extends to cover the midbrain and the anterior portion of the hindbrain.

HH stage 14 (50–53 hours) (Fig. 3.3h) This stage is characterized by the presence of approximately 22 somites. Separation of the prosencephalon into telencephalon and diencephalon begins, as indicated by a dorsal wall constriction. A broad cervical

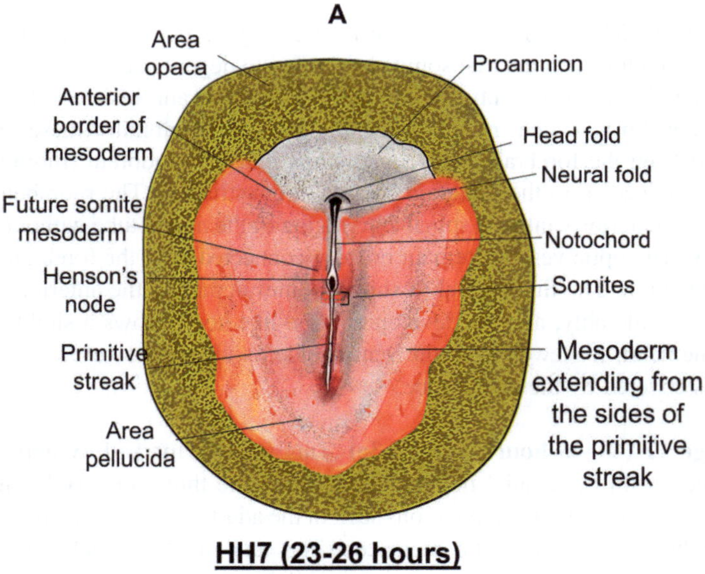

Fig. 3.3 Diagrammatic representation of the whole mount of chick embryos from approximately 23 to 53 hours of incubation. During these stages (HH 7 to HH 14), the number of somites plays a key role in defining embryonic development. (The diagrams have been prepared based on information gathered from Bellairs and Osmond (2005), Gilbert (2000), Hamburger and Hamilton (1992), Lillie (1919), Patten (1920). Created in https://BioRender.com and further refined using Infinite Painter (Android application).)

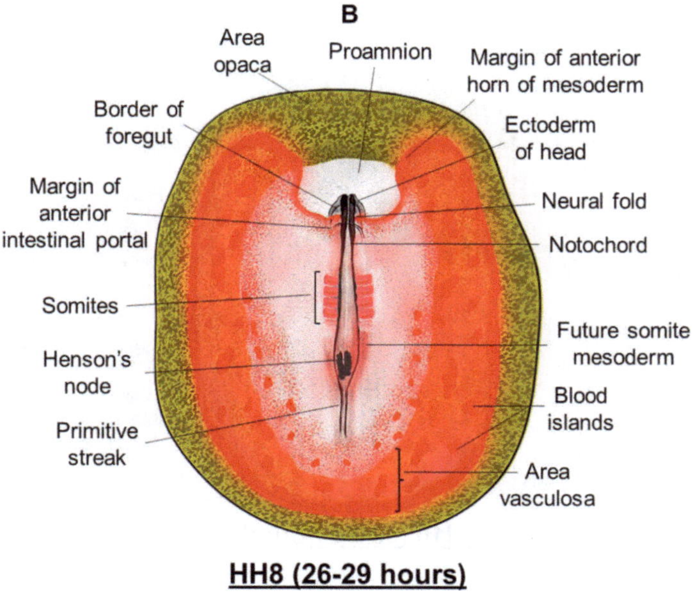

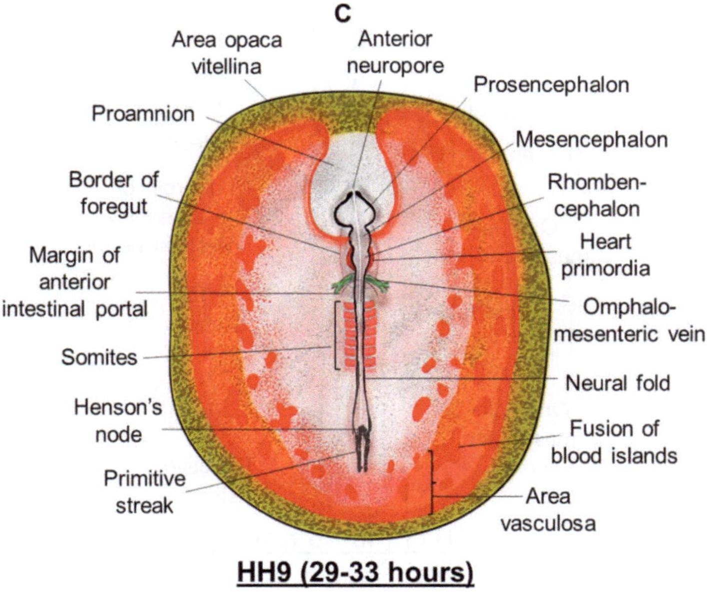

Fig. 3.3 (continued)

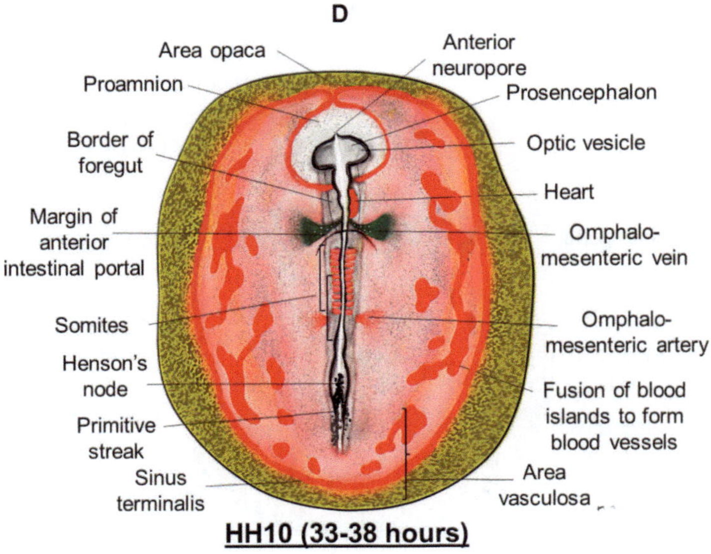

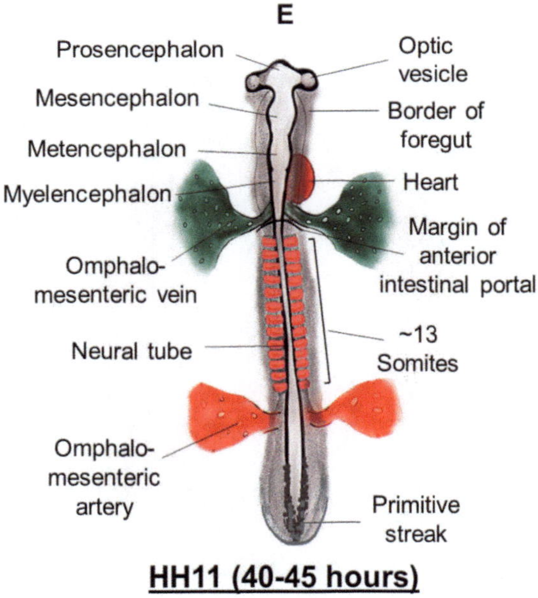

Fig. 3.3 (continued)

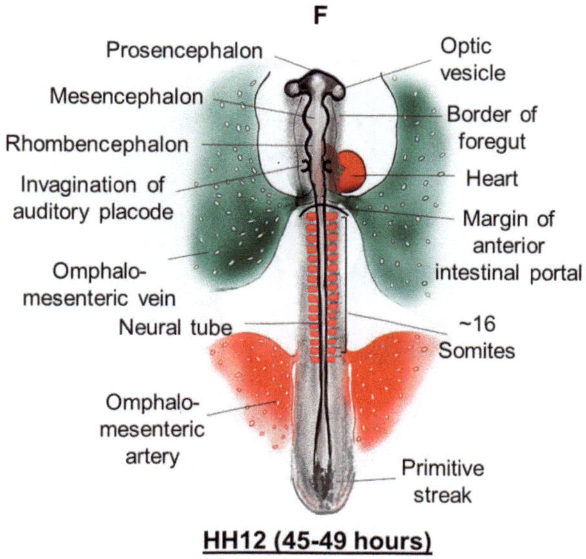

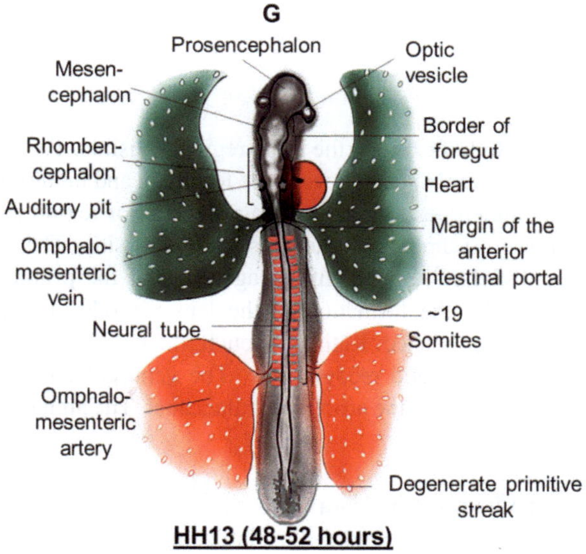

Fig. 3.3 (continued)

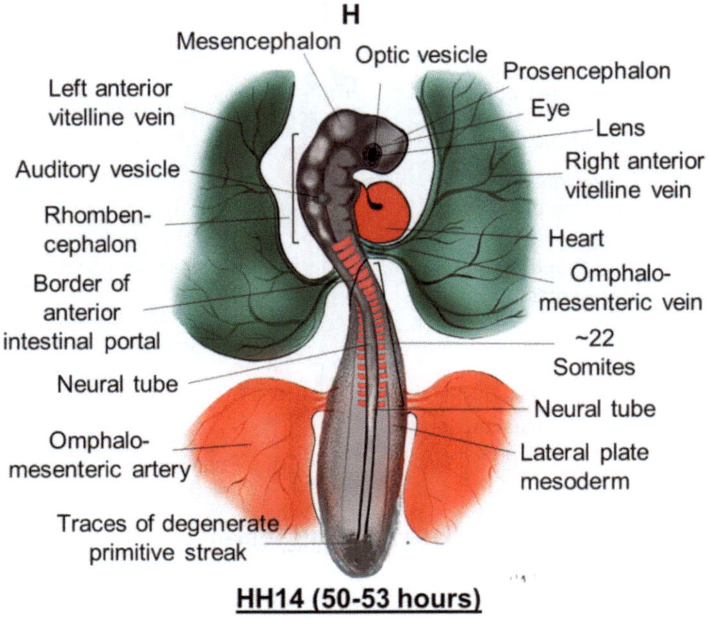

Fig. 3.3 (continued)

flexure is present. A slight flexure in the trunk region is visible. The cranial flexure at this stage forms a right-angle bend between the forebrain and hindbrain axis. Around this time, the head undergoes torsion, rotating the anterior region from the prosencephalon to somites 7–9 through 90 degrees to the left side. The first and second visceral arches and clefts are apparent. Invagination of the primary optic vesicle begins, accompanied by the formation of the lens placode. The opening of the auditory pit becomes constricted. Rathke's pouch, the primordium of the adenohypophysis, becomes distinct and recognizable. The ventricular loop of the heart is positioned ventrally relative to the atrioventricular canal. The amnion now extends to the level of somites 7–10.

HH stage 15 (50–53 hours) (Fig. 3.4a) Limb primordia remain flat and inconspicuous. The fore limbs develop around the level of somites 17–19, while the hind limb develops around the level of somites 26–32. The amnion extends to around the level of somite 14. Approximately 24–27 somites are present. The forebrain and hindbrain axis form an acute angle with nearly parallel ventral contours. The midbrain forms the anterior-most part of the embryo. The trunk is well-defined, with rotation (turning of the embryo on its left side) reaching the level of somite 13. The third visceral arch and cleft are distinct, and the optic cup is fully formed.

HH stage 16 (51–56 hours) (Fig. 3.4b) This stage is characterized by 26–28 somites. Limb primordia become visible as a thickened or elevated ridge in the forelimb, while the hindlimb remains flattened. Amnion reaches somite 18. Flexures deepen. The tail bud is short and straight. Visceral arches are visible, and the forebrain elongates with distinct brain region constrictions.

HH stage 17 (52–64 hours) (Fig. 3.4c, d) This stage is characterized by 29–32 somites. The fore and hind limb buds are of approximately equal size and appear as distinct swellings. Amnion coverage varies, extending either to somite 26 or fully to the tail. The cranial flexure remains as in HH stage 16, while the cervical flexure becomes more sharply bent. A distinct trunk flexure is visible at the brachial level, and rotation extends to somite 18. The tail bud, containing unsegmented mesoderm, bends ventrally. The epiphysis appears as a distinct knob, and the nasal pits become evident.

HH stage 18 (65–69 hours) (Fig. 3.4e, f) Limb buds enlarged compared to the previous stage, with hindlimb buds slightly larger than wing buds. The length (anterior-posterior dimension along the body-wall) of the wing bud is six times or more than its width (distance from body-wall to apex). Approximately 30–36 somites are present, extending beyond the leg buds. The amnion is typically closed, though an occasional oval opening may remain in the lumbar region. At the cervical flexure, the medulla's axis forms roughly a right angle with the posterior trunk's axis. The trunk flexure is shifted to the lumbar region, and rotation now extends to the posterior body. The mid-region of the body is slightly concave dorsally, as the embryo remains broadly attached to the yolk at this level. The tail bud is turned to the right at about 90° to the posterior trunk axis.

HH stage 19 (68–72 hours) (Fig. 3.4g, h) Limb buds enlarge further, with leg buds slightly larger and thicker than wing buds. The length of the wing bud is approximately four to six times its width. Approximately 37–40 somites are present, extending into the tail; the tail tip, directed forward, remains unsegmented. In the cervical flexure, the medulla axis forms an acute angle with the trunk axis. Trunk flexure is absent or nearly absent due to body rotation, leaving the posterior trunk contour straight to the tail base. The tail bud is curved, tip pointing forward. All four visceral arches (I, II, III, and IV) and clefts are apparent. The maxillary and mandibular processes of the first visceral arch are equal in length. The fourth cleft is seen as a dorsal slit with a shallow ventral groove. The allantois is a small, non-vesicular pocket of variable size concealed by the hind limb bud. Eyes remain unpigmented.

HH stage 20 (70–72 hours) (Fig. 3.4i, j) Limb buds are further enlarged, with hind limb buds markedly larger than wing buds. The ratio of the wing bud length is 3–3.9 times more than its width, and the leg bud length is 3–2.3 times more than its width. There are typically 40–43 somites, with the tail tip still unsegmented. Cervical flexure is more pronounced than in stage 19, and the tail bend extends forward into the lumbosacral region. Mid-trunk contour is straight, and the body rotation up to the level immediately posterior to the heart is completed. The maxillary and mandibular processes of the first visceral arch are distinct and about the same size. The second arch projects over the surface, while the fourth arch is smaller and less prominent than the third. The allantois is vesicular and variable in size. Eye pigment appears as a faint gray hue.

HH stage 21 (78 hours) (Fig. 3.4k, l) Both wing- and leg-buds are enlarged and slightly asymmetrical, with their proximodistal axis directed posteriorly. Posterior contours are steeper than anterior, meeting the baseline at ~90°. Wing buds are 2.3–2.7 times longer than their width; leg buds are 2.0–2.5 times longer than their width. Somites number 43–44; the extreme tail tip remains unsegmented. Dorsal trunk contour is straight or slightly bent. The maxillary process of the first visceral arch is longer than the mandibular process, reaching approximately mid-eye. The second arch projects over the surface, overlapping the third arch ventrally. The fourth arch is distinct, with the fourth cleft visible as a slit. The allantois is usually larger than in stage 20, sometimes extending up to the head. Eye pigmentation is faint.

HH stage 22 (78 hours) (Fig. 3.4m, n) The limb buds are elongated and point caudad, with their bases remaining nearly parallel. In the wing bud, the length is typically 1.5–2 times or more than its width, while in the leg bud, the length is approximately 1.3–1.8 times its width. The somites now extend to the tip of the tail. Flexures show little change from the previous stage, and the dorsal trunk contour appears straight to almost straight. The visceral arches remain essentially unchanged from HH stage 21; the maxillary process is still prominent, and the fourth visceral cleft is visible as a slit. The allantois sometimes extends as far as the head or even overlaps the forebrain. Pigmentation of the eye becomes evident at this stage.

HH stage 23 (78–96 hours) (Fig. 3.4o, p) The limb buds retain a similar overall shape to those in HH stage 22 but have increased in length, particularly at the proximal base, where the anterior and posterior contours remain parallel. In both wing and leg buds, the length now approaches or equals the width. The maxillary process is further elongated. The first to fourth visceral arches and their associated clefts remain distinct, appearing as protuberances or slits or narrow pits. The dorsal contour, extending from the hindbrain to the tail, forms a continuous curved line. The embryo remains attached to the yolk by an elongated, slender stalk, which contributes to the convex dorsal flexure of the body.

HH stage 24 (96 hours) (Fig. 3.4q, r) Both forelimbs and hindlimbs are distinctly longer than they are wide. In the wing bud, the digital plate is still indistinct, whereas in the leg bud, the toe plate is demarcated, though individual toes are not yet visible. At this stage, cranial ganglia begin forming from aggregating neural crest cells. The fifth (trigeminal) cranial nerve, or Gasserian ganglion, is the largest and most distinct. Anterior to the auditory vesicle lies a neural crest cell mass, representing the primordium of the seventh and eighth nerve ganglia. The mandibular arch and

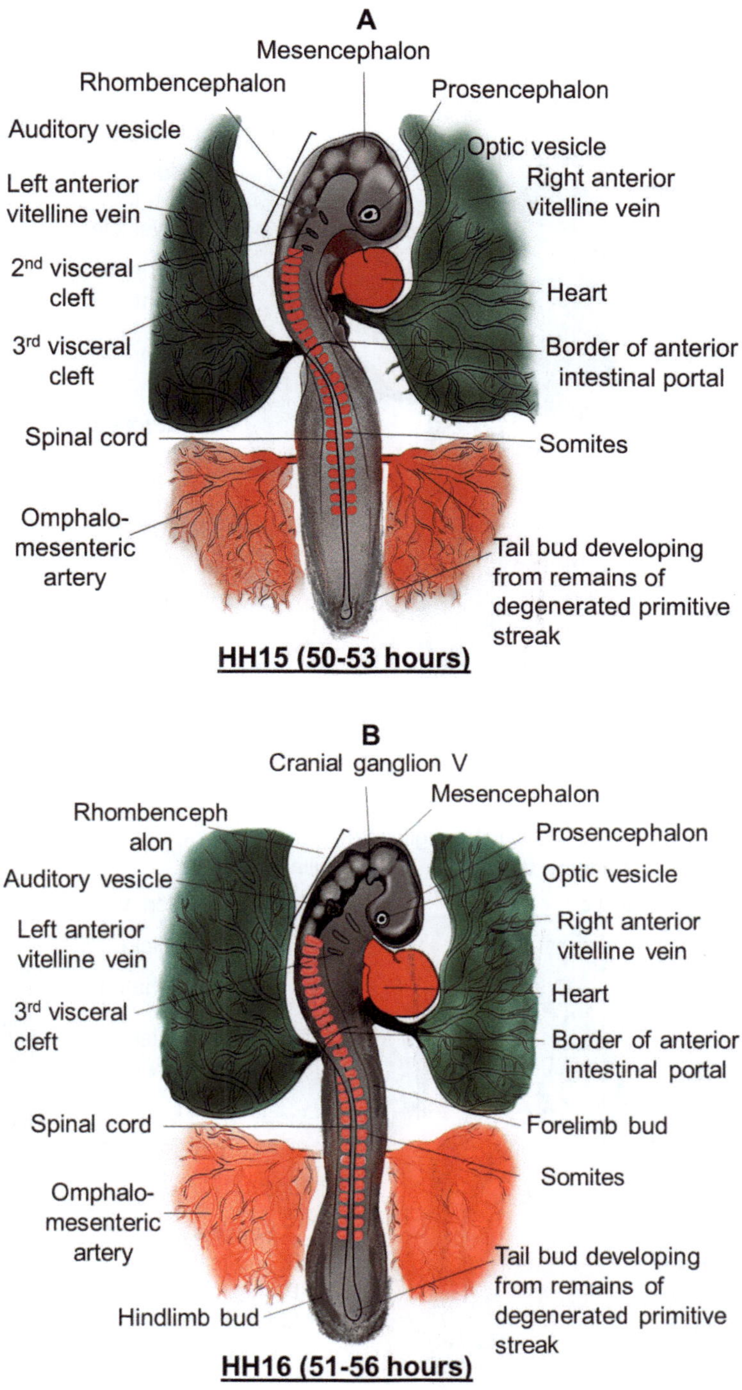

Fig. 3.4 Diagrammatic representation of the whole mount of chick embryos from approximately 50 to 96 hours of incubation. During these stages (HH 15 to HH 24), limb bud formation, visceral arch development, eye development, and organogenesis play a crucial role in defining embryonic development. (The diagrams have been prepared based on information gathered from Bellairs and Osmond (2005), Gilbert (2000), Hamburger and Hamilton (1992), Lillie (1919), Patten (1920). Created in https://BioRender.com and further refined using Infinite Painter (Android application).)

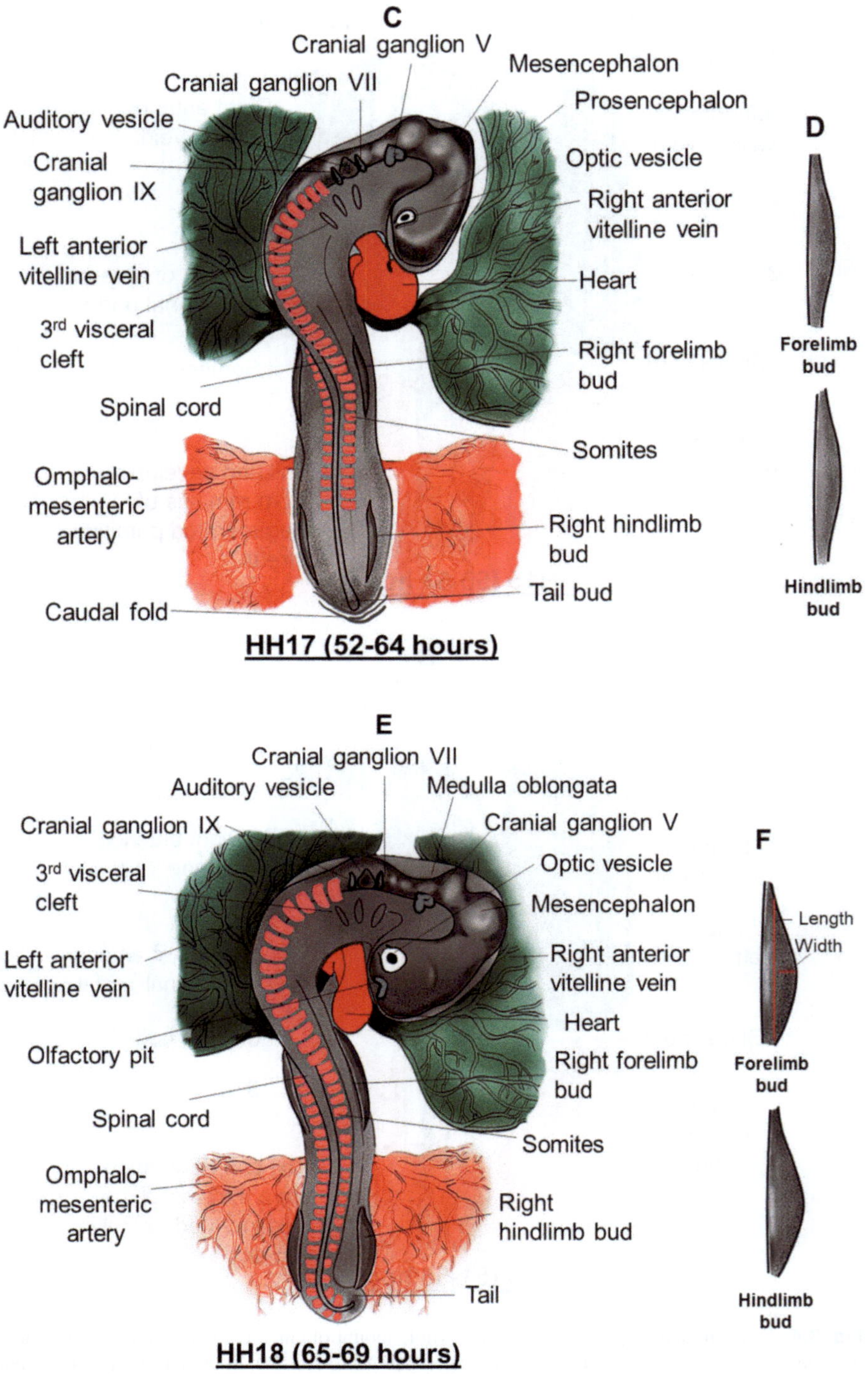

Fig. 3.4 (continued)

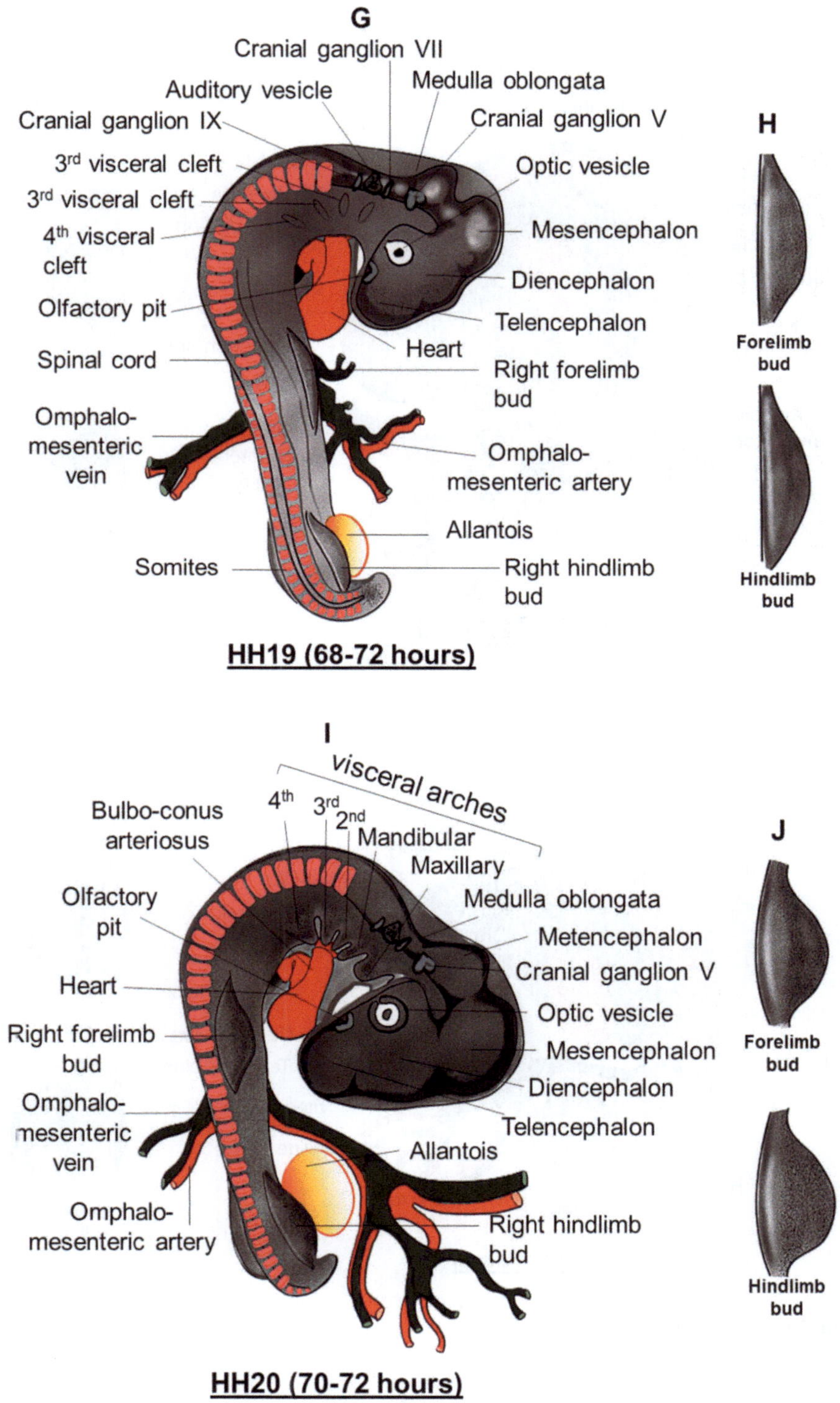

Fig. 3.4 (continued)

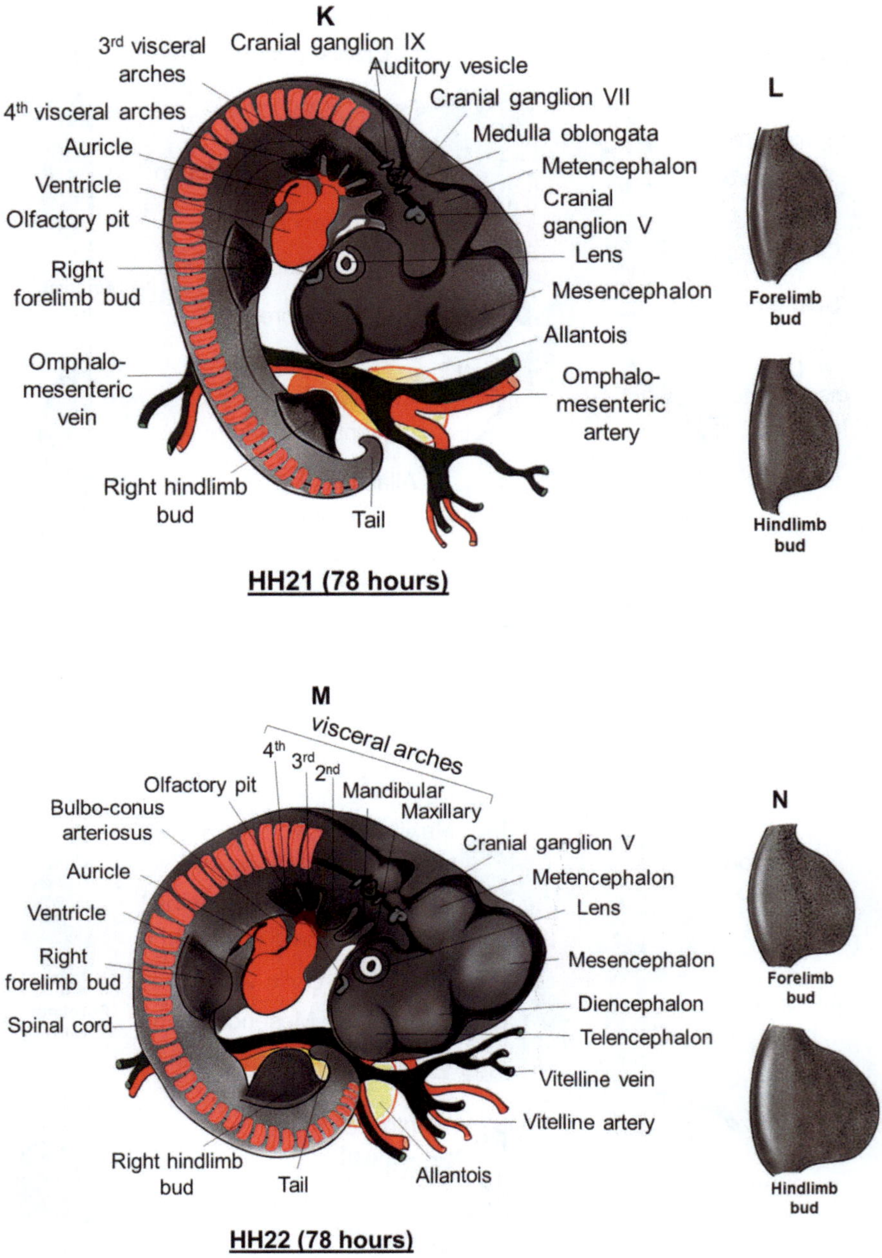

Fig. 3.4 (continued)

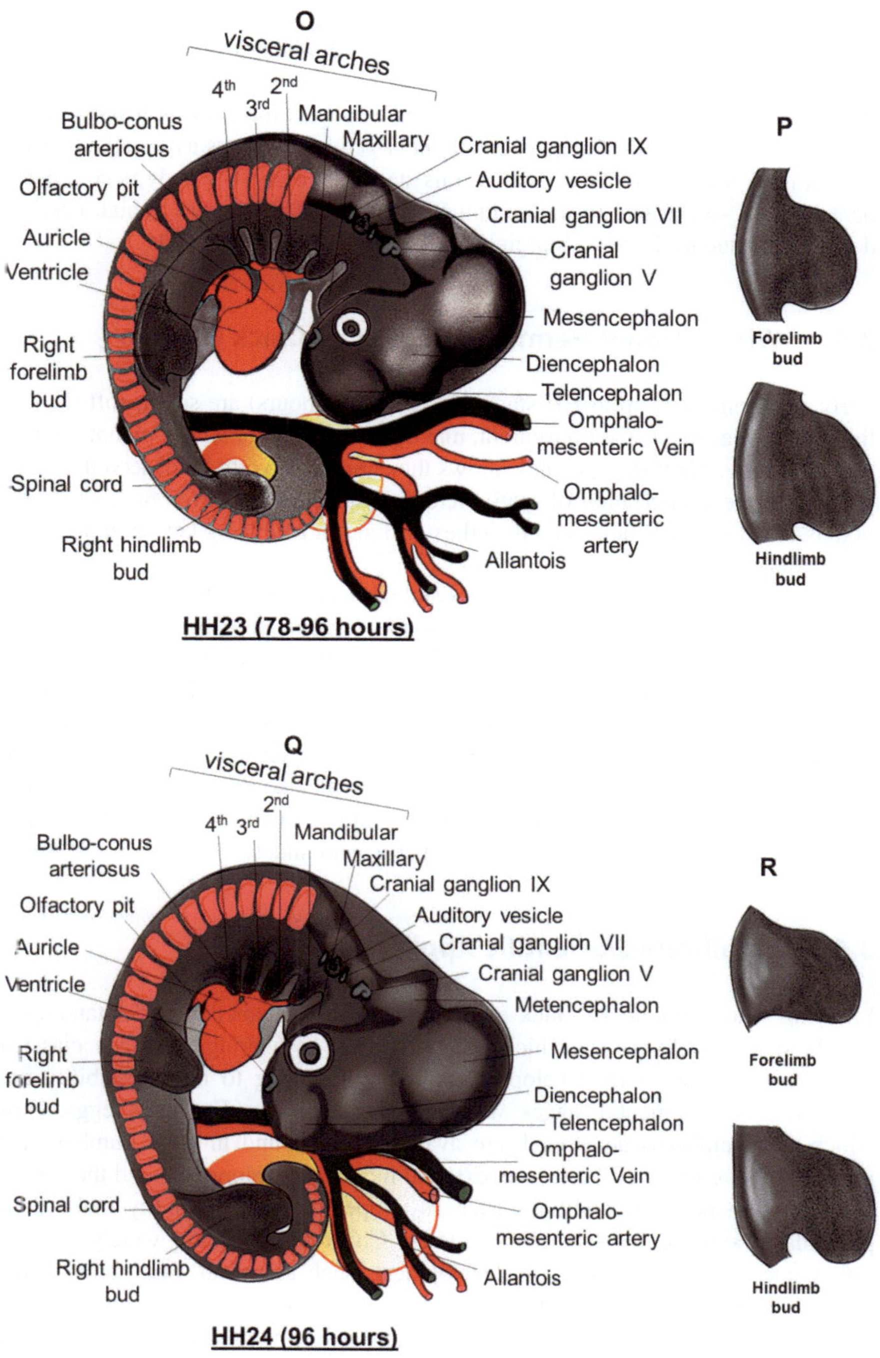

Fig. 3.4 (continued)

the second arch show clear differentiation. The third and fourth arches are buried beneath the surface, with the fourth visceral cleft diminished to a shallow, pit-like depression. The rotation of the entire embryo body to one side (left side) through 90 degrees is complete. The embryo's body now lies entirely on its left side over the yolk. The allantois at this stage is greatly enlarged and appears as a stalked vesicle projecting from the umbilical area. At this stage, the embryo's body is completely bent into a C-shape due to the continued progress of the body (cranial, cervical, dorsal, and caudal) flexures, bringing the head and tail close together.

3.5 Whole Mount Permanent Slides of Chick Embryo

Early chick embryos up to HH stage 21–22 (72–84 hours) are small, soft (lacking limb bones/placodes), and transparent, making them ideal for whole-mount preparation and study. Their transparency allows the entire embryo to be observed at once, facilitating the visualization of spatial relationships between developing structures, organs, and tissues. Whole-mount studies also aid in analyzing key morphogenetic events during embryogenesis.

Permanent slides of chick developmental stages are generally prepared through standard procedures. The blastoderm, or the region containing the developing embryo, is excised and placed in a wash glass with saline or directly on a slide with saline added. After carefully removing excess tissues with a soft brush, the embryo is fixed, stained with reagents such as borax carmine and eosin, dehydrated through graded alcohols, dealcoholized with clearing agents such as xylene, and finally mounted in a medium like DPX (Dibutylphthalate Polystyrene Xylene) and covered with a coverslip for long-term preservation. Figure 3.5a–z, shows microscopic images of the chick embryos from permanent slides.

3.6 Significance/Scientific Application

Studying whole mounts of chick embryos offers a fundamental understanding of vertebrate embryology. The chick embryo serves as a bridge between classical embryology and modern developmental genetics, owing to its accessibility and conserved developmental stages with higher vertebrates. Fertilized eggs, from which chick embryos developed, are available year-round in large numbers at a relatively low cost. Their large size, external mode of development, and the ease of controlling incubation conditions such as temperature and humidity make them particularly suitable for laboratory studies. Chick embryos are widely used to investigate embryonic and organ development, such as neural, limb, and cardiac morphogenesis. They also provide a well-established model for assessing the teratogenic and toxic effects of drugs, viruses, and environmental compounds on the developing embryo. Owing to these advantages, chick embryos remain an indispensable experimental system in developmental biology and toxicological research. Table 3.1 provides representative examples of research studies employing chick embryos.

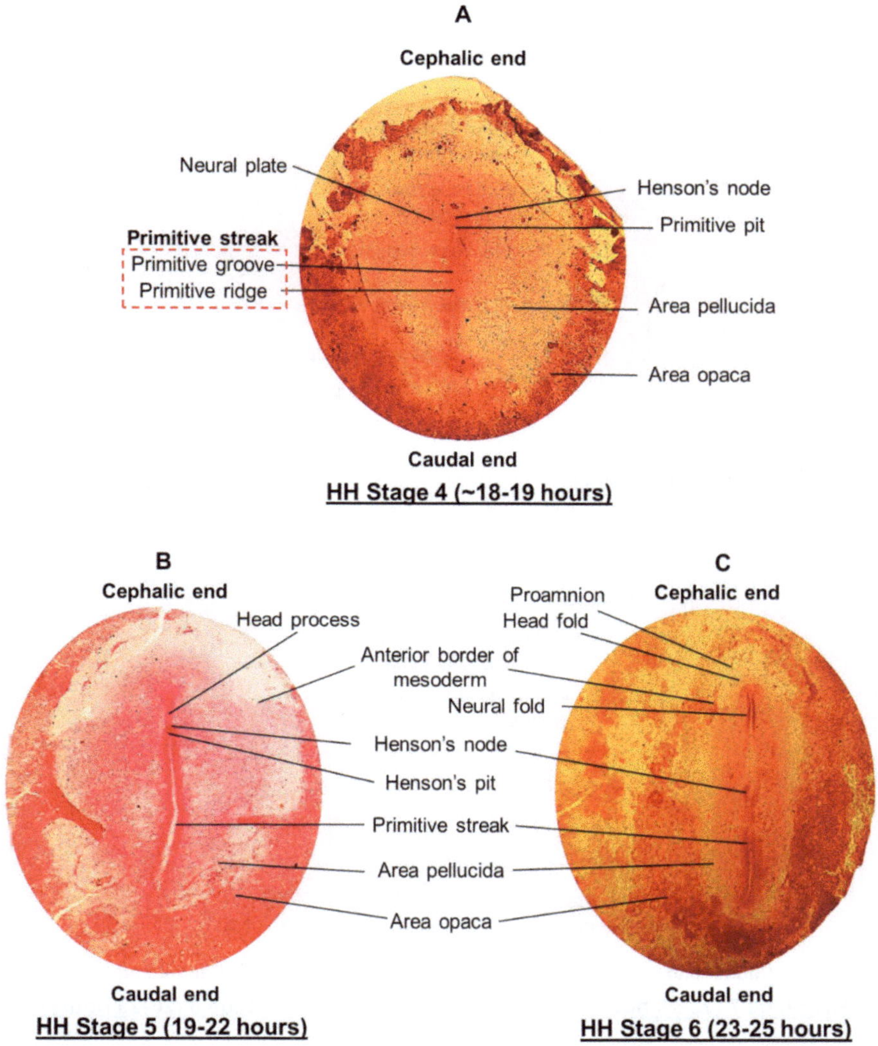

Fig. 3.5 Images of the permanent slides of the whole mount of chick embryos. (Source: Department of Zoology, Daulat Ram College, University of Delhi)

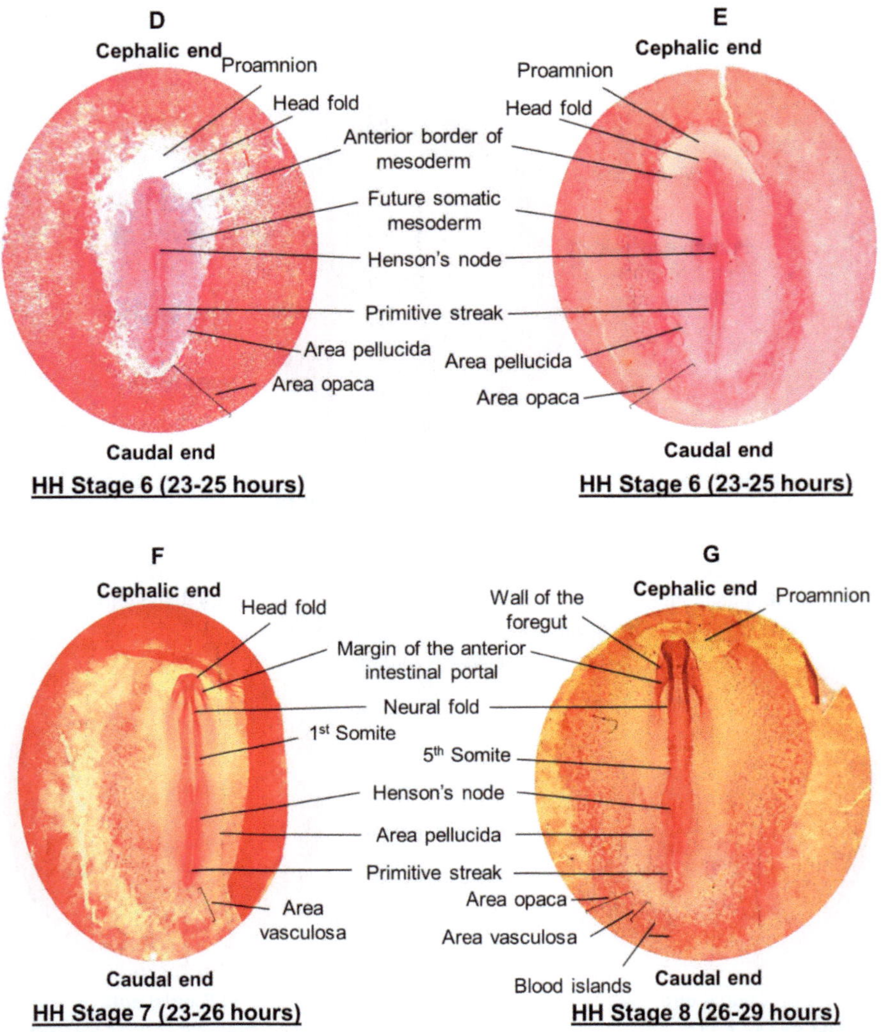

Fig. 3.5 (continued)

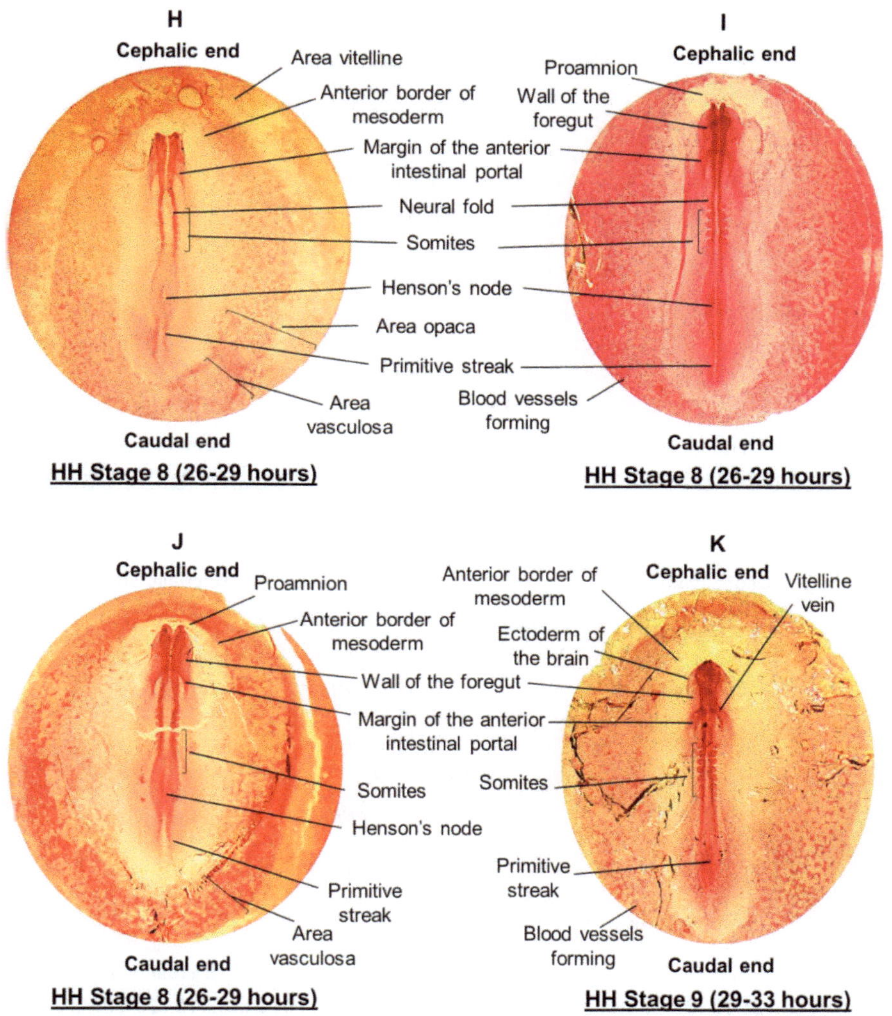

Fig. 3.5 (continued)

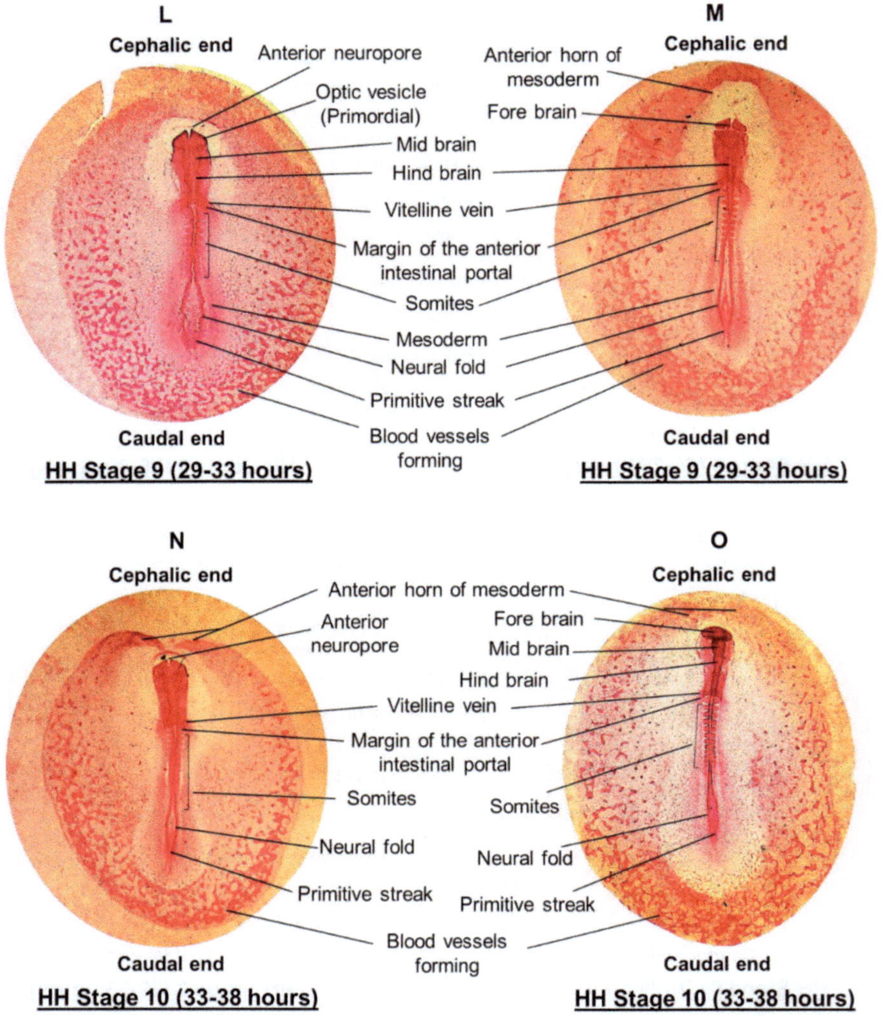

Fig. 3.5 (continued)

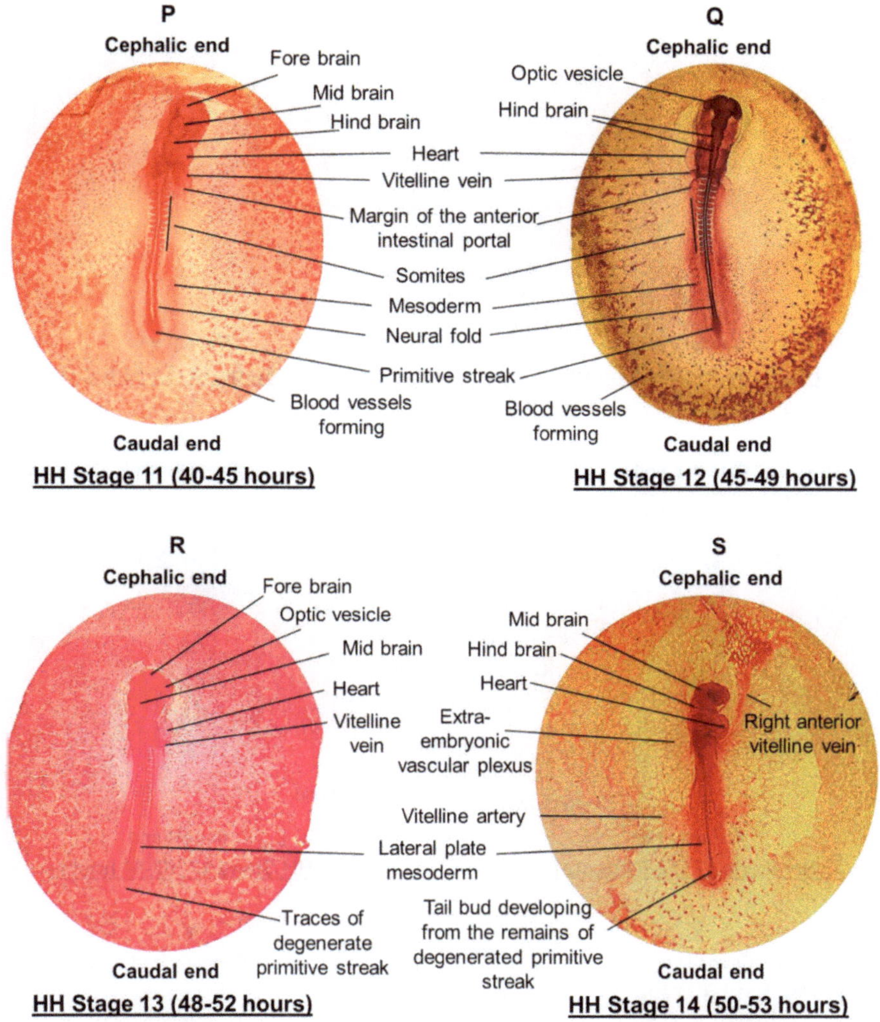

Fig. 3.5 (continued)

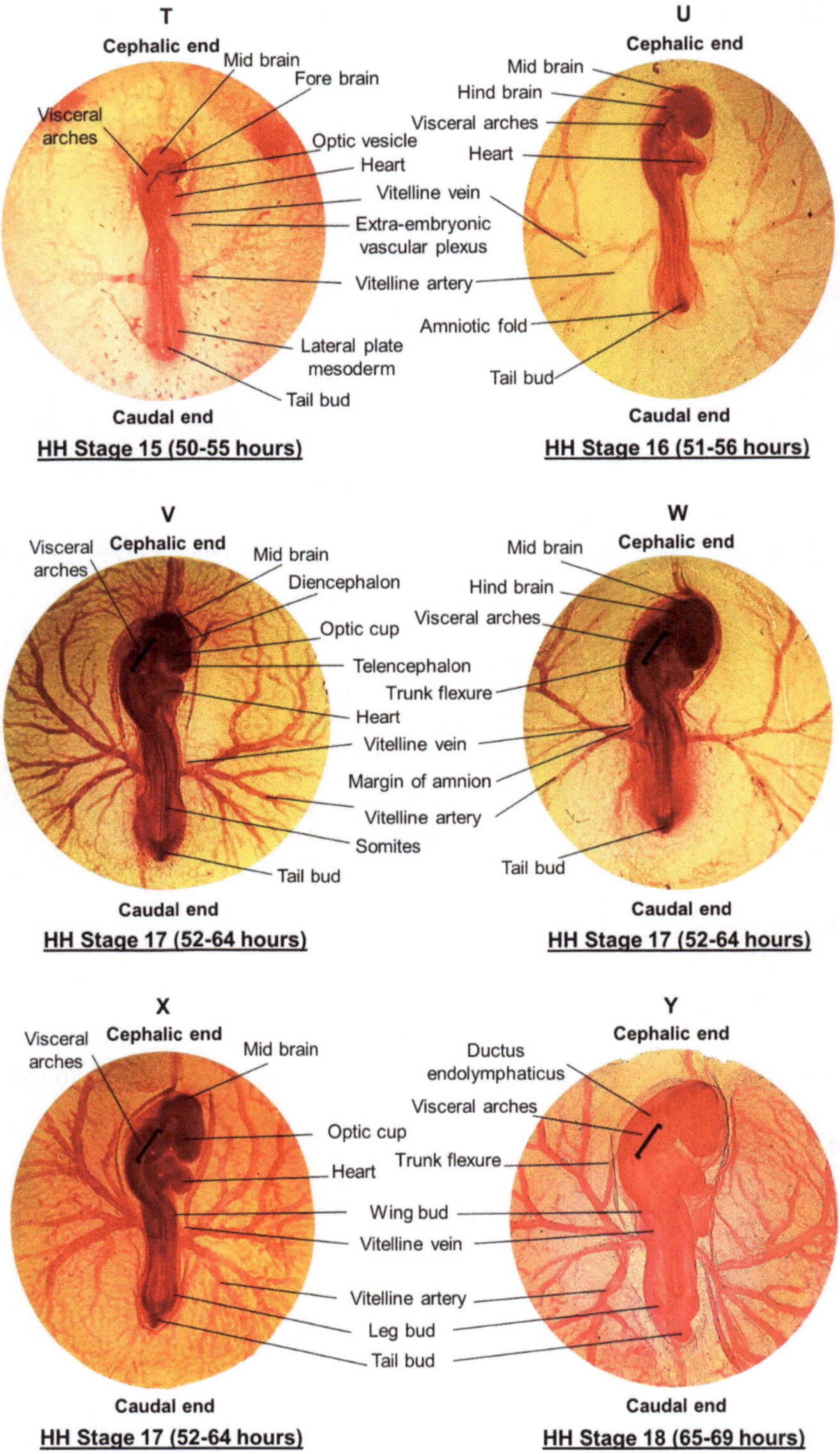

Fig. 3.5 (continued)

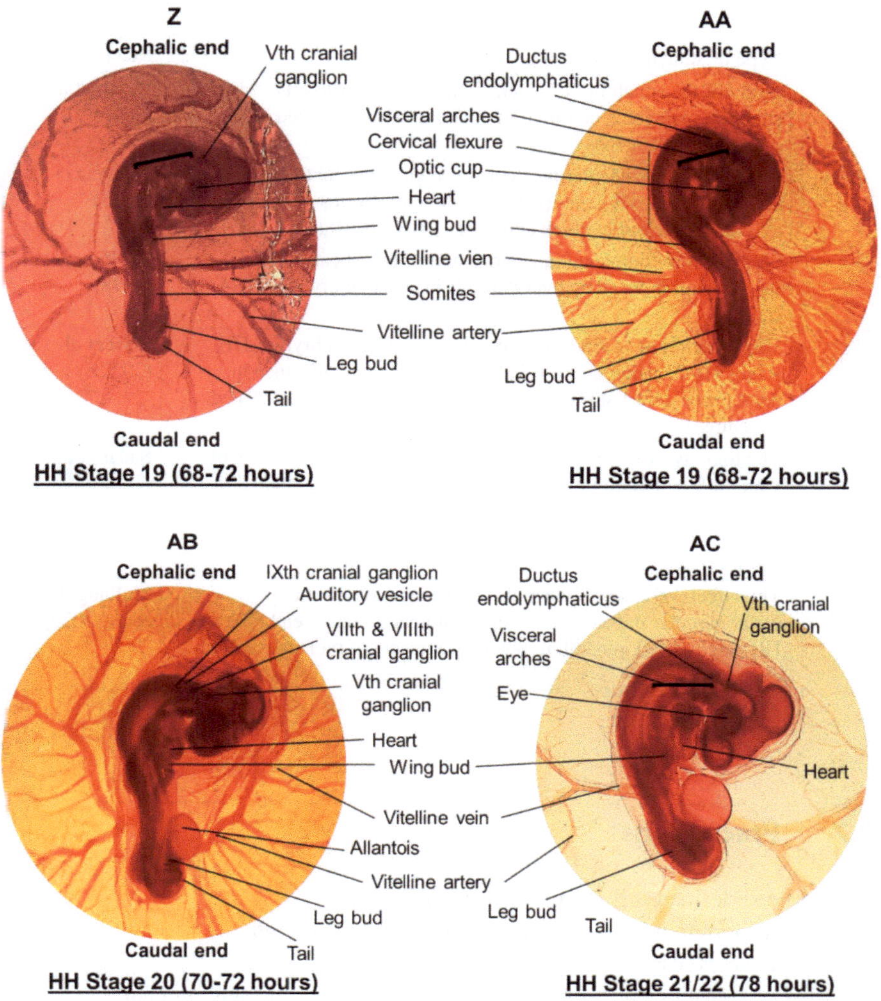

Fig. 3.5 (continued)

Table 3.1 Representative studies highlighting the significance of chick embryos

S. no.	Type of study/research	Reference
1.	Study of the suitability of chick embryos at HH stage 12–13 for studying melanoma cell migration and invasion induced by embryonic oncogenes	Busch et al. (2013)
2.	Evaluation of molecular mechanisms regulating the timely activation of HoxB genes (key regulators of anterior-posterior body axis specification) in chick embryos of HH7–14 stages	Gouveia et al. (2015)
3.	Study of developmental deviations and cellular heat shock mechanisms induced by hyperthermia in chick embryos of HH4–10 stages	Buckiova et al. (1998)
4.	Visualization of secondary neurulation in chick embryo of HH stages 10–22 using micro-CT, highlighting its significance in neural tube defect studies, to serve as a reference for associated structures	Choi et al. (2022)
5.	The toxicity of lambda-cyhalothrin and the protective effects of vitamin E and extra virgin olive oil in embryos were studied in golden black chick embryos. The test agents were administered to embryos at about 24 hours of incubation, corresponding to embryos of ~HH6 stage	Sadaf et al. (2022)
6.	The effects of alcohol (ethanol), a teratogen, on normal cardiovascular development were examined in chick embryos at HH stages 4–10, modelling potential impacts of prenatal alcohol exposure	McLaughlin et al. (2025)

3.7 Precautions

- Handle slides carefully to prevent damage.
- Use appropriate microscope settings for clear resolution.
- Do not use excessive light or heat on prepared slides.
- Label observations accurately and match them with the staging chart.

3.8 Quick Review Questions

3.8.1 Multiple Choice Questions

1. How many pairs of somites are typically present in a chick embryo at HH stage 7?
   (a) 1–3
   (b) 4–6
   (c) 7–9
   (d) 10–13
   Answer: a) 1–3

2. Which of the following structures is first clearly visible at about 13 hours of chick development (HH stage 4)?
   (a) Limb buds
   (b) Primitive streak
   (c) Somites
   (d) Optic vesicles
   Answer: b) Primitive streak
3. At HH stage 10 (33 hours), how many somite pairs are typically present?
   (a) 1–2
   (b) 10–13
   (c) 20–25
   (d) 30–35
   Answer: b) 10–13
4. Which of the following best describes the embryo at HH stage 17 (52–56 hours)?
   (a) Limb buds absent, only primitive streak visible
   (b) Limb buds just appearing, heart looped
   (c) Eye pigmentation fully developed, maxillary process elongates
   (d) Tail somites extend fully to the tip
   Answer: b) Limb buds just appearing, heart looped
5. The first organ system to become functional in the chick embryo is the:
   (a) Nervous system
   (b) Circulatory system
   (c) Respiratory system
   (d) Musculoskeletal system
   Answer: b) Circulatory system
6. Which of the following is characteristic of HH stage 20 (72 hours)?
   (a) Fully formed forelimbs
   (b) Maxillary process enlarges, optic vesicle pigmentation begins
   (c) Heart is a straight tube
   (d) Allantois absent
   Answer: b) Maxillary process enlarges, optic vesicle pigmentation begins
7. What is the amnion in chick embryo development?
   (a) An extraembryonic membrane enclosing the embryo in a fluid-filled cavity
   (b) A vascular membrane responsible for gas exchange
   (c) The embryonic structure forming the yolk sac
   (d) A thickened area of ectoderm giving rise to the nervous system
   Answer: a) An extraembryonic membrane enclosing the embryo in a fluid-filled cavity

8. What does the Hamburger–Hamilton (HH) staging system describe?
   (a) The nutritional requirements of chick embryos during incubation
   (b) The chronological time in hours of chick development
   (c) The morphological stages of chick embryonic development from fertilization to hatching
   (d) The order of somite formation in vertebrates

   Answer: c) The morphological stages of chick embryonic development from fertilization to hatching
9. What is the area vasculosa in the chick embryo?
   (a) The clear central region of the blastoderm without blood vessels
   (b) The peripheral avascular region surrounding the yolk
   (c) The vascularized region of the area opaca involved in nutrient absorption and circulation
   (d) The thickened area of epiblast giving rise to mesoderm

   Answer: c) The vascularized region of the area opaca involved in nutrient absorption and circulation
10. What are visceral arches in the chick embryo?
    (a) Transient embryonic structures that contribute to the formation of the face, neck, and pharyngeal organs
    (b) Paired outgrowths of mesoderm that develop into limb buds
    (c) Neural folds that give rise to cranial nerves
    (d) Extraembryonic folds that form the amnion and chorion

    Answer: a) Transient embryonic structures that contribute to the formation of the face, neck, and pharyngeal organs

3.8.2 Short Answer Questions

1. Define the primitive streak in chick embryonic development.
2. State the approximate number of somite pairs present at HH stage 10 (33–38 hours).
3. Differentiate between the proamnion and the amnion in chick embryonic development.
4. What is a somite in chick embryonic development?
5. Define visceral arches and name the arches visible between HH stages 20–24.
6. Explain the significance of whole-mount preparations in studying chick embryonic development.
7. Why are chick embryos widely used as a model system in vertebrate embryology?
8. What is the most significant feature used to differentiate and identify chick embryo stages in the Hamburger–Hamilton (HH) staging system?
9. Are somite numbers the most reliable feature for identifying chick embryos between HH stages 15 and 24? Explain?
10. State the function of Hensen's node during chick embryonic development.

Glossary

Allantois An extraembryonic membrane that develops as an outgrowth from the embryo's hindgut. It aids in gas exchange and storage of nitrogenous wastes during early development.

Anterior intestinal portal The anterior infolding of the splanchnopleure (lateral plate mesoderm fused with endoderm) that, with the caudal intestinal portal, meets the midgut (open to the yolk sac) to form the primitive foregut and hindgut.

Aortic arch Paired, curved embryonic blood vessels that arise from the truncus arteriosus of the developing heart and connect to the paired dorsal aortae. They appear within the pharyngeal arches and undergo remodeling to form the major arteries of the head, neck, and thorax.

Auditory (otic) placode A thickened, specialized region of surface ectoderm located on either side of the developing hindbrain in the embryo. It invaginates to form the otic pit, which subsequently pinches off to give rise to the otic vesicle.

Auditory (otic) vesicle A paired, spherical epithelial sac that develops from the thickened ectodermal otic placode located on either side of the hindbrain. This vesicle gives rise to the membranous labyrinth of the inner ear, including the cochlear and vestibular structures, which are essential for hearing and balance.

Area opaca The outer region of the blastoderm where the cells remain firmly attached to the yolk, giving it a darker and more opaque appearance.

Area pellucida The translucent central region of the blastoderm, separated from the yolk by the segmentation cavity. Owing to this separation, it appears translucent.

Area vasculosa The proximal region of the area opaca that becomes three-layered, consisting of ectoderm, mesoderm, and germ wall, and is characterized by the presence of blood islands.

Area vitelline The distal peripheral region of the area opaca, composed of only two layers, the ectoderm and endoderm, and lacks mesoderm.

Blood island Cell masses situated on the germ wall that give rise to the first blood corpuscles (erythrocytes) and blood vessels.

Bulbo-conus arteriosus A transient embryonic heart region between the primitive ventricle and truncus arteriosus, contributing to the outflow tract, right ventricle, ventricular outflow tracts, and bases of the great arteries.

Caudal fold The posterior embryonic fold formed by upward and lateral bending of the tail region, undercutting the posterior end of the embryo from the extraembryonic membranes, and contributing to the posterior body wall.

Caudal flexure A ventral bend at the posterior end of the developing embryo, near the tail region of the neural tube.

Cervical flexure A ventral bend in the embryonic neural tube at the junction between the developing hindbrain (rhombencephalon) and spinal cord.

Cranial flexure A ventral bend in the embryonic neural tube at the level of the midbrain (mesencephalon). First major flexure to appear during brain development.

Cranial ganglion A group of nerve cell bodies associated with cranial nerves that relay sensory signals or control autonomic functions in the head and neck.

Cranial ganglion V (trigeminal ganglion) A sensory ganglion that transmits pain, touch, and temperature signals from the face, oral cavity, and craniofacial areas to the brain.

Cranial ganglion VIII (vestibulocochlear ganglion) A sensory ganglion derived from the otic placode, transmitting signals for hearing (cochlear) and balance (vestibular) from the inner ear to the brain.

Cranial ganglion IX (glossopharyngeal ganglion) A sensory ganglion associated with the glossopharyngeal nerve, contributing to taste sensation, visceral sensory input, and innervation of the pharyngeal region.

Cranial ganglion X (vagal ganglion) A paired sensory and autonomic ganglion associated with the vagus nerve, involved in transmitting visceral sensory information, regulating autonomic functions, and innervating the pharynx, larynx, and thoracoabdominal organs.

Diencephalon The caudal portion of the forebrain, located between the anterior forebrain (telencephalon) and the midbrain, giving rise to structures like the hypothalamus and thalamus.

Epiblast The outermost cell layer of the early embryo (blastoderm), which generates ectoderm, mesoderm, and endoderm during gastrulation, and also forms extraembryonic structures such as the amnion.

Germwall The inner zone of the area opaca, consisting of cells partially freed from the yolk, often containing yolk granules, and marking the boundary between the area opaca and area pellucida.

Lateral plate mesoderm The lateral plate mesoderm, lateral to the intermediate mesoderm, splits into somatic mesoderm forming body wall and limbs, and splanchnic mesoderm forming the gut wall and associated organs.

Lateral body fold (lateral limiting sulci) Paired folds along the embryo's sides that deepen during development, undercutting the body, separating it from the yolk, and forming the lateral boundaries.

Leg bud A small lateral protrusion on the embryo marking the future hindlimb, composed of mesenchyme covered externally by ectoderm.

Margin of amnion The edge of the amniotic membrane.

Mesencephalon (midbrain) The middle brain region between the forebrain (prosencephalon) and hindbrain (rhombencephalon), responsible for vision, hearing, and motor coordination.

Metencephalon The anterior region of the hindbrain that develops into the pons and cerebellum, essential for motor control, balance, and coordination.

Myelencephalon The caudal region of the hindbrain that develops into the medulla oblongata, crucial for regulating vital functions such as breathing and heart rate.

Neck of Latebra Part through which the nucleus moves from the central yolk core towards the animal pole.

Neural fold Raised ridge along the edges of the neural plate, which later converges and fuses to form the neural tube, the precursor of the brain and spinal cord.

Neuropore A temporary opening at either end of the developing neural tube. The anterior (cranial) neuropore closes to form the brain, and the posterior (caudal) neuropore closes to form the spinal cord.

Notochordal process A rod-like structure of mesodermal cells that extends cranially from the primitive node (Hensen's node) along the midline. It serves as a precursor to the notochord.

Pharyngeal/visceral arches Paired, segmental transient structures that appear in the vertebrate embryonic pharyngeal region in a cranial-to-caudal sequence as bulges. They give rise to specific structures of the face, neck, and pharynx.

Pharyngeal/visceral arch I (mandibular arch) The first of the pharyngeal arches contributing to the formation of the lower jaw, maxilla, muscles of mastication, and associated skeletal elements. Its core consists of mesenchyme cells and neural crest cells, covered externally by ectoderm and internally by endoderm.

Pharyngeal/visceral arch II (hyoid arch) The second pharyngeal arch, contributing to the formation of the hyoid bone, muscles of facial expression, and associated skeletal elements. Like other arches, its core consists of mesenchyme cells and neural crest cells, covered externally by ectoderm and internally by endoderm.

Pharyngeal/visceral arch III It is the third pharyngeal arch contributing to the formation of parts of the hyoid apparatus and thymus, as well as associated muscles and arteries. Its core consists of mesenchyme cells and neural crest cells, covered externally by ectoderm and internally by endoderm.

Pharyngeal/visceral arch IV It is the fourth pharyngeal arch. It contributes to the formation of structures in the larynx, pharynx, and associated arteries, including parts of the aortic arch. Its core consists of mesenchyme cells and neural crest cells, covered externally by ectoderm and internally by endoderm.

Pharyngeal/visceral clefts Indentations or grooves that appear between arches externally during vertebrate embryonic development.

Pharyngeal/visceral cleft I The first groove between the first and second pharyngeal arches on the external surface. It contributes to the formation of the external auditory meatus (ear canal).

Pharyngeal/visceral cleft II The groove between the second and third pharyngeal arches on the external surface. It is largely overgrown by the second arch.

Pharyngeal/visceral cleft III The groove between the third and fourth pharyngeal arches on the external surface.

Pharyngeal/visceral cleft IV The groove between the fourth and fifth pharyngeal arches on the external surface.

Primitive streak The primitive streak is a midline thickening on the blastoderm that marks the future anterior-posterior body axis of the embryo and is the site of epiblast cell ingression during gastrulation.

Proamnion A mesoblast-free area located in the cephalic region, in the front and beneath the developing head of the early embryo. It plays a role in the formation of extraembryonic membranes, particularly the amnion and chorion, during early development.

Prosencephalon (forebrain) It is the anterior part of the brain. It gives rise to the diencephalon (thalamus and hypothalamus) and telencephalon (cerebral hemispheres), playing key roles in sensory integration, cognition, and hormonal regulation.

Omphalomesenteric vein (vitelline vein) Blood vessel that drains nutrient-rich blood from the yolk sac into the sinus venosus of the heart. It plays a crucial role in early embryonic circulation before the formation of definitive veins.

Omphalomesenteric artery (vitelline artery) It is a paired vessel that carries oxygenated blood from the dorsal aorta to the yolk sac, supplying nutrients to the developing embryo during early development.

Optic cup The optic cup is a double-layered, cup-shaped structure formed by the invagination of the optic vesicle. It gives rise to the retina, with the inner layer forming the neural retina and the outer layer forming the pigmented retina.

Sinus venosus A thin-walled, paired chamber that collects venous blood from the body and extraembryonic vessels (like the vitelline and umbilical veins) and channels it into the primitive atrium of the heart. It plays a key role in early embryonic circulation.

Somites Paired, segmented blocks of paraxial mesoderm (band of mesodermal tissue located on either side of the neural tube and notochord in the developing embryo) located alongside the neural tube. They give rise to vertebrae, skeletal muscles, and dermis, establishing the segmental organization of the body.

Tail bud A mass of undifferentiated mesenchymal cells at the posterior end. It contributes to the formation of the caudal vertebrae, spinal cord, and tail structures during late embryonic development.

Telencephalon It is the anterior part of the forebrain in vertebrate embryos, which develops into the cerebral hemispheres, including structures involved in cognition, sensory processing, and voluntary motor control.

Vena (sinus) terminalis The distinct peripheral boundary or a sinus-like blood vessel that marks the outer margin of the area vasculosa, separating the vascularized area vasculosa from the non-vascular area vitellina of the area opaca.

Wing bud A small, lateral protrusion on the body wall of a developing embryo that marks the future site of the forelimb. It consists of mesenchyme covered by ectoderm and gives rise to the wing structures.

References

Bellairs R, Osmond M (2005) Atlas of chick development. Elsevier

Buckiova D, Kubinova L, Soukup A, Jelinek R, Brown NA (1998) Hyperthermia in the chick embryo: HSP and possible mechanisms of developmental defects. Int J Dev Biol 42(5): 737–740. https://www.ncbi.nlm.nih.gov/pubmed/9712529

Busch C, Krochmann J, Drews U (2013) The chick embryo as an experimental system for melanoma cell invasion. PLoS One 8(1):e53970. https://doi.org/10.1371/journal.pone.0053970

Choi S, Kim KH, Kim SK, Wang KC, Lee JY (2022) Three-dimensional visualization of secondary neurulation in chick embryos using microCT. Dev Dyn 251(5):885–896. https://doi.org/10.1002/dvdy.441

Gilbert SF (2000) Developmental biology, 6th edn. Sinauer Associates, Sunderland (MA)
Gouveia A, Marcelino HM, Goncalves L, Palmeirim I, Andrade RP (2015) Patterning in time and space: HoxB cluster gene expression in the developing chick embryo. Cell Cycle 14(1): 135–145. https://doi.org/10.4161/15384101.2014.972868
Hamburger V, Hamilton HL (1992) A series of normal stages in the development of the chick embryo. 1951. Dev Dyn 195(4):231–272. https://doi.org/10.1002/aja.1001950404
Lillie FR (1919) The development of the chick: an introduction to embryology. H. Holt
McLaughlin D, Keane A, Quinn J, Tipping N (2025) The chick embryo model as an educational tool to explore the effect of alcohol on cardiovascular development. Adv Physiol Educ 49(2): 438–460. https://doi.org/10.1152/advan.00195.2024
Patten BM (1920) The early embryology of the chick. P. Blakiston's son & Company
Sadaf K, Ahmad KR, Ahmad SN, Kanwal U, Suleman S, Aslam Z, Inayat I, Siddique S, Kanwal MA (2022) Mitigating properties of vitamin E and olive oil on the teratomorphogenic impacts of lambda-cyhalothrin in developing chick embryos. Toxicol Rep 9:1901–1905. https://doi.org/10.1016/j.toxrep.2022.10.008

In Vivo Study of Chick Embryo Development by Windowing and Candling Methods

4

Anna Senrung and Alok Chandra Bharti

Abstract

The study of chick embryos has been fundamental to developmental biology and embryology and has recently emerged as a preferred alternative to rodent models. As an experimental system, the chick embryo helps reduce the use of higher vertebrate models, serving as an essential bridge between in vitro systems and mammalian models in biomedical research. Their suitability lies in the year-round availability of fertilized chicken eggs at a low cost (approximately Rs. 25–30 per egg in India) and their external development, which provides a manipulable, cost-effective system for in vivo experimentation. Chick embryos are particularly valued for their extraembryonic membranes, especially the chorioallantoic membrane (CAM), which serves as a vital model for angiogenesis and related studies. This practical exercise describes the maintenance of chick embryos within their natural eggshell environment, allowing observation and analysis through windowing and candling techniques. These methods facilitate the study of sequential developmental stages, physiological processes, and early morphogenesis while minimizing ethical concerns associated with higher vertebrate models.

Supplementary Information The online version contains supplementary material available at https://doi.org/10.1007/978-981-95-7269-4_4.

A. Senrung
Department of Zoology, Daulat Ram College, University of Delhi, New Delhi, India

A. C. Bharti (✉)
Molecular Oncology Lab., Department of Zoology, University of Delhi, New Delhi, India

A. C. Bharti et al. (eds.), *Textbook of Developmental Biology*,
https://doi.org/10.1007/978-981-95-7269-4_4

Keywords

Chick embryo · Windowing method · Candling method

Aim To study the sequential developmental changes in chick embryonic growth through direct observation within the egg at different stages of incubation using windowing and candling techniques.

Learning Objectives
Upon completing the experiment, students will be able to:

- Define and differentiate between the windowing and candling methods used in chick embryo culture
- Successfully culture chick embryos utilizing both windowing and candling techniques
- Identify and classify the various stages of chick embryo development based on morphological features

4.1 Introduction

Chick embryos serve as an invaluable model system in developmental biology, vertebrate embryology, and modern biomedical research. Their study provides fundamental insights into the processes of growth, differentiation, and morphogenesis in vertebrates. Observing chick embryo development in real time within the natural shell environment allows a deeper understanding of embryonic stages, survival, and growth patterns, contributing to improved experimental outcomes and management practices.

Handling and culturing live chick embryos offer a practical learning experience that goes beyond theoretical study or observation of preserved specimens, slides, or images. Such hands-on experimentation helps validate and reinforce concepts described in textbooks and literature.

The in vivo approach is the most natural and economical method for studying chick embryo development. The term in vivo refers to studies or experiments conducted within a living organism, distinguishing them from in vitro studies performed in artificial environments like test tubes or culture dishes. In vivo chick embryo studies maintain the developing embryo within its natural setting, the eggshell, preserving physiological conditions essential for normal development.

Among in vivo techniques, the windowing and candling methods are particularly significant. The windowing method enables direct visualization, manipulation, and experimental intervention on the developing embryo and is widely employed in biomedical research such as drug discovery and testing (Bahadir et al. 2024; Senrung et al. 2023a). In contrast, the candling method provides a non-invasive approach to monitor embryonic development and viability during incubation. It is commonly used in modern avian management to observe the interior of the egg, assess embryo growth, and track hatching progress. Candling also facilitates early detection of dead or degenerated embryos or eggs, reducing the risk of contamination or infection from fouled eggs within the incubator (Hall et al. 2023). Additionally, it allows safe withdrawal of vitelline or chorioallantoic blood samples for health and genetic analyses of the developing embryo by enabling the location of blood vessels (Hall et al. 2023). Together, these complementary techniques enable continuous and detailed observation of embryogenesis while preserving the embryo's natural growth environment.

For experimental or educational purposes, chick embryos can be cultured in ovo (within the shell) through windowing or ex ovo (outside the shell) using shell-less culture methods, depending on the objectives of the study.

This chapter outlines the procedure for culturing chick embryos and analyzing their morphological development through the windowing technique, along with overall visualization using the candling method.

4.2 Requirements

Fertilized eggs of white Leghorn, humidified chamber (REMI CHM-10 PLUS) or BOD, Mini Electric Drill, parafilm strip, plastic tray, spirit lamp, steel blade, match stick, camera, egg visualization platform, source of light/illuminator, 90% and 70% alcohol.

4.3 Methodology

4.3.1 Maintenance of Fertilized Eggs

Proper maintenance of fertilized eggs is crucial for successful embryonic development and experimental outcomes.

(i) Obtain pathogen-free fertilized eggs (white Leghorn, *Gallus gallus*) from a reliable farm source.
(ii) Wipe the eggs with 70% alcohol to sterilize the surface.
(iii) Designate the day the fertilized eggs are set for incubation as embryonic development day 0 (EDD-0).
(iv) Place the fertilized eggs inside a humidity chamber (e.g., REMI CHM-10 PLUS)/BOD.
(v) Set the humidity chamber to a temperature of 37 °C ± 1 and adjust the relative humidity (RH) to 80% ± 10 or ≥60% at least 3 h before incubation.

Fig. 4.1 Eggs laid horizontally in egg trays kept in humidity chamber

(vi) Wipe the egg trays with 70% alcohol using a muslin cloth. Allow the alcohol to dry completely before placing the trays in the humidity chamber.

(vii) Number the eggs sequentially (1, 2, ...) using a pencil for easy handling, identification, and recording.

(viii) Lay the eggs horizontally (Fig. 4.1) or with the pointed end facing downwards depending on the need of the study in the prepared egg trays placed within the humidity chamber.

(ix) Rotate the eggs at least every 3–4 h during the daytime (e.g., around 9:30–10 am, 2 pm, and 5:30–6 pm) to prevent the vitelline membrane from sticking to the eggshell.

Practical Tips: Maintenance of Fertilized Eggs

Rotation of the egg is essential and should be performed at least once daily. Failure to rotate can cause blood vessels to adhere to the inner eggshell membrane. Subsequent rotation may rupture these vessels, disrupting the embryo during albumin withdrawal for windowing culture method and potentially leading to embryo death or impaired development in chick embryo cultures used for candling-based visualization.

4.3.2 Windowing Method

The windowing method involves creating an opening in the eggshell to directly access and observe the developing chick embryo. This technique is particularly valuable for in vivo studies, as it allows for direct manipulation of the embryo and its surrounding environment. Researchers can introduce substances, perform microsurgery, or observe developmental processes such as organ formation and vascular development. The procedure typically involves carefully removing a section of the shell and sealing the opening with a transparent film (parafilm) or glass coverslip, ensuring the embryo remains in a stable environment. The windowing method has revolutionized embryological research by facilitating detailed morphological and physiological studies.

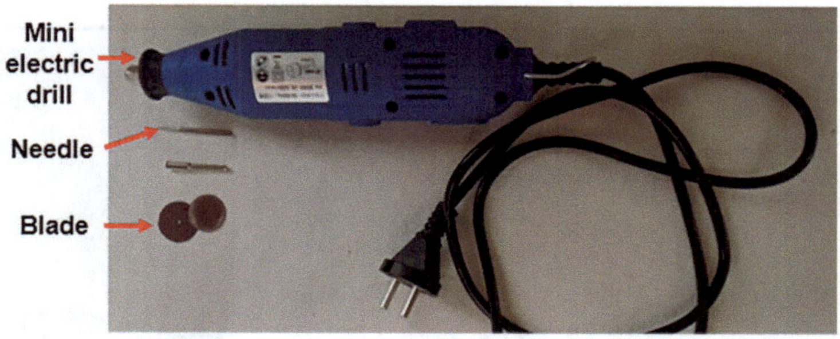

Fig. 4.2 230 V LX134MO 130 W Mini Electric Drill

Procedure for Windowing Culture

(Note: The windowing procedure should ideally be performed within the first 3 days of embryonic development day (EDD) post-incubation.)

(i) *Preparation*: Keep the eggs in the same horizontal position for about 2–5 min before beginning the window-making procedure. This positions the embryos at the top of the egg, facilitating setup and experimentation without harming them.

(ii) *Initial hole creation*: Using a Mini Electric Drill (Fig. 4.2) fixed with a needle (0.5–1 mm thick) or using scissors, create a small hole on the upper-facing surface of the narrower end of the egg, about 0.5 cm from the tip (Fig. 4.3Aa–c). After drilling and making the hole, ensure the eggs are not rotated thereafter.

(iii) *Albumin removal*: Using a sterilized 5 ml Dispovan syringe with a 24-gauge needle, slowly insert the needle through the hole downward toward the bottom of the egg. Angle/tilt the needle toward the center at approximately 45° angle and draw up 3–4 mL of albumin (egg white) through the hole (Fig. 4.3Ad).

(iv) *Seal the hole*: Seal the hole with parafilm to prevent drying of the egg contents after albumin removal and place it back into the humidity chamber (Fig. 4.3Ae–i).

(v) *Window creation*: On the following day, using the Mini Electric Drill fixed with a blade or using scissor, create a window in the middle region of the egg on the same side where the albumin removal hole was made. The window will enable easy access and visualization of the embryo.

(vi) *Window size*: The initial window, following the withdrawal of 4 ml of albumin, typically should be approximately 1.5–2 cm × 2–4 cm (short axis × long axis) to avoid damaging the developing embryo (Fig. 4.3Ba–f). The window size can be further enlarged as needed once the embryo's position is identified and exposed.

(vii) *Seal the window*: Seal the window with parafilm by gently melting the edges of the parafilm onto the eggshell around the window opening using a warm/

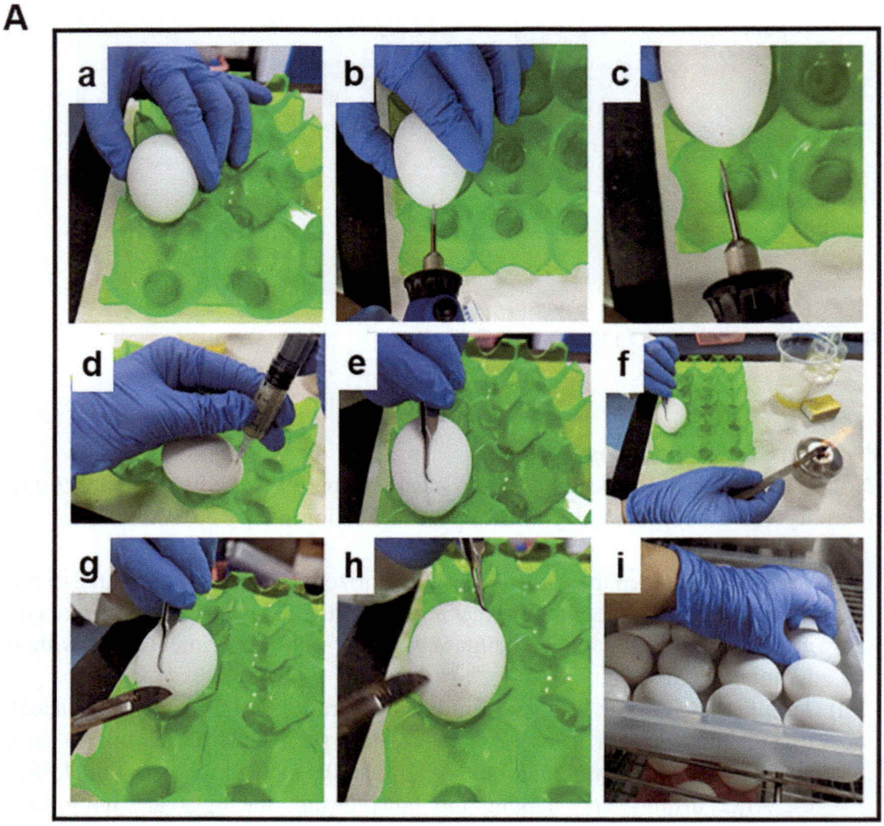

Fig. 4.3 Steps to windowing method. (**A**) Albumin removal from the egg. (**B**) Window making (windowing)

heated blade (Fig. 4.3Bg–j). This sealing step is crucial to maintain the humidity and prevent desiccation of the embryo.

(viii) *Incubation*: After sealing, carefully place the egg into the humidity chamber and place it on an egg tray (Fig. 4.3Bk–l). Ensure that the tray containing the culture dishes is positioned securely within the chamber to maintain stable culture conditions.

Practical Tips: Procedure for Windowing Culture

- After withdrawing the albumin from the egg, the egg should not be rotated.
- The window in the eggshell should not be made immediately after albumin withdrawal; the egg should be kept undisturbed overnight before creating the window.

(continued)

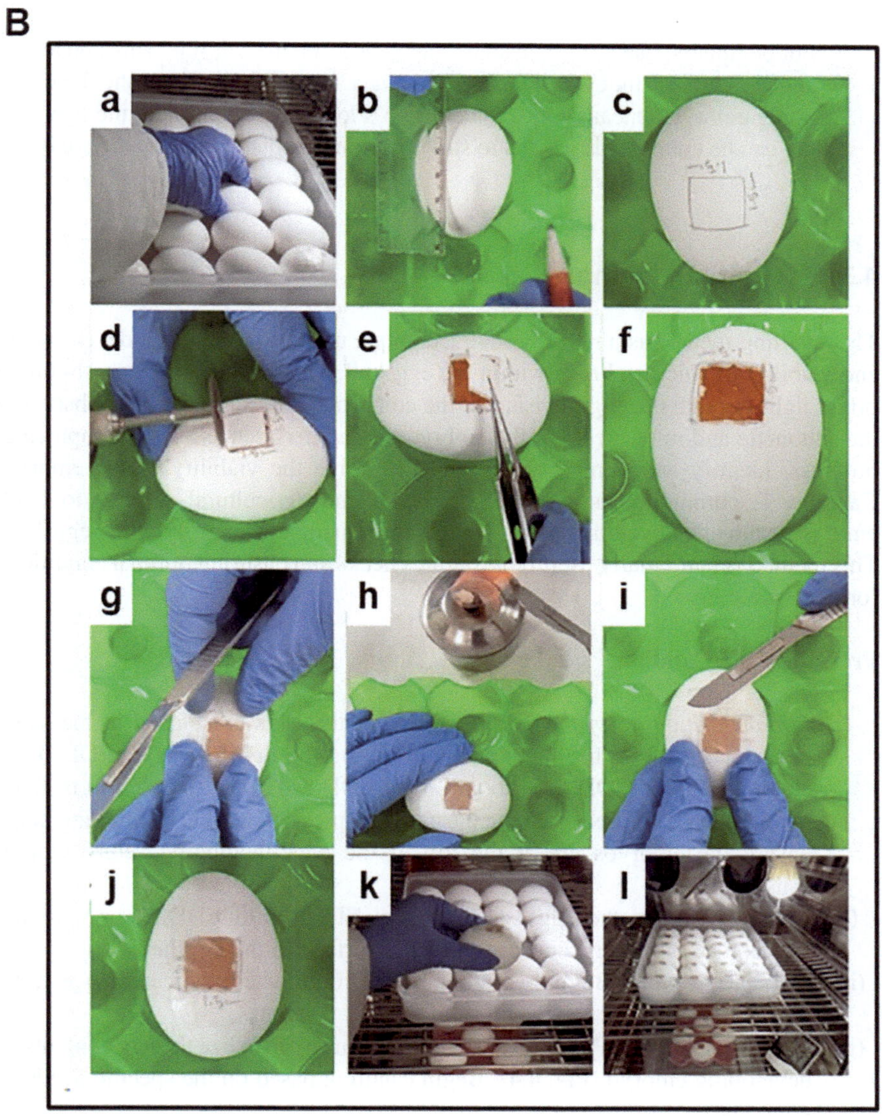

Fig. 4.3 (continued)

- The volume of albumin withdrawn should not exceed 4 ml, as excessive removal may result in abnormal or incomplete embryonic development.

(continued)

- For beginners, mark a square area of about 1.5 cm on each side with a pencil and create the window within this boundary to avoid damaging the developing embryo beneath. Once the embryo's location is identified, the opening can be enlarged as needed to suit the requirements of the specific study.

4.3.3 Candling Method

The candling method is a non-invasive technique used to monitor the development and viability of chick embryos within the eggshell. This method involves shining a bright light through the egg to visualize the embryo and its structures. By observing the silhouette and movements of the embryo, researchers can assess developmental progress, identify any abnormalities, and determine the viability of the embryo. Candling is commonly used in both research and agricultural settings to track embryo health and development without compromising the integrity of the eggshell. This method is particularly useful for routine checks and ensuring optimal conditions for embryonic growth.

Procedure for Candling

(i) *Initial positioning of the egg:* Ensure that the eggs are placed in the same horizontal position for about 2–5 min during the first few days (EDD 0–4), about 30 min on EDD 5–7, about 2 h on EDD 8–12, and over 5 h from EDD 13 to EDD 20 before beginning the candling procedure. This positioning helps ensure the embryos are at the top of the egg, facilitating setup without causing harm.

(ii) *Rotation:* Rotate the eggs starting from EDD1 until EDD20, or until the appearance of the first crack in the outer surface of the eggshell.

(iii) *Darken the room:* Begin by darkening the room where the candling is to be performed to ensure the visibility of the egg contents.

(iv) *Visualize contents:* Place the eggs one at a time to visualize the egg contents or developing embryo (Fig. 4.4). Begin candling based on the specific needs of your study.

(v) *Candling device:* Use a candling device, which may be a customized setup (Fig. 4.4), a specialized candling lamp, or simply a strong light source such as a flashlight.

(vi) *Handling the egg:* Hold or place the egg securely but gently, ensuring that its position is maintained. Depending on the setup, position the egg either horizontally (Fig. 4.4) or with the pointed end facing downwards to prevent disturbing the embryo. Marking the eggs to maintain consistent positioning will enhance the reliability of the process for observers.

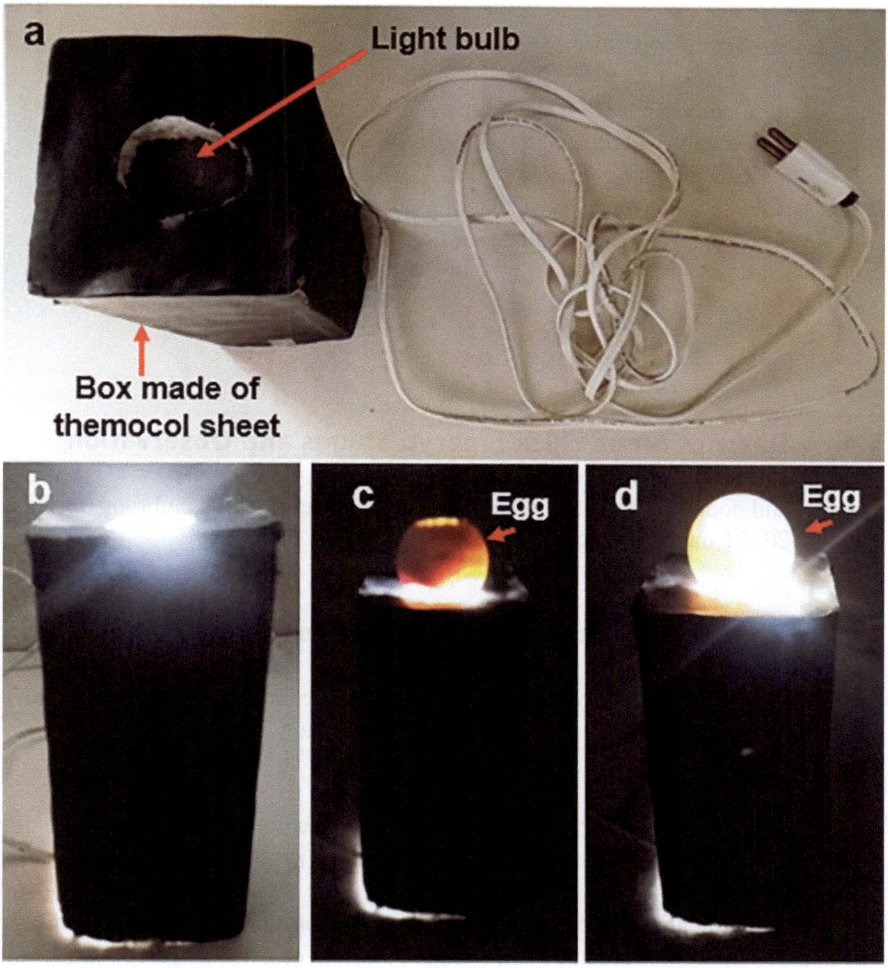

Fig. 4.4 Candling method with a customized illuminator

(vii) *Illuminate the egg:* Position the light source against the egg in the desired position and gently illuminate the shell to observe the contents within.

(viii) *Observation:* Look for indications of embryo viability or development progression, such as veins, shadows, or movement within the egg.

(ix) *Record observations:* Note the observations for each egg, paying attention to any abnormalities or notable signs of development.

Practical Tips: Procedure for Candling

- The opening on the egg visualization platform for the light source should be appropriately sized to snugly fit the egg, ensuring that the candling light is directed exclusively through the egg under observation for optimal visualization and imaging.
- Before starting embryo visualization and imaging, or on the first day of observation, mark for visualization surface to ensure consistent positioning. This allows uniform monitoring of gradual developmental changes in the embryos.

4.4 Description of Chick Embryo Stages and Observation

Observe and document the daily developmental changes in chick embryos from day 1 to day 20–21 until the chick emerges/hatches.

4.4.1 Timeline of Chick Embryo Development Day-Wise

The events in chick embryonic development can be broadly divided into three stages: preoviposition, the period between oviposition and incubation, and the incubation period.

Preoviposition Chick embryo development begins with fertilization, where the fusion of the oocyte and sperm forms a single-cell zygote in the infundibulum, occurring about 15–40 min post-oviposition. The zygote (fertilized oocyte) undergoes a sequential process of division starting approximately 5 h after fertilization and continuing for about 20 h, forming a blastula or a multilayered blastoderm (germinal disc or blastodisc) of ~60,000 cells, visible as a whitish spot on the top surface of a large yolk. Typically, about 24–25 h after fertilization, the eggs are laid at the blastula stage.

Period Between Oviposition and Incubation Upon being laid (oviposition), the egg cools, which halts cell division and renders the embryonic life inactive. Further development of the chick embryo resumes only when the egg is incubated under optimal conditions of temperature and humidity.

Incubation Period Upon incubating the fertilized eggs at 37 °C and a relative humidity of 80% (or greater than 60%), the embryo development or morphogenetic movement resumes after about 4–5 h. Below are the specific developmental features of the chick embryo on a day-by-day basis till hatching. The embryonic developmental changes below are documented according to the Hamburger-Hamilton (HH) staging system (Hamburger and Hamilton 1992).

Day 1 (HH 4–12): By 18–19 h, the definitive primitive streak is formed, averaging 1.88 mm in length. Around 19–22 h, the notochord or head process begins to appear and extends anteriorly. By approximately 23–25 h, the head-fold forms.

Day 2 (HH 13–20): At approximately 26 h, neural folds in the head region become noticeable. By 29–33 h, the development of the heart begins and the primary optic vesicles are established. Between 33 and 38 h, the first cranial flexure appears, the heart begins to bend to the right, and the three primary brain vesicles become distinctly evident. By 40–45 h, the bases of the optic vesicles are constricted, and the head begins to turn to the left side. At 45–49 h, the heart assumes a somewhat S-shaped form, the primary optic vesicles and stalk are defined, and 13–16 somites have formed.

Beyond 50–55 h of incubation, chick embryos exhibit prominent developmental features, including the emergence of limb buds, the formation of visceral arches, and the appearance of several other externally visible structures.

Day 3 (HH 20–23) (Fig. 4.5a)*:* Limb buds are formed, with leg buds being larger. Four visceral arches are developed, and nasal pits are evident. Approximately 37–43 somites have formed, and the amnion begins to develop. An avascular allantois appears, and by 70–72 h, eye pigmentation becomes visible as a faint grayish hue.

Day 4 (HH 24–26): Limb buds (wing and leg buds) are noticeably longer than they are wide. The wing and toe plates are distinct, but the digits have not yet formed. Four visceral arches are present, with the third and fourth clefts forming small pits. Other externally visible features include the optic vesicles, the expanding allantois, and clearer demarcations of the developing brain regions.

Day 5 (HH 27–28) (Fig. 4.5b)*:* The limbs are substantially lengthened compared to day 4, with distinct elbow and knee joints. The first three toes are demarcated, and beak formation is just beginning. The third and fourth small visceral pits become invisible at this stage. The optic, expanding allantois, neck elongation, and brain region demarcations are also prominently visible.

Day 6 (HH 29–30): The wing is bent at the elbow, with digits 1, 2, and 3 separated by shallow grooves. Webbing is present between toes 2, 3, and 4. The rudiment of the fifth toe is evident. The limbs exhibit a pointed contour as the second digit and the third toe are distinctly longer than the others. The beak outgrowth is prominent, but the egg-tooth has not yet appeared. The mandibular process of the first visceral and the second arch become fused, and the dorsal auditory meatus becomes distinct. The optic vesicles, expanding allantois, and regional brain demarcations remain clearly visible.

Day 7 (HH 31–33): Feather germs have developed, and the neck as well as all digits and toes are conspicuously lengthened. Indication of webbing is present between the first and second digits. The rudiment of the fifth toe disappears. Differentiation of the digits and toes becomes significant. The beak at this stage is more prominent than the previous, and egg-tooth is distinct as a slight protuberance.

Day 8 (HH 34–35) (Fig. 4.5c)*:* The differential growth of the second digit and third toe is prominent, and the neck continues to lengthen. Toe phalanges become

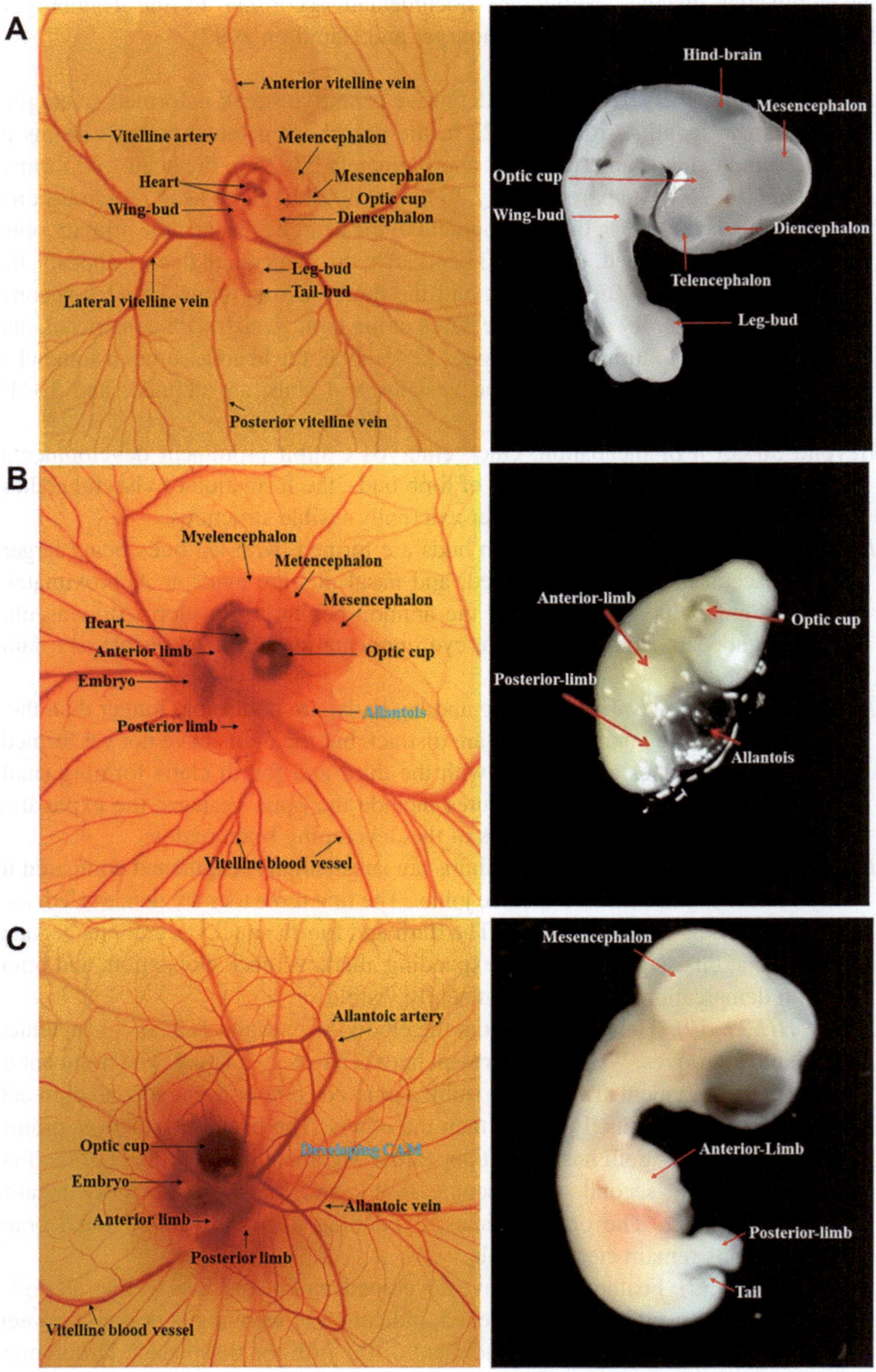

Fig. 4.5 Representative close-up images of few developmental stages of the chick embryo cultured using polystyrene weigh boat. Chick embryo at EDD3 (**a**), EDD5 (**b**), EDD8 (**c**), EDD10 (**d**), EDD12 (**e**), EDD14 (**f**). *Left panel*: images in situ; *Right panel*: dissected images of the corresponding developmental stage. (Images **c**–**f** are adapted from (Senrung et al. 2023a) with permission)

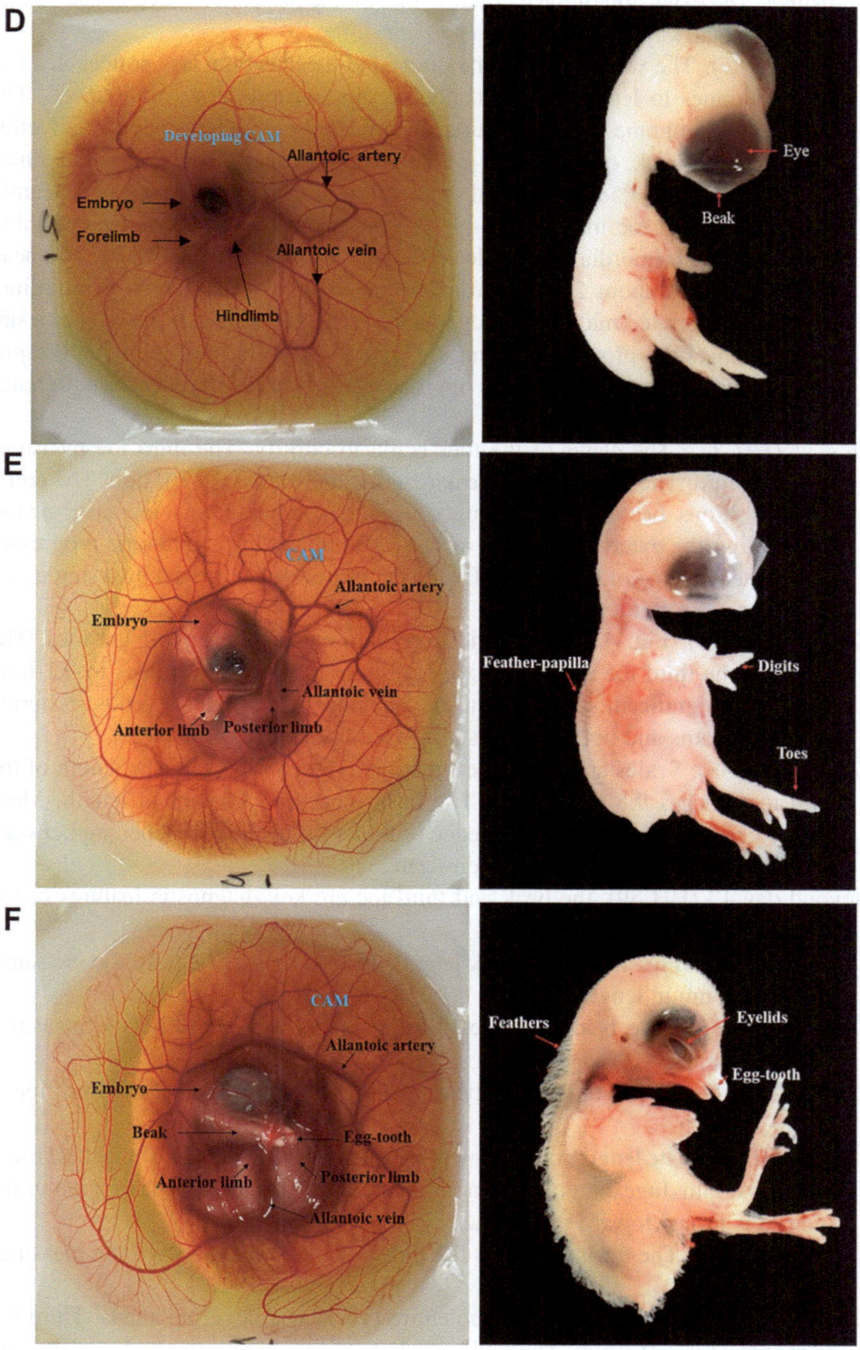

Fig. 4.5 (continued)

distinct. The extension of the nictitating membrane between the eyelid and scleral papillae is halfway complete.

Day 9 (HH 35): The webs between the digits and toes become less noticeable. The beak continues to lengthen, and feather germs become more pronounced. The nictitating membrane extends further toward the outer scleral papillae. Eyelids extend toward beak, overgrowing eyeball, forming an ellipsoidal circumference.

Day 10 (HH 36) (Fig. 4.5d)*:* The distal segments of the wing and leg are significantly longer. The third toe measures 5.4 ± 0.3 mm from tip to mid metatarsal joint. Tapering claw primordia are visible on the toe tips and on wing digit 1. The beak length measures about 2.5 mm from the nostril angle. The comb primordium appears as a serrated mid-dorsal ridge on the beak, and the nostril is seen as a slit. Flight feathers are pronounced, and the coverts at the base of the wing are just visible. The nictitating membrane approaches the cornea. Eyelid circumference forms a narrowing ellipse.

Day 11 (HH 37): The claws of the toes begin to cornify. The third toe measures 7.4 ± 0.3 mm. The length of the beak reaches approximately 3.0 mm. The comb with serrations becomes increasingly visible. Feather germs increase, nearly surrounding the external auditory meatus. The nictitating membrane progresses to the anterior edge of the cornea. Eyelid circumference bordered by a single row of feather germ primordia.

Day 12 (HH 38) (Fig. 4.5e)*:* Cornification continues at the tips of the toes. The length of the third toe measures 8.4 ± 0.3 mm. The beak lengthens to 3.1 mm. There is a significant reduction in the opening between the eyelids. External auditory meatus and upper eyelid surrounded by feather germs.

Day 13 (HH 39): Scales appear on the superior surface of the leg. The length of the third toe reaches 9.8 ± 0.3 mm. The beak measures 3.5 mm in length. More feather germs are added along the edge of the lower eyelid. The opening between the eyelids is condensed to a thin crescent.

Beyond day 13 (HH 39), the beak and third toe are key diagnostic features of the developing chick embryos.

Day 14 (HH 40) (Fig. 4.5f)*:* The beak lengthens to 4.0 mm. The third toe measures 12.7 ± 0.5 mm in length.

Day 15 (HH 41): The beak measures 4.5 mm in length. The third toe measures 14.9 ± 0.8 mm.

Day 16 (HH 42): The length of the beak is 4.8 mm. The length of the third toe is 16.7 ± 0.8 mm.

Day 17 (HH 43): The length of the beak is 5.0 mm, while the length of the third toe is 18.6 ± 0.8 mm. Ingestion of the amniotic fluid is initiated. The beak region of the embryo is turned toward the air sac.

Day 18 (HH 44): The beak measures 5.7 mm in length, and the third toe measures 20.4 ± 0.8 mm.

Day 19 (HH 45): Beak shortens as the entire peridermal covering sheds. Third toe measures 21.4 ± 0.8 mm. The yolk sac starts to retreat into the body cavity, and the blood vessels of the CAM are significantly reduced.

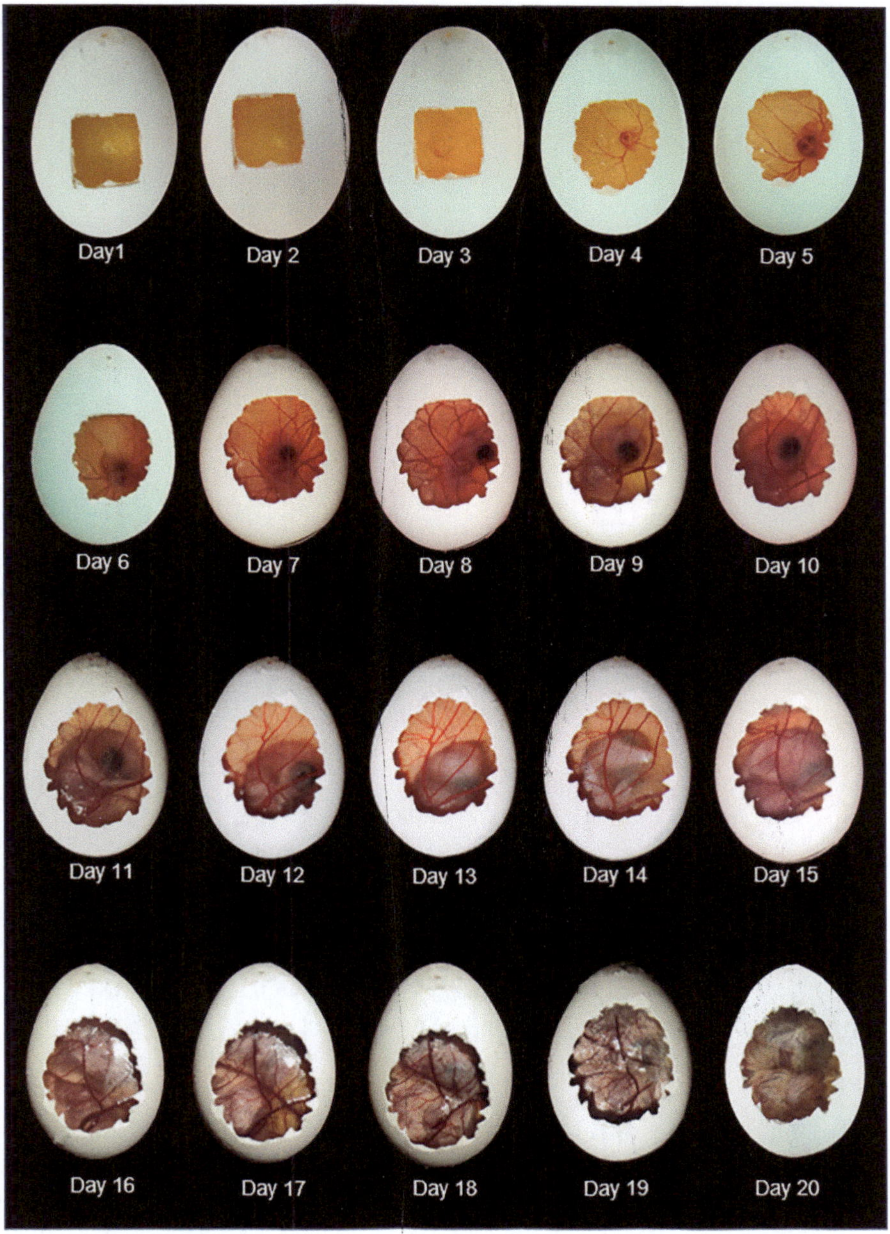

Fig. 4.6 Timeline of chick embryo developmental stages day-wise by windowing method. (Video 4.1)

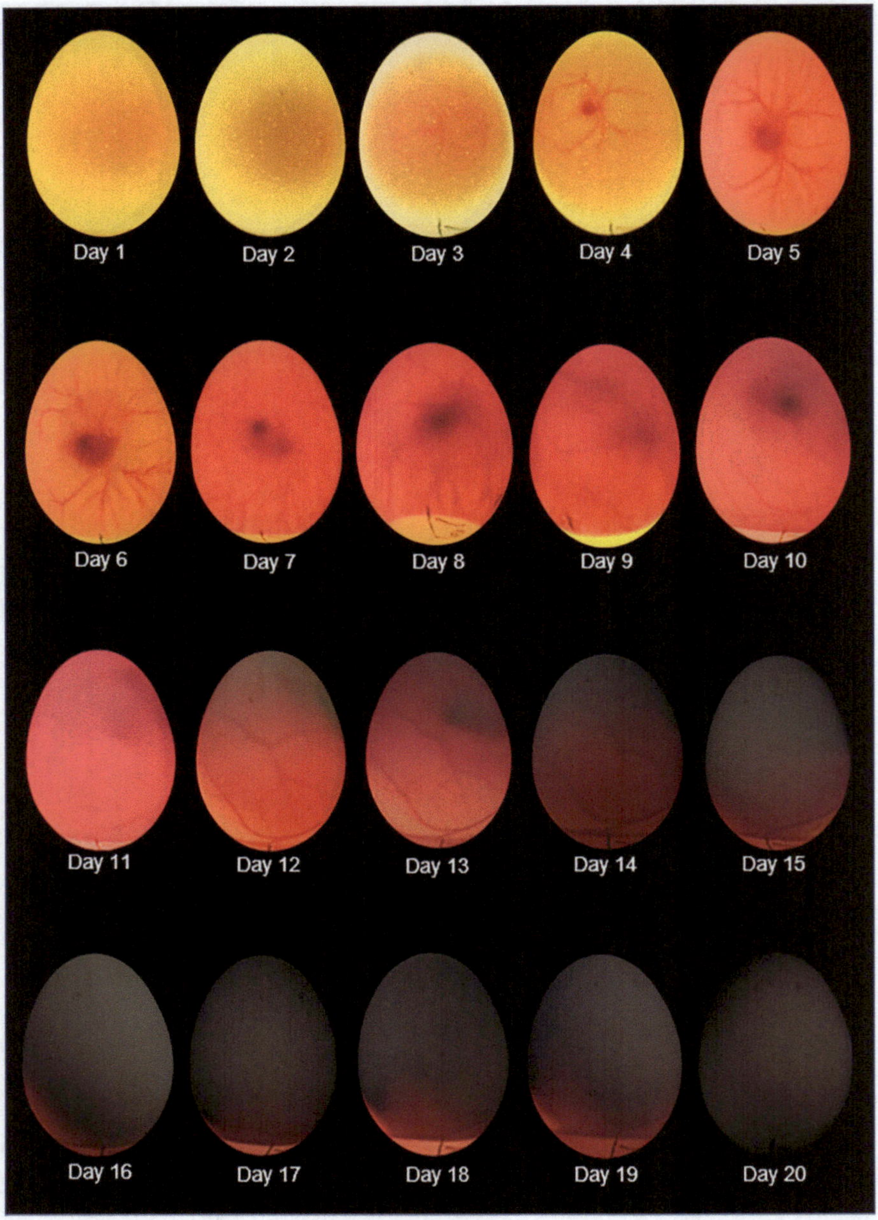

Fig. 4.7 Timeline of chick embryo developmental stages day-wise by candling method. (Video 4.2)

Day 20 (HH 45–46): The yolk sac is fully withdrawn into the body cavity, and the space within the egg is completely occupied by the embryo.
Day 21 (HH 46): The chick hatches.

These significant developmental milestones can be effectively studied through windowing (Fig. 4.6) or candling methods (Fig. 4.7). Initially, during the first two days of incubation, the embryo appears as a whitish tissue that is clearly visible through the windowing method. In contrast, candling reveals a deeper coloration at the blastoderm site, intensified by EDD2 compared to EDD1. As the embryo continues to develop, the blood vasculature becomes more pronounced, visible through candling along with the heartbeat in the early stages and subsequent embryo movements. The windowing method allows for the observation of additional developmental features such as brain differentiation, head region bend, body flexure contour, allantois formation, limb and digit elongation, beak development, feather germs, and reduction in yolk content.

4.4.2 Pipping and Hatching

Pipping refers to the short, high-pitched sound or the initial pecking noise made by a young chick as it begins breaking through the eggshell using its egg tooth. Hatching is the broader process in which the young fully emerge from the egg. Pipping marks the first stage of hatching, typically occurring about 24 h before the chick fully emerges. The chick uses its egg tooth to make an initial peck, and with the help of strong neck muscles, it gradually widens the crack until the shell is sufficiently broken for it to push through (Fig. 4.8).

Practical Tips: Pipping and Hatching

- In the windowing method of chick embryo culture, when the open area is sealed with parafilm, it is advisable to loosen the parafilm after the first pipping if it occurs within the sealed area, as breaking through the parafilm can be more difficult than the natural process of cracking the eggshell for chick emergence.
- As the eggshell opening widens at the pipping site, closely monitor the emerging embryo. Newly hatched chicks may wander unpredictably and can fall from trays or platforms, especially if these are positioned above the ground or lower surfaces of the incubator.

4.4.3 Eggs Gone Wrong

In in vivo chick embryo development studies using the windowing method, the embryo is easily visualized directly through the opening/window, allowing for the quick identification of problematic eggs, such as those that are non-viable, dead, or contain abnormal embryos. In contrast, the candling method does not provide direct visualization but is effective in determining the viability of the developing embryo.

Fig. 4.8 Chick embryo pipping and hatching. (Videos 4.1 and 4.2)

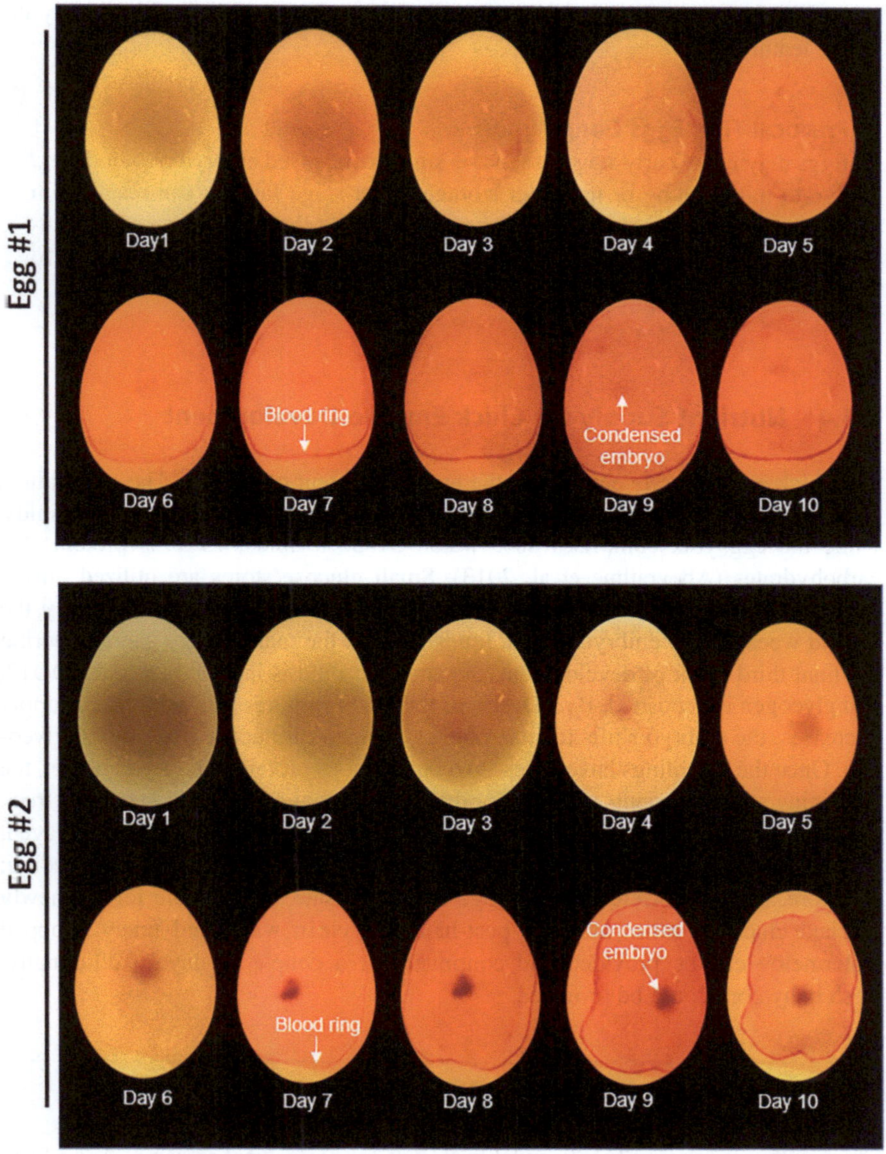

Fig. 4.9 Timeline of chick embryo development day 1–10 in eggs gone wrong by candling method. (Video 4.3)

Key indicators for identifying problematic eggs through candling include the presence of a blood ring, the absence of a network of blood vasculature, and lack of embryo growth and movement. The blood ring is typically formed as blood moves away from a dead embryo. In the first few days of development (up to about EDD4), it may be difficult to identify weak or non-viable embryos (Fig. 4.9). However, as the developmental days progress, abnormal embryos become easier to identify by the above features.

Practical Tips: Eggs Gone Wrong
Eggs may fail at early stages, seen as a small, condensed embryo or blood ring, due to mishandling or improper storage. Later-stage failures can result from physical fluctuations, prolonged exposure outside the incubator, or contamination during visualization. Such embryos should be promptly removed to protect the healthy developing embryos.

4.4.4 Nutrient Supply for Chick Embryo Development

The chick embryos rely entirely on the nutrient content of the egg. The egg white's composition includes 88% water, 11% protein, 1% carbohydrates, ash, and lipids, while the egg yolk comprises 50% water, 31–35% lipids, 15–17% protein, 1% carbohydrates (Abeyrathne et al. 2013). Small glucose stores are utilized during the first week to maintain the nutrient supply to the developing embryo. By the second week, as the embryo grows, lipids become the main energy source. During the final third week of development, the amniotic fluid is ingested around EDD17, and glycogen is deposited. By EDD19, as the CAM ruptures and the oxygen supply decreases, the embryo shifts to anaerobic catabolism of glucose from stored glycogen. Once the hatchlings have access to an exogenous protein-carbohydrate diet, the remaining stored nutrients are used for the development of the gastrointestinal tract. To prevent mortality or developmental abnormalities, feeding should not be delayed. Once the hatchling dries and stabilizes after emergence, it typically begins feeding; therefore, the food offered at this stage should be safe and suitable for the newly hatched chicks. On the first day post-hatching, clean water and finely chopped watermelon can be provided and continued for a couple of days. Additionally, fresh baby corn could be provided.

4.5 Significance/Research Applications

Chick embryos serve as highly suitable models for biomedical research. Apart from having a body system comparable to higher vertebrates, fertilized eggs are readily available throughout the year, easy to handle, and highly economical. Notably, fertilized eggs can be stored under cold conditions for several days without compromising viability. In the present study, fertilized eggs were stored for up to one week without any observable reduction in viability. The highly vascularized extraembryonic membrane of the chick embryo, the CAM, provides an excellent model for studying angiogenesis and the growth of tumor xenografts. The windowing and candling techniques described in this chapter enable direct

Table 4.1 Representative studies on the application of chick embryos cultured through the windowing and candling visualization method in biomedical research

S. no.	Type of condition/disorder	Type of study/research	Reference
1.	Glioblastoma (GBM) tumor microenvironment	Assessment of potential phytochemicals as antiangiogenic agents against GBM was conducted using the chick embryo model through the windowing method	Senrung et al. (2023b)
2.	Colorectal liver metastases	Examination of the molecular profile of desmoplastic liver metastasis in a chemo-naive patient using the chick embryo chorioallantoic membrane (CAM) model via the windowing method	Ceausu et al. (2021)
3.	GBM	Assessment of the anti-tumorigenic and anti-neoangiogenic potential of Sema3C using the chick embryo CAM model through the windowing method	Valiulyte et al. (2019)
4.	Ovarian carcinoma	Serving as a metastatic model system for the screening of drugs using chick embryo CAM model through the windowing method	Lokman et al. (2012)
5.	Lewis lung carcinoma	Development of chick embryo as a fluorescent imageable metastatic cancer model system for screening drugs	Bobek et al. (2004)
6.	Embryonic toxicity of omega-3 acids (docosahexaenoic and eicosapentaenoic acids)	The embryonic toxicity of omega-3 acids was evaluated using chick embryos maintained within their natural eggshell environment. The candling method was employed to visualize embryonic growth prior to the administration of the test compounds	Salari et al. (2021)
7.	Teratogenic potential of ondansetron	The teratogenic potential of ondansetron was investigated in chick embryos over an 18-day incubation period, during which embryo survival was monitored through candling	Bokariya et al. (2016)

visualization and manipulation of the chick embryo and provide a reliable biomedical model for testing therapeutic agents introduced through openings in the eggshell. Table 4.1 summarizes representative studies that employ these techniques in biomedical research.

4.6 Precautions

- The eggs should undergo sterilization before incubation to eliminate living contaminants and any potential non-living elements present on the eggshell's surface.
- Set the BOD incubator to the required conditions a day before incubating the fertilized eggs to ensure it is fully functional.
- Ensure that the eggs remain inside the humidity chamber under the specified conditions of 37 ± 1 °C and relative humidity of 80 ± 10% (≥60 °C) throughout the incubation period. Conduct all culture and experimental procedures within this controlled environment to maintain optimal conditions for embryo development.
- Carefully cut and remove the eggshell to expose the underlying membrane and embryo. Ensure that the cut area does not exceed 2 cm × 2 cm to avoid damaging the developing embryo beneath.
- When sealing the window, the blade should be heated to approximately 60 °C to slightly melt the parafilm and ensure a tight seal. Avoid excessive heating, as it can cause the parafilm to completely melt.

4.7 Quick Review Questions

4.7.1 Multiple Choice Questions

1. The "windowing" method in chick embryo study is used to:
   (a) Remove the yolk from the egg
   (b) Observe the embryo directly through a small opening in the shell
   (c) Count the number of yolk layers
   (d) Measure shell thickness
   Answer: b) Observe the embryo directly through a small opening in the shell
2. Candling is a technique used to:
   (a) Measure the embryo's heart rate
   (b) Visualize the embryo inside the egg using light
   (c) Incubate eggs at a constant temperature
   (d) Remove the shell for microscopic observation
   Answer: b) Visualize the embryo inside the egg using light
3. Define the blastocoel.
   (a) The opening of the archenteron
   (b) A solid mass of cells formed after cleavage
   (c) A fluid-filled cavity within the blastula
   (d) The space between the yolk and the embryo
   Answer: c) A fluid-filled cavity within the blastula
4. Significance of the gray crescent in symmetry formation.
   (a) It determines the future anterior end of the embryo
   (b) It marks the site of sperm entry

(c) It establishes radial symmetry in the embryo
(d) It indicates the future dorsal side and helps establish bilateral symmetry
Answer: d) It indicates the future dorsal side and helps establish bilateral symmetry

5. The process of formation of the three germ layers during frog embryonic development is called:
(a) Cleavage
(b) Gastrulation
(c) Organogenesis
(d) Neurulation
Answer: b) Gastrulation

6. The most prominent feature of a frog embryo at the neurula stage is:
(a) Formation of gill arches
(b) Appearance of the neural tube
(c) Closure of the blastopore
(d) Formation of limb buds
Answer: b) Appearance of the neural tube

7. In which hemisphere is the yolk in a frog egg predominantly concentrated?
(a) Animal hemisphere
(b) Vegetal hemisphere
(c) Both equally
(d) None of the above
Answer: b) Vegetal hemisphere

8. In the late gastrula stage of the frog, which structure indicates the future digestive tract?
(a) Notochord
(b) Neural groove
(c) Archenteron
(d) Blastocoel
Answer: c) Archenteron

9. In the early stages of practical embryology, whole mounts are preferred because:
(a) Sections are not feasible in early embryos
(b) Early embryos are small and transparent
(c) Whole mounts give a 3D view of development
(d) All of the above
Answer: d) All of the above

10. The purpose of studying sections of frog embryonic development stages is to:
(a) Observe pigmentation only
(b) Examine internal organ formation
(c) Identify the blastopore position
(d) Avoid using whole mounts
Answer: b) Examine internal organ formation

4.7.2 Short Answer Questions

1. Why do we clean fertilized eggs with 70% alcohol when bringing them to the laboratory from the farm?
2. Why is it necessary to rotate eggs during incubation?
3. Define the term candling in the context of chick embryo observation.
4. Highlight the significance of the candling method.
5. Mention the ideal incubation temperature and humidity required for chick embryo development.
6. What developmental features can be observed in a chick embryo on day 3 of incubation?
7. Name the extraembryonic membranes formed during chick embryo development.
8. What is the function of the chorioallantoic membrane (CAM) in chick embryos?
9. Explain how in vivo observation differs from in vitro culture in embryology.
10. Write two applications of the windowing method in biomedical or developmental research.

Glossary

Albumin The major water-soluble protein component of egg white, constituting approximately 60% of its total protein content.

Amnion A thin, elastic membrane forms the amniotic sac filled with fluid, protecting the developing embryo and regulating temperature.

Amniotic fluid A clear fluid surrounding the developing embryo or fetus within the amniotic sac, providing cushioning, temperature regulation, hydration, and free movement essential for normal growth and development.

Anaerobic catabolism Breakdown of organic molecules in the absence of oxygen to release energy.

Avian Relating to birds.

Blastoderm (germinal disc or blastodisc) A cell layer formed on the yolk surface during early embryonic development in birds and other oviparous animals, giving rise to embryonic tissues and organs.

Blastula An early embryonic stage following cleavage, consisting of a hollow sphere of cells surrounding a fluid-filled cavity called the blastocoel.

Blood ring A circular accumulation of blood between the yolk and the shell membrane, indicating that the embryo has stopped developing.

Candling techniques Method used to examine the contents of avian or reptilian eggs by shining a light through the eggshell. This allows visualization of the developing embryo, blood vessels, yolk, and other structures without breaking the egg.

Chorioallantoic membrane (CAM) Vascularized membrane in amniote eggs, formed by fusion of chorion and allantois, provides oxygen and calcium to the developing embryo.

Cornea The transparent, dome-shaped front part of the eye covering the iris and pupil; it focuses light onto the retina and protects against dust and microbes.

Cranial flexure A primary bend in the developing brain between the midbrain and hindbrain, causing the forebrain to curve ventrally and aiding brain orientation.

Desmoplastic liver metastasis A type of liver metastasis characterized by a dense fibrotic (desmoplastic) stromal reaction surrounding tumor cells.

Egg albumin The egg white contains protein and essential nutrients like niacin, riboflavin, and minerals.

Egg tooth A temporary, sharp projection on the beaks or snouts of hatchlings, enabling them to break free through eggshells.

Embryonic development day (EDD) Refers to the specific day of an embryo's growth, counted from the day the egg is laid/fertilized or incubated.

Ex ovo Culturing or development of an embryo outside the eggshell.

Feather germs Small epidermal and dermal projections appearing on the skin of developing avian embryos, representing the earliest stage of feather formation.

Hamburger–Hamilton (HH) staging system A standardized system that classifies chick embryo development from fertilization to hatching, based on specific morphological features instead of chronological age.

In ovo Procedures or experiments conducted within the egg, particularly before hatching.

Incubation The process of maintaining eggs under controlled environmental conditions to support the development and hatching of embryos. It typically involves regulating temperature, humidity, ventilation, and sometimes egg rotation to mimic the natural conditions provided by a brooding parent.

Limb buds Small protrusions that appear on the sides of a developing embryo, representing the early stages of limb formation.

Morphogenetic movement Coordinated cell movements during embryogenesis that shape the embryo, forming germ layers and establishing body plan organization.

Nictitating membrane A transparent third eyelid in many vertebrates; protects, moistens, and cleans the eye while maintaining vision.

Notochord A flexible, rod-like mesodermal structure in the early embryo is vital for nervous system development and body axis alignment.

Optic vesicles Paired outpouchings from the forebrain during embryonic development that form the eyes.

Oviposition A process by which a female animal lays eggs or the act of depositing fertilized or unfertilized eggs outside the body.

Pipping Process by which a hatching bird embryo breaks through the eggshell using its egg tooth, marking the beginning of emergence from the egg.

Preoviposition The period between mating and egg-laying in oviparous animals.

Primary brain vesicles Three primary swellings of the embryonic neural tube, forebrain, midbrain, and hindbrain, which form early and develop into major brain regions.

Primitive streak Temporary structure emerging 14 days after fertilization; guides embryonic development through epithelial-mesenchymal transition, forming mesoderm and definitive endoderm.

Real time Immediate processing or response to events as they occur, allowing immediate visualization and analysis without delay.

Scleral papillae Temporary conjunctival epithelial thickenings in chick embryos' eyes during incubation.

Somites Mesodermal segments in vertebrate embryos; form the vertebral column, muscles, and dermis.

Teratogenic Any agent or factor that can disturb the development of an embryo or fetus, leading to congenital malformations or birth defects.

Tumor xenografts The blood vessels that carry deoxygenated blood from the heart to the gills for oxygenation.

Visceral arches Cartilages or bones that support the pharyngeal region and connect jaws to the skull.

Vitelline membrane A thin, transparent membrane surrounding the egg yolk, separating it from the albumen, and facilitating nutrient regulation and gas exchange for the developing embryo.

Webbing Membrane between digits (fingers or toes) supporting limb structure during development.

Windowing techniques Method used to create a small opening (or window) in the eggshell to directly observe and manipulate a developing embryo without fully opening or destroying the egg.

Yolk The yellow center of an egg, rich in cholesterol and essential nutrients, providing valuable health benefits.

Yolk sac An extraembryonic membrane that surrounds the yolk of an egg and provides essential nutrients to the developing embryo during early growth.

References

Abeyrathne ENS, Lee H, Ahn DU (2013) Egg white proteins and their potential use in food processing or as nutraceutical and pharmaceutical agents—a review. Poult Sci 92(12): 3292–3299

Bahadir B, Ozgural O, Mammadkhanli O, Aktan ES, Tas DO, Bahadir EA, Unlu MA (2024) Investigation of the effect of Tenoxicam on neural tube defect using Chick embryo model. Turk Neurosurg 34(5):827–832. https://doi.org/10.5137/1019-5149.JTN.45237-23.2

Bobek V, Plachy J, Pinterova D, Kolostova K, Boubelik M, Jiang P, Yang M, Hoffman RM (2004) Development of a green fluorescent protein metastatic-cancer chick-embryo drug-screen model. Clin Exp Metastasis 21(4):347–352. https://doi.org/10.1023/b:clin.0000046138.58210.31

Bokariya P, Kothari R, Gujar V, Shende M, Doshi MA (2016) Evaluation of teratogenic potential of ondansetron on developing Chick embryo. Indian J Anat 5(3). https://doi.org/10.21088/ija.2320.0022.5316.22

Ceausu RA, Ciolofan A, Blidisel A, Cretu OM, Cimpean AM, Raica M (2021) Liver metastatic colorectal tumor cells change their phenotype during consecutive passages on Chick embryo Chorioallantoic membrane: lessons from the lab to the clinic. In Vivo 35(5):2711–2718. https://doi.org/10.21873/invivo.12555

Hall CA, Potvin DA, Conroy GC (2023) A new candling procedure for thick and opaque eggs and its application to avian conservation management. Zoo Biol 42(2):296–307

Hamburger V, Hamilton HL (1992) A series of normal stages in the development of the chick embryo. 1951. Dev Dyn 195(4):231–272. https://doi.org/10.1002/aja.1001950404

Lokman NA, Elder ASF, Ricciardelli C, Oehler MK (2012) Chick chorioallantoic membrane (CAM) assay as an in vivo model to study the effect of newly identified molecules on ovarian cancer invasion and metastasis. Int J Mol Sci 13(8):9959–9970. https://doi.org/10.3390/ijms13089959

Salari Z, Tavakkoli H, Khosravi A, Karamad E, Salarkia E, Ansari M, Dabiri S, Mortazaeizdeh A, Mosallanejad SS, Sharifi F (2021) Embryo-toxicity of docosahexaenoic and eicosapentaenoic acids: in vivo and in silico investigations using the chick embryo model. Biomed Pharmacother 136:111218

Senrung A, Tripathi T, Janjua D, Yadav SK, Chhokar A, Aggarwal N, Yadav J, Chaudhary A, Joshi U, Sethi P, Bharti AC (2023a) Chick chorioallantoic membrane: a valuable 3D in vivo model for screening nanoformulations for tumor antiangiogenic therapeutics. Int J Dev Biol 67(4):115–135. https://doi.org/10.1387/ijdb.230198pb

Senrung A, Tripathi T, Yadav J, Janjua D, Chaudhary A, Chhokar A, Aggarwal N, Joshi U, Goswami N, Bharti AC (2023b) In vivo antiangiogenic effect of nimbolide, trans-chalcone and piperine for use against glioblastoma. BMC Cancer 23(1):1173. https://doi.org/10.1186/s12885-023-11625-4

Valiulyte I, Curkunaviciute R, Ribokaite L, Kazlauskas A, Vaitkeviciute M, Skauminas K, Valanciute A (2019) The anti-tumorigenic activity of Sema3C in the Chick embryo Chorioallantoic membrane model. Int J Mol Sci 20(22). https://doi.org/10.3390/ijms20225672

Study of Indirect Development and Metamorphosis by Rearing Any One Insect

5

Jyotsna Singh and Sanjiv Mullick

Abstract

The rearing of tobacco caterpillar, *Spodoptera litura,* in laboratory conditions is an ideal method for studying indirect development and holometabolous metamorphosis. The study envisages the developmental stages and life cycle of this agricultural pest moth under lab conditions, incorporating photographs and short videos alongside well-labeled schematic diagrams. Moreover, it elaborates on the laboratory rearing methods for these insects, encompassing the preparation of artificial diets in case the natural food source (castor leaves) is not available. These methods provide valuable insights into the growth and development of *S. litura* and help maintain a perennial healthy culture of this insect pest, which is a prerequisite for conducting any further experiments for its management and control in the agricultural setup.

Keywords

Spodoptera litura · Indirect development · Metamorphosis · Castor

Supplementary Information The online version contains supplementary material available at https://doi.org/10.1007/978-981-95-7269-4_5.

J. Singh (✉)
Department of Zoology, Daulat Ram College, University of Delhi, Delhi, India
e-mail: jyotsnasingh@dr.du.ac.in

S. Mullick
Department of Zoology, Dyal Singh College, University of Delhi, Delhi, India

A. C. Bharti et al. (eds.), *Textbook of Developmental Biology*,
https://doi.org/10.1007/978-981-95-7269-4_5

Aim

To study indirect development and metamorphosis in insects through the rearing of *Spodoptera litura* in laboratory conditions.

Learning Objectives

Upon completing the experiment, students will be able to:

- Define and differentiate between direct and indirect development.
- Define metamorphosis and classify its types in insects.
- Successfully culture *Spodoptera litura* in lab conditions using natural food and also know about the preparation of artificial diet.
- Identify the various larval instars of *S. litura* and its other developmental stages based on morphological features.

5.1 Introduction

Development is the progressive series of morphological, physiological, and biochemical changes through which an organism advances from the embryonic stage to a fully mature state, attaining higher levels of structural and functional organization. Based on the nature of post-embryonic changes, development may be categorized as:

- *Direct development*—in which the juvenile stage closely resembles the adult stage in form, with no radical change in body plan.
- *Indirect development*—in this type of development, the organism undergoes one or more *morphologically distinct larval stages*, each adapted to specific ecological niches, before attaining the adult form.

Further differentiation of direct and indirect development is given in Table 5.1.

Table 5.1 Comparison between direct & indirect development

Feature	Direct development	Indirect development
Resemblance of young to adult	Juveniles resemble miniature adults	Juveniles (larvae) are morphologically different from adults
Larval stage	Absent	Present (one or more larval stages)
Metamorphosis	Absent	Present (mild to complete structural transformation)
Ecological niche	Same for young and adult	Different for larva and adult
Energy allocation	Growth and reproduction occur in similar habitats	Larva mainly for growth and dispersal, adult for reproduction
Examples	Earthworm, reptile, human	Frog, butterfly, mosquito, starfish

5.1.1 Indirect Development

Indirect development is characterized by a complex life cycle comprising discrete stages that exhibit significant differences in morphology, physiology, and behavior. The larval stages are typically specialized for growth and dispersal, whereas the adult stage is specialized for reproduction. Such differentiation allows for reduced intra-specific competition by exploiting different resources and habitats during ontogeny.

5.1.2 Metamorphosis

Metamorphosis refers to the process of extensive, often abrupt, structural, and physiological reorganization that transforms a larva into a sexually mature adult. It involves coordinated cellular proliferation, differentiation, and apoptosis, regulated by endocrine and genetic factors. In insects, metamorphosis occurs via shedding of the external cuticle, referred to as molting or ecdysis, facilitated by the ecdysone hormone. The shed cuticular skin is called the exuviae, which marks the progression to the next stage of metamorphosis. The stage between each molt or ecdysis is called an instar. The neonate larva or nymph post hatching or intermediate molt is referred to as the first instar. The insect then molts into the second instar, and this process continues until the last molt, when the adult or imago emerges. The intervals between each molt are referred to as stages or stadia.

In insects, metamorphosis can be further divided into three categories, as follows:

Ametabolous Insects showing complete lack of metamorphosis or zero metamorphosis, with the larval form undergoing a highly variable number of molts to form the adult. It mostly includes primitive, wingless insects like silverfish, springtails, etc. In this type, both juvenile and adult forms are found in the same ecological niche.

Hemimetabolous This includes insects with incomplete development, also referred to as direct development, where the immature larval form or the nymph resembles the mature adult form but lacks wings and fully developed genitalia. For example, insects from the order Hemiptera (true bugs such as bed bugs, cicadas, etc.), Orthoptera (grasshoppers), and Isoptera (termites).

Holometabolous (Fig. 5.1) Insects belonging to the orders Lepidoptera (moths and butterflies), Coleoptera (beetles), Hymenoptera (bees and wasps), Diptera (housefly, mosquito), Neuroptera, and Trichoptera undergo complete transformation from worm-like larval form with soft, cylindrical body to winged, hard, sclerotized adult forms with an intermediate non-feeding pupal stage. This pupal stage sets the holometabolous insects distinctly apart from hemimetabolous or ametabolous insects.

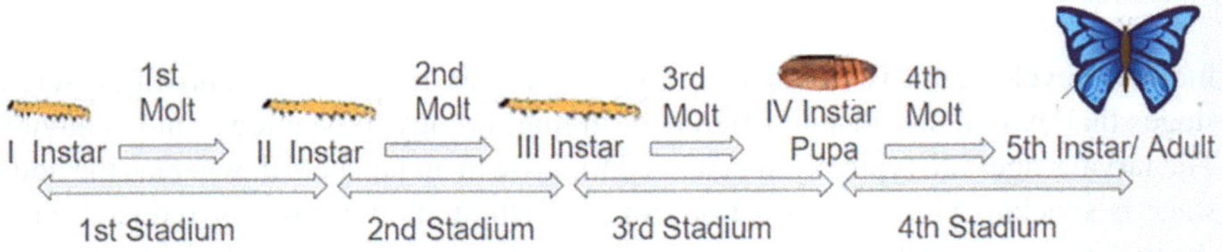

Fig. 5.1 Schematic representation of holometabolous metamorphosis in insects

Metamorphosis is a hallmark of holometabolous insects (e.g., Lepidoptera, Diptera, Coleoptera) and amphibians (e.g., Anura, Urodela). It enables the organism to occupy distinct ecological niches during its life cycle, thereby optimizing survival and reproduction.

The best way to understand indirect development is through studying the life cycle of any insect that shows the transition from the egg to the larval, pupal, and adult stages, clearly highlighting the morphological changes.

One such insect displaying indirect development and complete metamorphosis is the tobacco caterpillar, *Spodoptera litura,* which is a polyphagous pest available all year round in the Indian sub-continent.

5.2 Requirements

Fresh castor leaves (*Ricinus communis*), oviposition cage made of plexiglass (l 25 cm × b 25 cm × h 25 cm), plastic jars (15 cm height ×10 cm diameter), small plastic bottle caps or diet cups, blotting sheet, B.O.D. incubator, cotton swabs, 10% sucrose solution in water, sodium hypochlorite solution (0.2%), formaldehyde solution, $KMnO_4$ powder, 0.05% sodium hypochlorite (NaOCl) solution, gram flour/ chickpea flour, yeast, casein, ascorbic acid, sorbic acid, methyl parahydroxybenzoate, streptomycin, cholesterol, wheat germ oil, tocopherol acetate, distilled water, agar powder, moth catching nets, adult moths both males and females caught from Delhi University area.

5.3 Insect Culture/Rearing

- *Spodoptera litura* culture is established from males and females captured during night hours below light.
- The whole culture is maintained in insectary conditions at standard laboratory conditions of 27 ± 2 °C, 60–65% R.H., and a photoperiod regime of 14D:10 L.
- Male and female moths are transferred to a mating cum oviposition cage (Fig. 5.2), made up of plexiglass (25 cm × 25 cm × 25 cm), having two circular windows on the side walls for ventilation and a front door for handling the moths. The side window can be fixed with a net cover, and the front door with a loose net cover.

Fig. 5.2 Pictograph showing oviposition cage setup

- The floor of the cage is lined with a moist blotting sheet.
- Diet cups with cotton swabs, soaked in 10% sucrose solution, are provided as the adult diet.
- A castor leaf with its peduncle dipped in water, fixed in a reagent bottle, is provided in the cage as an oviposition substrate.
- Approximately 10 pairs of males and females per cage are introduced for mating. Adults should be transferred to a fresh cage daily.
- After successful mating, the eggs are collected daily, sterilized with sodium hypochlorite solution (0.2%), and rinsed with formaldehyde solution.
- The eggs are then washed thoroughly with distilled water, dried at room temperature, and kept in small jars for emergence.
- The larvae are reared on castor leaves (*Ricinus communis*), which are one of its natural host plants and are the most suitable for its growth and development (Bhalani 1989).
- Castor leaves are collected and washed with water containing a pinch of $KMnO_4$ powder to prevent any infection and then washed again with normal water.
- Larvae are reared in plastic jars (15 cm height ×10 cm diameter) (Fig. 5.3).
- On hatching, the neonate larvae are transferred to the plastic jars and provided with young, tender castor leaves. The bottom of the jars is lined with a blotting sheet.
- Larvae of the same stage/instar are kept in one jar, and the number of larvae in each jar is regulated in a way to avoid overcrowding for their proper growth and development.
- Larvae are transferred to fresh jars and provided with fresh leaves every day until pupation.
- During pupation, larvae usually bore inside the blotting sheet and pupate.

Fig. 5.3 Pictograph showing rearing jars used for maintaining *S. litura* culture

Fig. 5.4 Pictograph showing quantified artificial diet kept in rearing jars for different instars

- Pupae are collected and washed with 0.05% sodium hypochlorite (NaOCl) solution to prevent any infection (Ignoffo and Dutkys 1963).
- Pupae were air-dried, sexed, and kept in separate jars for the emergence of adult moths.

5.4 Preparation of Artificial Diet

When the natural host plant of the insect is too scarce or completely absent or seasonal, then the best way to rear the insect in the laboratory is by using an artificial diet (completely devoid of natural host plant) (Fig. 5.4) or a semi-synthetic diet (with some percent of host plant mixed in the artificial diet).

Table 5.2 Ingredients for preparing an artificial diet for Noctuidae moths

Constituents	Quantity
Part I	
Gram flour or chickpea flour	84 g
Yeast	25 g
Casein	10 g
Formaldehyde solution (10%)	1 ml
Part II	
Ascorbic acid	5 g
Sorbic acid	1 g
Methyl parahydroxybenzoate	2 g
Streptomycin	0.2 g
Cholesterol	0.2 g
Wheat germ oil and tocopherol acetate (10:1)	0.6 g
Part III	
Distilled water	575 ml
Agar	11 g

Artificial diet was prepared as per the ingredients in Table 5.2 for the moths belonging to family Noctuidae. Required quantities of Part I ingredients (Table 5.1) can be weighed and made into a homogenous paste in water. Similarly, ingredients of Part II can be weighed separately and kept in a beaker. Balance water (Part III) should be boiled in a 2000 ml beaker. When water starts boiling, agar powder must be gradually added and mixed vigorously. This mixture then should be poured into an airtight storage box and left to cool down. Once set, this diet can be stored in a cool place (refrigerator).

A semi-synthetic diet for *S. litura* can be prepared by a similar method as the artificial diet but by adding castor leaf powder. The freshly excised leaves of the castor plant could be dried at room temperature and ground to a fine powder in an electric grinder. Twenty percent of the chickpea seed powder in the artificial diet can be replaced by this dried leaf powder. All the other ingredients and their proportion would remain the same.

Once the diet is ready, a similar method of rearing as the previously mentioned procedure for rearing *S. litura* on castor leaves can be followed for rearing the insect on this artificial/semi-synthetic diet.

5.5 Result/Observation

Record all the changes in morphological features of all instars as the development progressed from egg stage to the adult emergence, paying attention to the exuviae shed after each molt or other notable signs of development such as increase in the width of head capsules during the larval stages.

The life cycle of *Spodoptera litura* would exhibit the following stages when reared on a proper diet under suitable lab conditions. In the present study, the images of the insect are those reared on the natural host castor plant.

5.5.1 Eggs (Fig. 5.5)

- *Site of egg laying*: Post mating, the female moths lay the eggs on the underside of castor leaves kept inside the mating cage, in clusters of 2–3 layers covered with hairy scales from their anal tuft. At times, eggs are also deposited on the cage walls or underneath the blotting sheet spread on the cage floor.
- *Morphological features*: Freshly laid eggs are round in shape and grayish white in color. The color changes to pale yellow in 1–2 days and dark brown prior to hatching due to the presence of black head condition of the emerging first instar.
- *Incubation period*: Usually 3–5 days, after which the first larval instar or L1 emerges.

5.5.2 First Instar Larvae

Also called the neonate larvae or L1 are tiny, pale green in color with a prominent shiny black head capsule (Fig. 5.6a). Their body usually measures about 1.4 to 1.6 mm in length, and the width of head capsules is 0.2–0.3 mm on an average. The larval duration for this instar is around 2–3 days. These larvae are usually leaf scrapers (Fig. 5.6b).

5.5.3 Second Instar Larvae or L2

The larval body becomes translucent green in color with a conspicuous black mark on the first abdominal segment (Fig. 5.7a). The head color changes to pale brown

Fig. 5.5 Photomicrographs of *S. litura* eggs covered with hairy scales on castor leaf

Fig. 5.6 Photomicrographs of neonate larvae of *S. litura*. (**a**) Emerging from egg mass. (**b**) Feeding on castor leaf (Video 5.1)

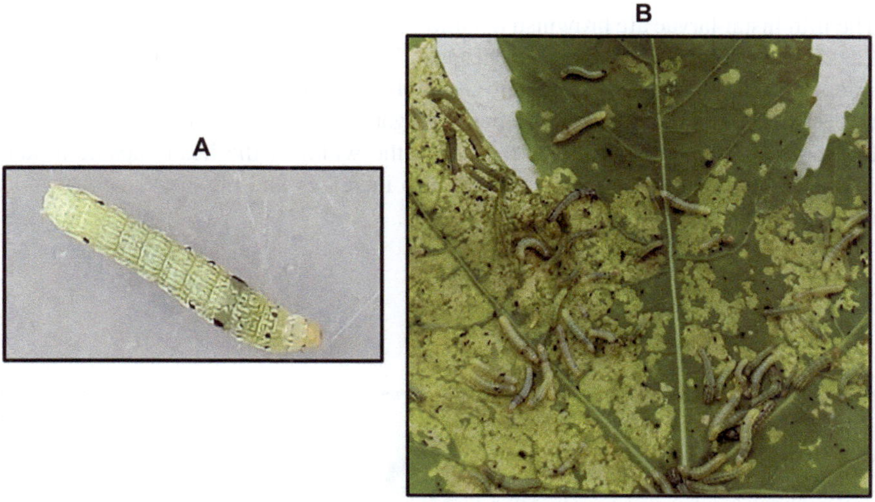

Fig. 5.7 Photomicrographs of (**a**) freshly molted L2 larvae of *S. litura*; (**b**) L2 larvae feeding on castor leaf (Video 5.2)

with a prominent triangular mark. Three longitudinal whitish lines may be present, one on the dorsal surface and one each on the dorso-lateral surface. The width of the head capsule ranges from 0.35 to 0.45 mm, while the body length varies from 3.5 to 4.5 mm. The average duration of the second instar is 2–3 days. Both first and second instar larvae feed by scraping the layer of tender leaves (Fig. 5.7b).

5.5.4 Third Instar Larvae or L3

This instar is normally blackish-green in color with three thin yellow lines running on the body, one dorsally in the center and two on the sides (Fig. 5.8). A row of black dots and small triangles begins to appear on the sides. Body length is about 9-12 mm, and the head capsules width is around 0.6–1.1 mm. This stage lasts for 3–4 days on an average.

5.5.5 Fourth Instar Larvae

The body color lightens and becomes brownish green at this stage, with the same three thin lines on the back as present in the third instar (Fig. 5.9). The black dots and triangle become more conspicuous at this stage. The head capsule starts to get darker in color from the sides (Fig. 5.10). L4 measures approx. 17–22 mm in length with head capsule width varying from 1.1 to 1.7 mm. Larval duration is about 3–5 days.

5.5.6 Fifth Instar Larvae

The fifth instar larvae are brownish or earthy in color with the same three thin yellow lines down the back, as in previous instars (Fig. 5.11). The row of black dots on its sides becomes more pronounced with a row of dark triangles. The head capsule starts to bear a V-mark. The larva becomes a gregarious feeder at this stage. The body length ranges from 26.5 to 29.5 mm, with the width of the head capsule around 1.5 to 1.9 mm. Overall, the body width also increases and the larvae appear more round in shape. This stage lasts for 4–5 days.

Fig. 5.8 Photomicrographs of L3 larvae of *S. litura* (Video 5.3)

Fig. 5.9 L4 stage of *S. litura*

Fig. 5.10 Pictomicrograph showing L4 larva preparing to molt. Note the segments coming closer as the body shrinks and the cuticle becomes darker in color

5.5.7 Sixth Instar Larvae (Fig. 5.12)

At this stage, the larval body is clearly divided into a head, thorax, and abdomen. The body color gets darker, and the larva looks grayish brown in color. The three yellow lines become dull, out of which the lateral lines begin to disintegrate. The head capsule has a conspicuous "V" mark on it. They continue to feed gregariously, requiring more leaves at this point. The body length of the sixth instar ranges from

Fig. 5.11 Pictomicrograph showing L5 stage of *S. litura*

Fig. 5.12 L6 larva with V mark on head capsule

38 to 40 mm, and the breadth of the head capsule is about 2.2 to 2.5 mm. After 3–4 days, it proceeds to the prepupa stage and stops feeding.

5.5.8 Prepupa

This stage lasts for 1–2 days. Once the larval development is completed, the final instar ceases to feed and move. Its body shrinks in size and becomes comma-shaped (Fig. 5.13). Body coloration also starts to darken and becomes black. The insect usually moves underneath the blotting paper lining the culture jar to take refuge prior to entering the prepupa stage. The body length is reduced to around 18–21 mm.

Fig. 5.13 Prepupa stage (comma-shaped)

Fig. 5.14 Pictomicrographs showing pupal stages of *S. litura* (Video 5.4)

5.5.9 Pupa

The freshly molted pupae are pale green in color, and after 4 hours it gradually changes to dark brown due to tanning of the cuticle (Fig. 5.14). The pupa though is an inactive stage that does not move or feed but is dynamically an active stage with major morphological changes taking place within. Pupa can move its abdomen on tactile signals, and this movement is also used to check the viability of this stage. Female pupae are larger than the male pupae. The pupae can be sexed according to the position of the gonopore located on the eighth abdominal sternum in case of the female pupae and on the 9th abdominal sternum in case of the male pupae. The pupal duration ranges from 5 to 7 days on an average.

5.5.10 Adult Moths

Both male and female moths (Figs. 5.15 and 5.16) are ash-brown in color with black compound eyes and filiform antennae on the head. Their thorax and abdomen are usually covered with brown-colored scales. The light brown forewings with prominent straw-colored markings are narrower than the off-white hindwings.

(a) *Males*: The male moths have shiny silver bands on the terminal side of forewings and a big prominent yellow or straw-colored oblique patch in the middle of the forewing. Their abdomen is thin and tapering toward the end.

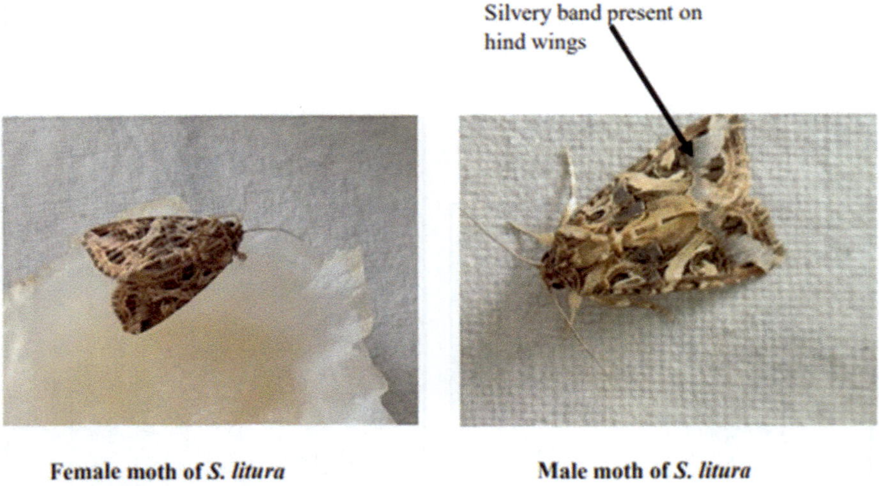

Fig. 5.15 Pictomicrograph showing adults of *S. litura*

Fig. 5.16 Male moth with thin pointed abdomen and female moth with stout blunt abdomen. (Source: https://commons.wikimedia.org/wiki/File:Spodoptera_litura_male.jpg; https://commons.wikimedia.org/wiki/File:Spodoptera_litura_female.jpg)

Fig. 5.17 Pictomicrograph showing oviposition cage used for rearing *S. litura*

(b) *Females*: The female is generally bigger in size than the male moth. The abdomen of a female is stout and blunt. It lacks the silver bands and oblique patch on the forewings. The wingspan of females is wider than that of males.

On emergence from the pupal case, the moths are transferred to the oviposition cage for mating (Fig. 5.17). Usually after 1–2 days, oviposition begins, and the eggs are transferred to small rearing jars or boxes for hatching.

5.6 Significance/Research Applications

Metamorphosis in insects is an important biological process that helps in understanding fundamental mechanisms of development.

- It highlights the hormonal and genetic control of transitions between larval, pupal, and adult stages, involving tissue remodeling, cell death, and organ formation.
- Insects like *Spodoptera litura* can act as key models to study holometabolous metamorphosis as well as direct development due to their short life cycle and easy handling in the lab conditions.
- Many genes and signaling pathways involved are conserved in vertebrates, including humans, providing valuable insights into human developmental biology and related diseases.

In summary, this study not only enhances our understanding of *S. litura* biology but also provides the essential methodological groundwork for a wide range of agricultural, ecological, and educational applications.

5.7 Precautions

- Keep temperature (27 ± 2 °C), relative humidity (60–65%), and photoperiod (14 L:10D) constant using a B.O.D. incubator or climate-controlled room.
- Avoid placing culture jars in direct sunlight or near fluctuating heat sources.
- The culture jars or containers must be changed every alternate day for the first 3 instars and almost every day for the later 3 instars as they are gregarious feeders.
- Early instars must be handled using a soft brush as they are delicate.
- Neonate larvae must be offered soft, tender leaves for feeding.
- Fresh castor leaves must be provided every day to the growing instars. They must be washed properly and treated with $KMNO_4$ solution to avoid any infections.
- Eggs and pupa must be sterilized properly (avoid prolonged exposure to prevent reduced hatchability or pupal mortality) before transferring to the rearing jars.
- The bottom of the jars containing insect larvae and the oviposition cages with adult moths must be lined with blotting sheets to avoid the extra moisture created due to water content of leaves, larval frass, excreta, and respiration.
- Maintain separate rearing containers for each instar and provide enough food to avoid cannibalism and disease transmission.
- Do not disturb larvae when they are preparing to molt (characterized by reduced movement, body contraction, loosening of head capsule). Excessive handling at this time may lead to injuries.
- Larvae and adults must be transferred to fresh jars/cages on a regular basis to avoid chances of contracting infections.
- Maintain detailed notes on instar durations, mortality, environmental conditions, and feeding behavior to ensure authenticity of the data.

5.8 Quick Review Questions

5.8.1 Multiple Choice Questions

1. Which of the following best describes indirect development?
   (a) The juvenile stage closely resembles the adult stage with no radical change in body plan.
   (b) The organism undergoes one or more morphologically distinct larval stages before attaining the adult form.
   (c) Growth and reproduction occur in similar habitats for both young and adult stages.
   (d) Metamorphosis is absent, and structural transformation is mild or incomplete.
   Answer: b) The organism undergoes one or more morphologically distinct larval stages before attaining the adult form.
2. What is the defining feature that sets holometabolous insects apart from hemimetabolous or ametabolous insects?

(a) They are typically wingless in their adult form.
(b) They possess an intermediate, non-feeding pupal stage.
(c) Their immature form (nymph) resembles the adult but lacks wings.
(d) They undergo a highly variable number of molts without significant transformation.

Answer: b) They possess an intermediate, non-feeding pupal stage

3. What is the primary natural host plant used for rearing *Spodoptera litura* in the present study?
(a) Wheat germ
(b) Chickpea flour
(c) Castor leaves
(d) Artificial diet

Answer: c) Castor leaves

4. Which insect order does *S. litura* belong to?
(a) Isoptera
(b) Diptera
(c) Lepidoptera
(d) Coleoptera

Answer: c) Lepidoptera

5. Which of these disinfectants is most commonly used for sterilising castor leaves?
(a) $KMNO_4$ solution
(b) Slaked lime
(c) Sodium hypochlorite solution
(d) Charcoal

Answer: a) $KMNO_4$ solution

6. Which of these disinfectants is most commonly used for sterilising the eggs and pupae of *S. litura*?
(a) Charcoal
(b) Sodium hypochlorite solution
(c) Slaked lime
(d) $KMNO_4$ solution

Answer: b) Sodium hypochlorite solution

7. Silverfish undergoes which type of metamorphosis?
(a) Holometabolous
(b) Hemimetabolous
(c) Ametabolous
(d) All of the above

Answer: c) Ametabolous

8. The artificial diet of *S. litura* contains ______ as chief ingredient.
(a) Chickpea flour
(b) Bajra flour
(c) Ragi flour
(d) Jowar flour

Answer: a) Chickpea flour.

9. Which of the following is a hemimetabolus insect ?
   (a) Termites
   (b) Grasshoppers
   (c) Butterflies
   (d) Bugs
   Answer: c) Butterflies
10. Which of the following is not part of the artificial diet of *S. litura*?
   (a) Methyl parahydroxybenzoate
   (b) $KMNO_4$ solution
   (c) Formaldehyde solution
   (d) Yeast
   Answer: b) $KMNO_4$ solution

5.8.2 Short Answer Questions

1. Define "metamorphosis" and explain the term "instar" in the context of insect development.
2. List three key morphological features used to distinguish the second instar larvae (L2) of *Spodoptera litura*.
3. Why is *Spodoptera litura* considered an ideal model for studying holometabolous metamorphosis?
4. Briefly describe the purpose of lining the bottom of culture jars and oviposition cages with blotting sheets when rearing *Spodoptera litura*.
5. Differentiate between hemimetabolous and holometabolous metamorphosis with suitable examples.
6. Give the characteristics of the pupal stage of holometabolous insects.
7. List the ingredients used for preparing an artificial diet for *S. litura*.
8. Give the key characteristics of differentiating the females of *S. litura* from its males.
9. How do the early instars (1st and 2nd) of *S. litura* differ from its later instars (3rd to 6th) in terms of their feeding behavior?
10. Enumerate the methods used to avoid infections in culture while rearing *S. litura* in lab conditions.

Glossary

Ametabolous Refers to insects that show a complete lack of metamorphosis, with the larval form undergoing a highly variable number of molts to form the adult; typically includes primitive, wingless insects.

Artificial diet A prepared food source used for insect rearing in laboratory conditions that is completely devoid of natural host plant material.

Cuticle The outermost layer of body covering found in insects.

Direct development A type of development where the juvenile stage closely resembles the adult stage in form, with no radical change in body plan.

Ecdysis The process of shedding the external cuticle in insects, facilitated by ecdysone hormone, also known as molting.

Ecdysone A hormone that facilitates the process of molting or ecdysis in insects.

Exuviae The shed cuticular skin that remains after an insect undergoes a molt.

Hemimetabolous Describes insects with incomplete development, where the immature larval form or nymph resembles the mature adult form but lacks wings and fully developed genitalia.

Holometabolous Describes insects that undergo complete transformation, including a worm-like larval form, an intermediate non-feeding pupal stage, and a winged, hard, sclerotized adult form.

Imago The adult, sexually mature insect that emerges after the final molt.

Indirect development A complex life cycle involving one or more morphologically distinct larval stages adapted to specific ecological niches, before attaining the adult form.

Insects Arthropods with three pairs of legs and a pair of sensory antennae.

Instar The developmental stage of an insect between each molt or ecdysis. For example, the neonate larva is the first instar.

Intraspecific competition The struggle between the individuals of the same species for limited resources of survival like water, food, shelter, and potential mates.

Larva A morphologically distinct juvenile stage in insects undergoing indirect development, typically specialized for growth and dispersal.

Lepidoptera One of the important insect orders under the sub-division Endopterygota that exhibit complete metamorphosis and internal wing development. It consists of moths and butterflies. Since the wings of insects under this order are scale-like, they are referred to as lepidoptera (lepid means scale-like in Latin and ptera are the wings).

Metamorphosis The process of extensive structural and physiological reorganization that transforms a larva into a sexually mature adult, involving cellular proliferation, differentiation, and apoptosis.

Morphology Branch of science dealing with the structure and form of organisms.

Molting The process by which an insect sheds its external cuticle to grow, also known as ecdysis.

Neonate larvae The newly hatched first instar larvae.

Oviposition Cage A specially designed cage used in insect rearing for adult moths to mate and lay eggs.

Polyphagous pest An agricultural pest that feeds on a wide variety of host plants.

Photoperiod An important environmental cue for all organisms highlighting the duration of light exposure for their normal growth and development.

Physiology Branch of science dealing with the functions of biological organisms.

Prepupa A transitional stage, typically lasting 1–2 days, where the final instar larva ceases feeding and movement, shrinks in size, and prepares for pupation.

Pupa An intermediate, non-feeding, and generally inactive stage in holometabolous insects, during which major morphological changes occur as the larva transforms into an adult.

Rearing The practice of culturing or breeding organisms, such as insects, under controlled laboratory conditions.

Sclerotized Thickened and darkened over a period of time, usually a body tissue that hardens gradually.

Semi-synthetic diet A laboratory diet for insect rearing that includes some percentage of the natural host plant material mixed with an artificial diet.

Stadia The intervals or periods of time between each molt or ecdysis in insects.

References

Bhalani PA (1989) Suitability of host plants for growth and development of leaf eating caterpillar *Spodoptera litura* (Fabr.). Indian J Ent 51(4):427–430

Ignoffo CM, Dutky SR (1963) The effect of sodium hypochlorite on the viability and infectivity of Bacillus and Beauveria spores and cabbage looper nuclear-polyhedrosis virus. J Insect Pathol 5(4):422–426

Study of the Developmental Stages and Life Cycle of *Drosophila*

6

Sonika Sharma and Madhu Singh

Abstract

The life cycle of *Drosophila melanogaster* is short and exhibits holometabolous development, making it a vital model in genetics and developmental biology. At an optimal temperature, the entire cycle from fertilized egg to adult completes in 10–14 days. The cycle includes four distinct stages: embryo (egg), larva (comprising three instars), pupa, and adult. The brief generation time, distinct morphological transitions, and high reproductive output emphasize *Drosophila*'s utility as a powerful experimental organism. In this chapter, we will learn about rearing and maintaining *Drosophila* in the laboratory. *Drosophila* is a very easy organism to rear and maintain in the lab with minimal requirements. We will also see an overview of the life cycle stages with distinct characteristics of each stage. Short life cycle, high fecundity, easy and low-cost maintenance, observable developmental stages, and genetic similarity to humans make it an important model organism for research.

Keywords

Drosophila melanogaster · Life cycle · Egg · Larvae · Pupa

Aim The chapter/study is to showcase the developmental stages and life cycle of *Drosophila* (an animal research model). Additionally, to provide detailed methods

S. Sharma · M. Singh (✉)
Drosophila Resource Center, Department of Zoology, Daulat Ram College, University of Delhi, New Delhi, India
e-mail: madhu@dr.du.ac.in

A. C. Bharti et al. (eds.), *Textbook of Developmental Biology*,
https://doi.org/10.1007/978-981-95-7269-4_6

for culturing *Drosophila*, including diet preparation, and furthermore their research applications.

Learning Objectives
After doing this practical, students will be able to:

- Identify *Drosophila* males and females
- Identify and differentiate between the larval stages
- Identify the pupal stages
- Prepare food for culturing the flies
- Rear and maintain flies in the laboratory

6.1 Introduction

The *Drosophila*, commonly known as the fruit fly, has played a vital role in biological research. *Drosophila* is not only a common home item but is also a well-studied creature in many scientific labs. The life cycle and development of *Drosophila melanogaster* is a thoroughly researched process that sheds light on the genetics and developmental aspects.

Drosophila melanogaster has served as a foundation for biological research, especially in the fields of developmental biology and genetics. Its high fecundity, short generation time and life cycle, ease of breeding in the laboratory, low cost management, and well-understood genetics make it an ideal model organism for studying development and life cycle processes. This chapter will explore *Drosophila*'s life cycle, developmental stages and rearing of *Drosophila* culture in the laboratory. *Drosophila melanogaster* has played a crucial role in advancing our understanding of developmental genetics. The identification of developmental genes in *Drosophila* has provided insights into the genetic mechanisms underlying the development of other animals, including humans. The homology of these genes across species has allowed researchers to study complex developmental processes in a relatively simple organism (Wolpert et al. 2019).

Drosophila melanogaster is a holometabolous insect. It undergoes complete metamorphosis showing transition through four distinct stages: egg, larva, pupa, and adult (Fig. 6.1). Identifying these stages and understanding their transition is indispensable for both academic and research purposes.

6.2 Rearing and Handling *Drosophila* in the Laboratory

Drosophila is easy to rear and maintain, and the following are the requirements for setting up a *Drosophila* lab (*FlyBase:New to Flies—FlyBase Wiki* 2013).

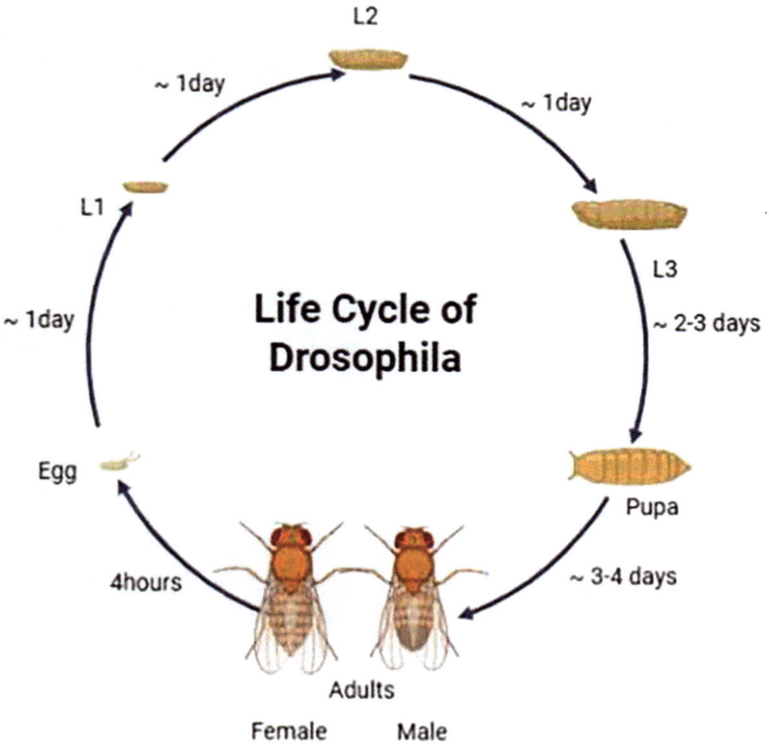

Fig. 6.1 Life cycle of *Drosophila* showing egg, larvae, pupa and adults. L1—First larval instar, L2—Second larval instar, L3—Third larval instar

6.2.1 Environment

Drosophila is easy to rear and requires only a BOD incubator (Fig. 6.2) with the temperature maintained between 20 and 25 °C. It is important to maintain the temperature because increased temperatures may lead to sterility, and lower temperatures are known to delay life span and viability. The flies are viable between 14 and 29 °C. Generally, higher temperatures lead to rapid development, shorter lifespan, and smaller adults. On the other hand, lower temperatures slow the development, increase life span, and produce larger adults. Beyond the viable range, there is a drastic reduction in viability (Petavy et al. 2001).

6.2.2 Culture Bottles and Vials

The flies are reared in glass milk bottles or glass vials (Figs. 6.3a and 6.3b). Non-absorbent cotton is used to plug these bottles and vials for rearing the flies. Flies should be transferred regularly to fresh bottles/vials to prevent overcrowding as

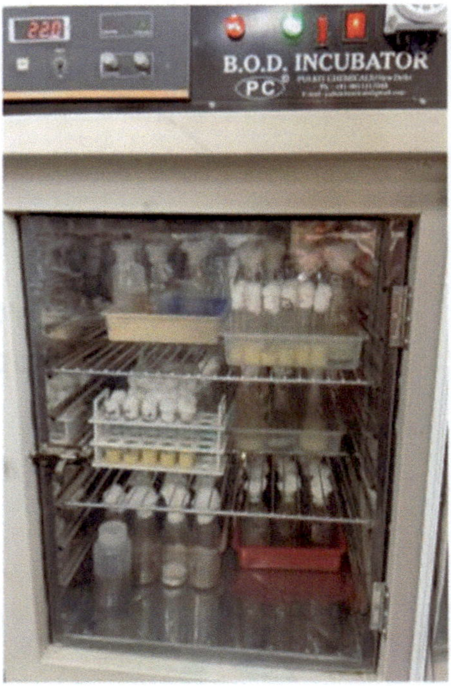

Fig. 6.2 Picture showing BOD Incubator for rearing flies

well as buildup of dead and decaying flies, which may lead to bacterial, fungal, or mite infections.

6.2.3 Essential Equipment

For handling the flies and larvae, a very simple setup is required, consisting of fine brushes, etherizer and re-etherizer, sponge pad, insulin syringes, and blunt forceps (Fig. 6.4). The flies are examined under stereo zoom microscopes (Fig. 6.5).

6.2.4 Food for *Drosophila*

Flies are reared on agar-based semi-solid food containing maize flour, sugar, and yeast. Different labs have their own food preparation protocols (Lakhotia 2021). In our lab, we follow this optimized protocol. The composition for 100 ml food is as follows.

(a) Maize flour: 6 g
(b) Sugar: 12.6 g
(c) Yeast: 3 g
(d) Agar-agar: 0.86 g

Fig. 6.3a Picture showing culture bottles with flies and larvae

Fig. 6.3b Picture showing culture vials with flies

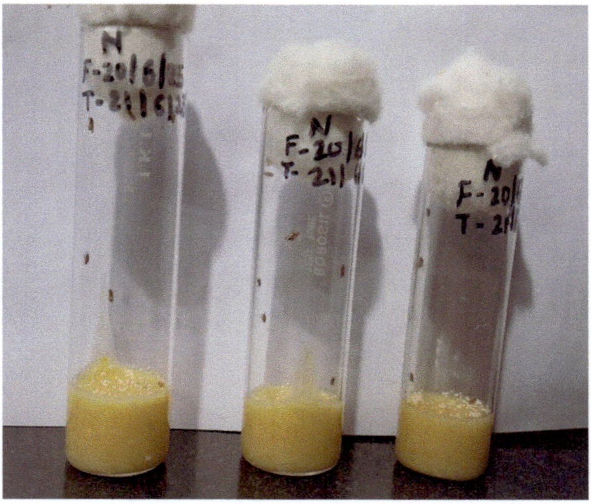

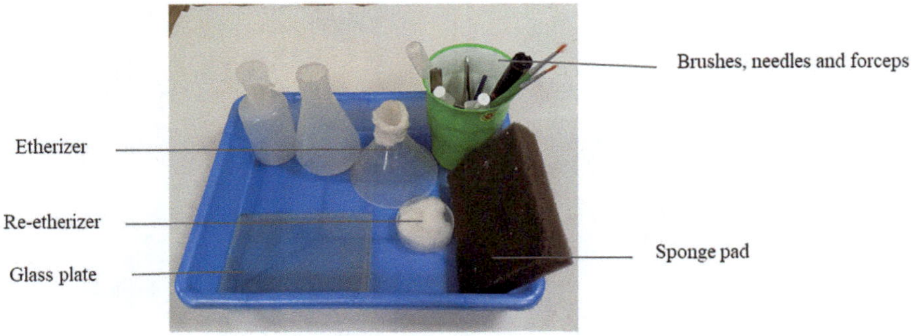

Fig. 6.4 Picture showing a tray with minor tools for handling *Drosophila*

Fig. 6.5 Picture showing stereo zoom binocular microscope used to observe stages of life cycle

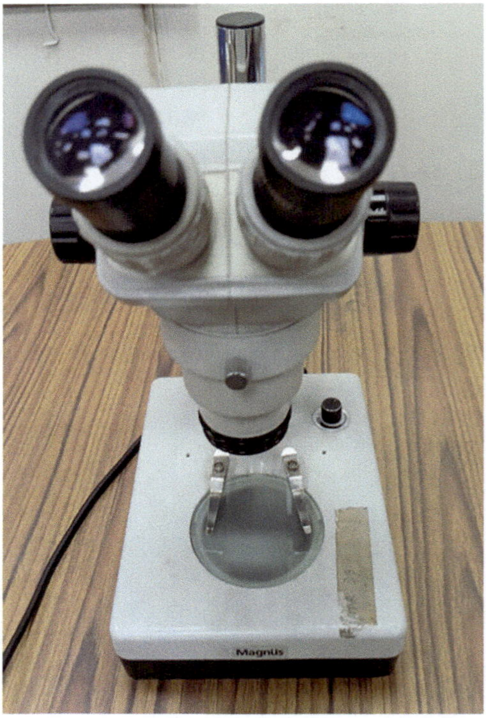

(e) Distilled water: 100 ml
(f) Preservatives: Propionic acid—143 μl, sodium benzoate—pinch

Take 72 ml water in a container and add the maize flour while stirring constantly to avoid forming lumps. When the consistency starts to thicken, add the sugar and stir till it melts. Next add the yeast and stir for some time. Heat the remaining 28 ml water in a glass beaker and add the agar till it starts frothing. Add this to the other container and mix well until the food starts to thicken. Remove from flame and add

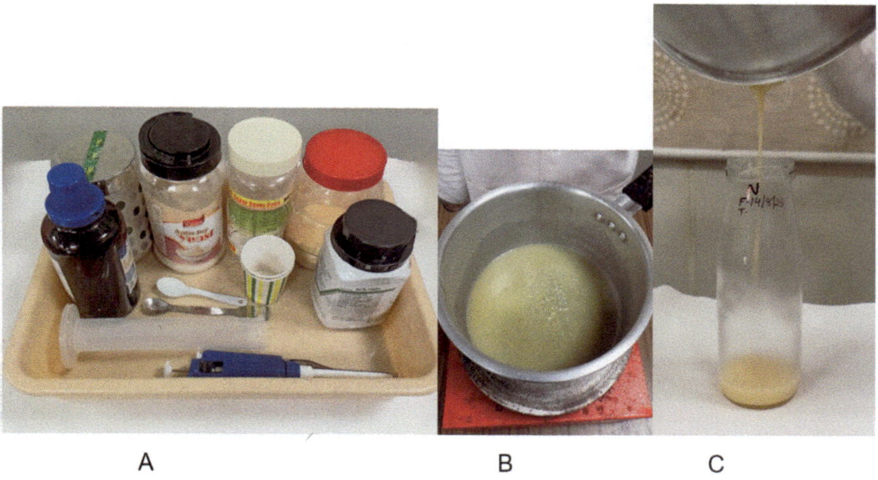

Fig. 6.6 Picture showing food preparation. (a) Tray with all components. (b) The food being prepared in a pan. (c) Food being poured into a culture bottle

the preservatives before pouring into the glass bottle. 25–30 ml of food is added to each glass bottle and 10–15 ml food is added to each glass vial. The bottles are covered with muslin cloth till the food solidifies. After solidification the bottles are sealed with cotton plugs and can now be used for transferring flies (Fig. 6.6).

6.2.5 Reculturing

Transfer of adults to fresh bottles should be done in around 3 weeks to avoid overcrowding and maintaining a healthy stock of flies. For transferring flies, the new bottle should be prepared with a funnel placed inside it. The old bottle is tapped gently a couple of times on a sponge pad to temporarily settle the flies near the food area before quickly opening the cotton plug and inverting it on the funnel placed on the new bottle. This setup is tapped on the sponge pad, and the flies will automatically move into the new bottle. The funnel is carefully removed, and the new bottle is sealed with a cotton plug.

Morgue: Flies that will no longer be used should be discarded in the morgue, which is a beaker or flask containing non-evaporating oil or soap, water, or 70% alcohol in an airtight bottle (Fig. 6.7).

6.2.6 Immobilizing Flies

To visualize and analyze the flies, they have to be immobilized using one of the following methods. Each one has its advantages and disadvantages, so the choice is based on the experimental requirements (Lakhotia 2021).

Fig. 6.7 Picture showing morgue flies that are no longer required are discarded in this bottle containing 70% alcohol

- Volatile anesthetics like ether: The flies are exposed briefly to a low dose of ether in an etherizer. An etherizer is made up of a conical flask or bottle with a funnel placed inside it. The tube of the funnel is wrapped with a thin layer of cotton tied with a thread. To anesthetize the flies, cotton on the funnel tube is moistened with ether and placed on the funnel. A bottle containing flies is briefly tapped on a foam and quickly inverted on the etherizer and tapped to transfer the flies. Soon the flies stop moving and can be transferred to the observation plate with the help of dry fine brushes for visualizing under a stereo zoom binocular microscope. If the flies start moving again, they can be re-anesthetized with the help of a re-etherizer made by sticking cotton on the inside of a small petri dish. The cotton is moistened with ether and inverted on the flies on the observation plate. This can be repeated if the observation is not complete, taking care that overexposure can kill the flies (Fig. 6.8).
- CO_2 exposure: The flies are exposed directly to CO_2 from a carbon dioxide chamber. This is a quicker and easier alternative with flies remaining immobile till exposure continues.
- Cold anesthesia: Chilling the flies on ice or cold environments is a chemical-free alternative for immobilizing the flies.

Fig. 6.8 Picture showing immobilization of flies using ether. The glass bottle with a funnel to whose stem cotton has been wrapped is the etherizer

6.2.7 Setting Crosses

The ideal female-to-male ratio for setting crosses is 3:1. Many genetic crosses are to be set between males and females of desired genotypes. *Drosophila* females can store sperm after mating and can use them throughout life, but this could hinder genetic crosses. So, it is desired to use unmated females in such genetic crosses. To do so, we may either collect unmated virgin adult females, or sexing can be done in the larval and pupal stages too.

- Females usually mate 4–6 h after eclosion, so if males and females are segregated within 2 h of emergence, we can collect unmated virgin females (Markstein 2019).
- Segregating mature pupa by observing their ventral side for the presence or absence of sex combs on the first tarsal segment of the prothoracic legs. Sex combs can be easily visualized through the pupal case to segregate male and female pupa into separate vials.
- Male and female flies can also be segregated during the larval stage by examining their gonads through the larval cuticle, which is semi-transparent. The male larvae have conspicuous gonads in the lower abdominal segments when observed under a microscope.

6.3 Overview of the *Drosophila* Life Cycle

Drosophila with four distinct phases completes its life cycle in around 10–14 days depending upon the temperature at which it is being reared. The life span of an adult *Drosophila* is around 80–100 days.

6.3.1 Courtship Behavior and Fertilization

In their life span, adult male and female *Drosophila* mate together to have internal fertilization. In the case of *Drosophila*, males initiate courtship rituals. Males dance around females while buzzing their wings and present her with the courtship songs. These songs are in the form of wing vibrations. Mating occurs only after females give an acceptance signal by slowing down, extruding her ovipositor, and spreading her wings. The length of time of mating depends on how many times the male partner has previously mated. During mating male is positioned on top of the female. Males release their sperm through penis into the female uterus. From the uterus, the sperms travel into the female seminal receptacle and spermatheca. Sperms are stored here for fertilization.

6.3.2 Oviposition

During oviposition, the eggs emerge from the female's ovipositor posterior end first. The female lays their fertilized eggs preferentially on a moist food surface in a humid atmosphere. Females can lay up to 50–100 eggs per day under optimal conditions. If the atmosphere is too dry, the female may have her food, but she won't oviposit. A female can lay eggs for most of the period of their lifespan. But with age, the number of eggs and their viability may decline.

6.3.3 Cleavage

After fertilization, an egg starts undergoing cleavage. In its initial days of laying eggs, a female can hold eggs within its uterus for a period of time as late as just prior to hatching of larva. Later on, around the 4th and 5th day, the uterus is cleared fastly by the female. So, embryos with an early stage of cleavage can be obtained by using 4–5-day-old culture of flies.

In *Drosophila* egg, the yolk is centrally concentrated and the cytoplasm is pushed to be placed toward the periphery. Such eggs are called centrolecithal eggs. Centrolecithal eggs have an unusual pattern of cleavage called superficial or peripheral cleavage. Nuclear division (karyokinesis) occurs many times before cytoplasmic division (cytokinesis) starts. During early cleavage (up to first 7), newly divided nuclei are placed centrally only. After this, they commence migrating to the periphery and still continue dividing. After about 5000 nuclei are lined up to the periphery (syncytial blastoderm), cytoplasmic division starts to enclose individual nuclei. This results in formation of an embryo with a peripheral layer of distinct cells with their individual nuclei. This is called cellular blastoderm. Cellularization of nuclei does not take place in a synchronized manner, rather it starts first at the posterior end. These newly formed, relatively large cells are called pole cells which descend to primordial germ cells during further embryonic development. These cells around the yolk are patterned along the anterior-posterior (A-P) and dorsal-ventral (D-V) axes by maternal and zygotic gene expression (Gilbert 2000).

6.3.4 Gastrulation

Following cleavage, gastrulation in *Drosophila* rapidly transforms the cellular blastoderm into a multi-layered embryo. *Drosophila* gastrulation's initial movements separate the presumed mesoderm, endoderm, and ectoderm. It begins with the invagination of the ventral furrow. The ventral furrow is created when the approximately 1000 cells that make up the embryo's ventral midline fold inward. Eventually, this furrow separates from the surface to form the embryo's ventral tube. Below the ventral ectoderm, it flattens to create a layer of mesodermal tissue.

At the anterior and posterior extremities of the ventral furrow, the prospective endoderm invaginates as two pockets. Together with the endoderm, the pole cells are internalized. The embryo now forms the cephalic furrow by bending (Sharma 2023).

Once the mesoderm and endoderm are internalized, the entire trunk region of the embryo, referred to as the germ band, dramatically elongates. The main cause of this extension is cell intercalation, in which cells rearrange themselves in the epithelial layer, trading neighbors to make the tissue longer along the anterior-posterior axis and narrower along the dorsal-ventral axis (Gilbert 2000).

The posterior structures are brought close to the head area by the germ band's posterior extension, which eventually wraps around the embryo's dorsal side.

On the dorsal side of the embryo, a noticeable transverse invagination known as the cephalic furrow develops, dividing the potential head area from the trunk.

Following germ band extension, the embryo experiences germ band retraction. This brings the posterior segments back to their caudal position. This is followed by further morphogenetic processes, including segmentation, organogenesis, and neurogenesis, ultimately leading to the formation of the larva.

6.3.5 Emergence of First Instar Larva (L1)

Usually between 22 and 24 h after fertilization (at 25 °C), the larva emerges from the egg. They are referred to as L1 or first instar larva. It is tiny, worm-like, and starts eating the culture medium in large quantities right away. They are called voracious feeders. The L1 stage is marked by moderate growth, mostly due to cell enlargement (hypertrophy) in numerous larval-specific organs and cell division to some extent. Certain tissues, like imaginal discs, which will ultimately develop into adult structures, start to expand initially as the overall size increases. The L1 stage lasts for about 24–25 h.

6.3.6 Molting of First Instar (L1) to Second Instar Larva (L2)

The first molt (ecdysis) signifies the change from L1 to L2. A new, bigger cuticle is secreted underneath after it sheds its old one, along with its mouthparts and spiracles. This makes it possible for the size to grow even more. Second instar larvae keep up their rapid growth and active feeding. Growth during this stage is greatly aided by both cell expansion and cell proliferation (mitosis). Rapid cell division also occurs in the Imaginal discs, although they do not differentiate. Usually, the L2 stage lasts for roughly 24 h.

6.3.7 Molting of Second Instar (L2) to Third Instar Larva (L3)

The second molt leads to change from L2 (Second instar) to L3 (third instar). L3 is the largest of all instars in dimensions. Of all instar durations, the third instar is the longest phase. Like L1 and L2, L3 continue their voracious feeding and exhibit exponential growth, reaching their maximum larval size. This phase duration and intense feeding are crucial for storing energy reserves required for future transformation and growth and building up imaginal discs, which are going to form adult tissue (Jürgens et al. 2024). Imaginal discs grow in size by extensive cell division, but are still undifferentiated. Toward the end of the L3 phase, the larvae stop feeding and move away from food sources looking for a comparatively drier place. These changes in behavior are induced by hormonal fluctuation to prepare larvae to undergo pupation.

6.3.8 Hormonal Regulation of Molting

It is a steroid hormone called ecdysone that drives the intricate process of molting in *Drosophila* and other insects. Ecdysone is produced and released by the prothoracic gland, an endocrine organ in larvae. Molting in *Drosophila* larva L1 and L2 is triggered by low-titer ecdysone pulses. During molts, the old cuticle, or exoskeleton, is shed (ecdysis), which permits the larva to enlarge and develop (Kannangara et al. 2021). Before the old cuticle is lost, a larger, new one is synthesized underneath it. By the conclusion of the L3 stage, there is a high-titer pulse of ecdysone. This high-titer pulse signals the end of larval feeding, the beginning of pupariation (the production of the pupal case from the hardened larval cuticle), and the start of the pupa's spectacular metamorphosis (Warren et al. 2006; Wegman et al. 2010).

6.3.9 Pupation

Pupation is a crucial stage of *Drosophila melanogaster's* holometabolous (full) metamorphosis. It signifies the abrupt change from the feeding and growing phase of larvae to the non-feeding, transforming stage that results in the adult fly. Large-scale cellular and tissue rearrangement occurs at this time. Some tissues being retained, some being destroyed, and some being constructed from specialized imaginal tissues.

The entire process of pupation can be conceptually divided into two main phases: pupariation and the subsequent pupal development (metamorphosis).

Pupariation is the initial event where a voracious feeder and highly active larva transition to an inactive and immobile pupa. Before pupariation, the L3 larvae, which have stopped eating, enter a wandering stage. In this stage, larvae look for a safer and drier place away from food sources. The purpose is to protect delicate pupae from rotten food and infecting pathogens. After this, the larvae immobilize themselves by shortening their body and making them round. The anterior spiracles evert for respiration. The immobilization is followed by hardening and darkening of larvae's old cuticles. It forms a barrel-shaped outer protective covering called puparium (Wegman et al. 2010). Shortly after puparium formation, apolysis takes place, in which larvae epidermis detaches itself from the old cuticle. Below, a delicate new "pupal cuticle" is secreted to envelop the growing pupa. Twelve hours after pupal formation, head eversion occurs. After head eversion, they are referred to as true pupa. Before this, they are considered as prepupa (Thomas B. Brody 2020).

After a true pupa is formed, the process called metamorphosis starts taking place. During metamorphosis, the larva transforms into an adult structure. Most of the larval-specific tissue and organs undergo apoptosis and histolysis. Their degraded cellular component is used to power the formation of adult structures. The imaginal discs start forming adult structures like wings, legs, eyes, etc. by differentiation and morphogenesis. While there are some systems and organs which undergo considerable remodeling but are not fully histolyzed, like the central nervous system, gonads, and portions of the intestine (Thomas B. Brody 2020). Through the translucent puparium, exterior changes are obvious as growth continues. The wings appear as folded structures, the compound eyes become pigmented, and the adult form as a whole becomes recognizable. The puparium gets darker with the progression of days. At 25 °C, the pupal stage normally lasts 4–5 days, though this greatly depends on the temperature.

6.3.10 Hormonal Regulation of Pupation

Ecdysone is the mastermind behind this whole metamorphic program, frequently working in tandem with the declining levels of juvenile hormone (JH). Pupariation is started by a high ecdysone pulse, and the distinct developmental events of adult differentiation are driven by subsequent pulses and various ecdysone titers throughout the pupal stage. In larvae, JH inhibits metamorphosis; in L3 larvae, its low or absent levels permit ecdysone to start pupation (Warren et al. 2006).

6.3.11 Eclosion

The pupal stage culminates with the emergence of the new adult fly (imago) from the puparium. This is called pupal eclosion. This is a hormone-controlled event. In addition to this, newly formed adult uses specialized mechanisms and internal pressure to crack open the pupal case. The newly born adult is soft, pallid, and has crumpled wings at first, but when it fills its body with air and hemolymph, it quickly hardens its cuticle and spreads its wings.

6.4 Key Identifying Features for Life Cycle Stages of *Drosophila*

Egg (Fig. 6.9):

- *Drosophila* eggs are tiny in shape measuring about 0.5 mm in length.
- They are oval in shape with an appearance of rice grain.
- They are off-white or pale yellow in color.
- Its outermost covering is made up of chorion, which is completely proteinaceous in composition and considered to be the most complex layer.
- Chorion is there to protect against mechanical damage and infection by pathogens.
- Inner to chorion there is hydrophobic wax layer followed by vitelline membrane as innermost layer (Sharma 2023).
- They are equipped with two projections on the anterior dorsal side called respiratory filament. They are for respiration and also referred to as breathing tubes.
- There is one more specialized opening on the anterior pole called micropyle. This is to allow entry of sperm for fertilization. Visible under microscope.
- Often laid on moist food surfaces in cluster or separately.
- Eggs are oriented in food with two respiratory filaments projected outward.

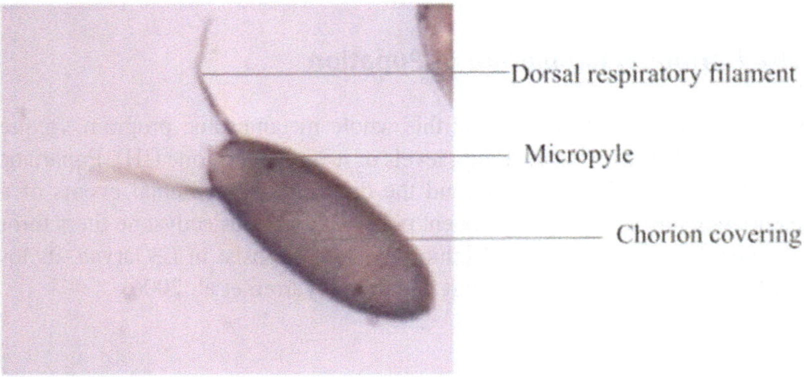

Fig. 6.9 Picture showing *Drosophila* egg with ventral side up

First instar larva (Fig. 6.10):

- The larval stage that hatches directly from egg is the first instar larva.
- They are very small in size, measuring about 0.5–1 mm.
- It is whitish or translucent in appearance.
- It is a voracious feeder, so generally found inside the food.
- Body is segmented but lack appendages and legs.
- Head region is less well defined as compared to other instar.
- Black color mouth parts are visible at the anterior side (help in localizing larvae inside the food). Mouth carries smaller hooks and less teeth as compared to later instar larvae.
- Larvae also carry anterior and posterior spiracles, but they are less distinct as compared to L2 and L3.
- They move inside food by peristaltic movement.
- They grow in size mainly by hypertrophy and to a lesser extent by hyperplasia.

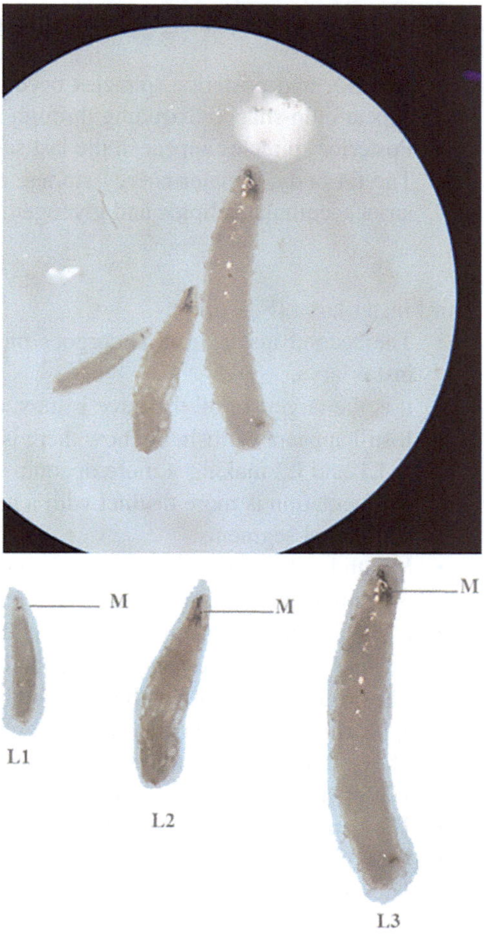

Fig. 6.10 Picture showing comparison of body length and mouth parts of all three larval instars of *Drosophila melanogaster*. L1: first instar Larva, L2: second instar, L3: third Instar larva, M: mouth

Second instar larva (Fig. 6.10):

- First instar larva undergoes its first molt to emerge as second instar larva.
- Second instar larva (L2) is larger in size as compared to L1. It ranges in size from 1.5 to 2.5 mm.
- Increase in size is contributed by both cell enlargement and cell proliferation.
- Appearance wise L2 is similar to L1 (worm-like and segmented) but appears thicker and longer.
- Color also remains whitish or translucent. Mostly appear inside food, being intense feeders.
- Mouthparts become more prominent and developed. Size of hooks and number of teeth increases.
- More strong mouthparts develop for efficient scraping of food.
- The larval airway system can be visualized clearly as a network of interconnected tubes organized in a hierarchical structure. Two longitudinal dorsal trunks are connected to the outside world via openings on the body surface called spiracles. Primary branches originate from the longitudinal dorsal trunk. The blind endings of the tracheal system are made from terminal branches.
- Anterior and posterior spiracles become more prominent. Anterior spiracles appear club-shaped extruding through the first thoracic segment.
- Posterior spiracles appear in the last segment as two prominent dots.
- The fat body, a major energy storage organ, continues to grow in size as the larva accumulates lipids and glycogen.

Third instar larva (Fig. 6.10):

- The second instar larva undergoes molting to change to the third and final instar larva.
- It is the largest of all the three instars, approximately 4–5 mm in length.
- It still appears whitish, but now there is more accumulation of fat as compared to L1 and L2 making it more opaque.
- Segmentation is more distinct with a head, three thoracic segments, and eight abdominal segments.
- Mouth hooks become more prominently large and robust, with more number of teeth as compared to L1 and L2.
- L3 is also characterized by intense and continuous feeding.
- The anterior spiracle is now larger and branched in shape.
- The posterior spiracles at the very caudal end are also large and clearly visible, appearing as two dark, orange-colored spots with distinct respiratory slits. They often protrude slightly from the body.
- The internal tracheal system is also more developed and more branched.
- They are highly active and intense feeders.
- Late third instar larvae are characterized by their wandering behavior.
- They stop feeding and move away from the food in order to prepare themselves to undergo pupation.

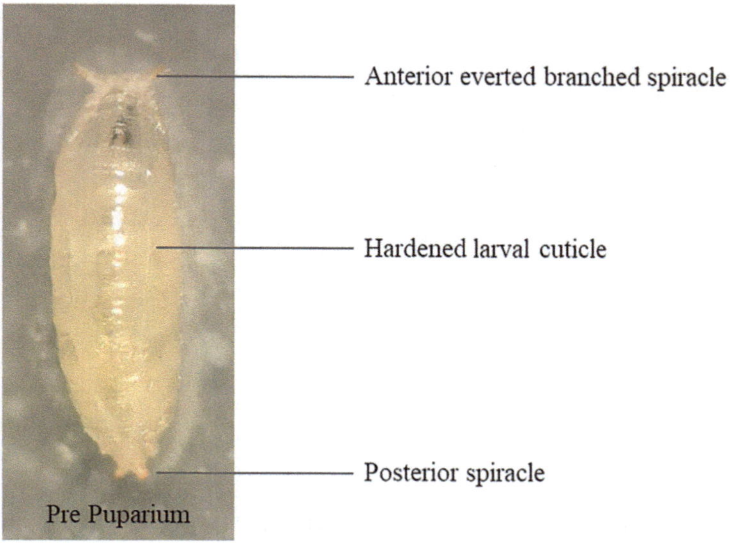

Fig. 6.11 Picture showing barrel-shaped prepuparium (off-white color)

Prepupal stage (Fig. 6.11):

- The transition phase that connects the end of the larval phase and the commencement of true pupal development.
- It begins when larva stops feeding, moving away from the food and anchoring themselves to the substrate, making them immobile.
- The larva (L3) cuticles harden, shorten, and start getting darker. Still they appear off-white in color.
- The long, worm-like L3 body shortens and becomes more barrel-shaped within the puparium.
- The anterior spiracles evert and spread out. They appear as two small, fan-like structures that are more exposed on the anterior end of the puparium. These are pivotal for respiration during pupation.
- The larva head region becomes less distinct because of its inward retraction.
- It lasts for 10–12 h at 25 °C.

Pupal stage (Figs. 6.12 and 6.13):

- It is an immobile and non-feeding stage of development.
- Pupa is enclosed within a barrel-shaped puparium, which gets harder and darker (due to pigmentation) with progression of time.
- The puparium changes its color from off-white to pale, pale to light brown, light to reddish brown, and ultimately reddish brown to black brown (Fig. 6.14).
- The everted anterior spiracle and posterior spiracle remain visible.

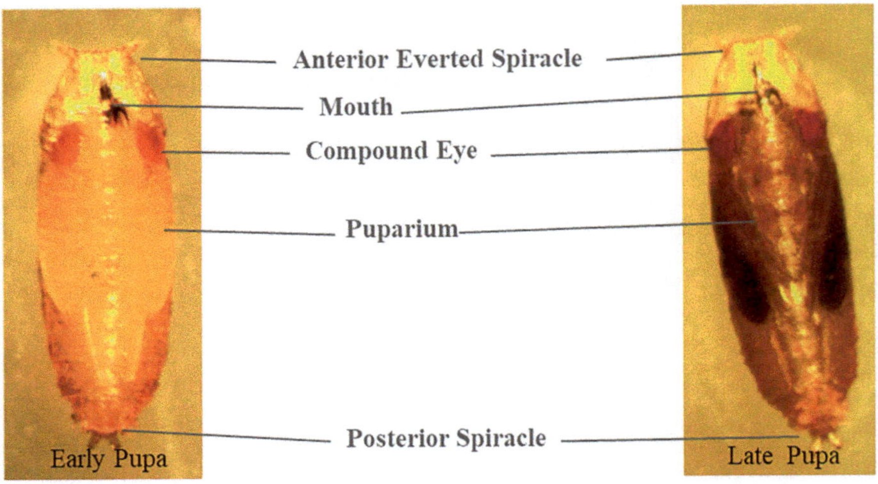

Fig. 6.12 Picture showing ventral side of early and late pupa

- In later stages of pupation, everted head and thoracic appendages can be visualized taking their adult position beneath the cuticle covering.
- Development of wings, eye, halter, and abdominal segment also become distinct with further development.
- It typically lasts for 3–4 days at 25 °C.

Adults (Figs. 6.15 and 6.16):

- Adults are small insects measuring in length about 2–4 mm. They are yellowish brown in color (wild-type).
- The body is divided into the head, thorax, and abdomen.
- The head is round and distinct. The head bears two very large, prominent, brick-red compound eyes on either side of the head. Each compound eye is composed of hundreds of individual light-sensing units called ommatidia, providing a wide field of vision.
- The wild type fly has red eyes which is due to specific pigments (ommochromes and drosopterins).
- The head also carries three smaller, simple eyes (ocelli), which are positioned on the top of the head, between the compound eyes. These are involved in light intensity detection and possibly photoperiodism.
- Short, stubby, and feathery or bristled antennas are located on the head. Each antenna has a basal segment, a pedicel (second segment), and a funicle (third segment) which bears a feathery bristle called the arista. Antennae are crucial for detecting odors, air currents, and vibrations.
- The fly has a sucking and sponging type of mouthpart adapted for sopping up liquids.

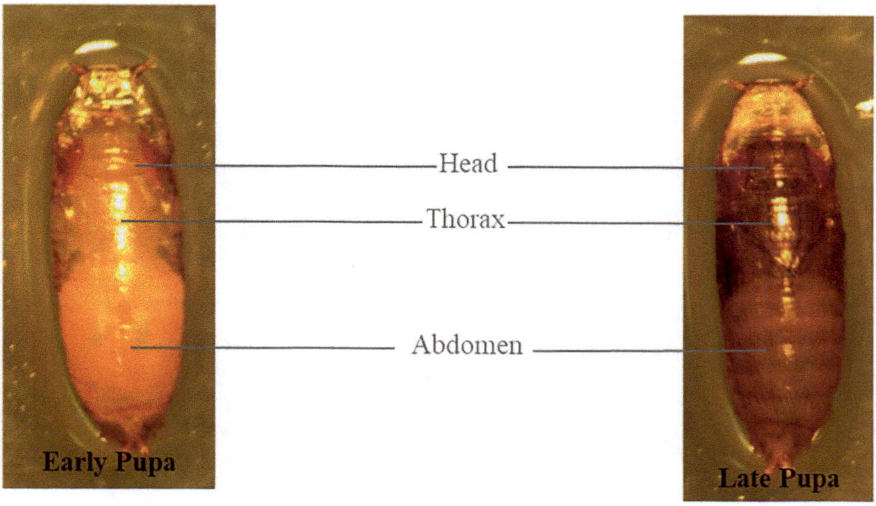

Fig. 6.13 Picture showing dorsal side of early and late pupa

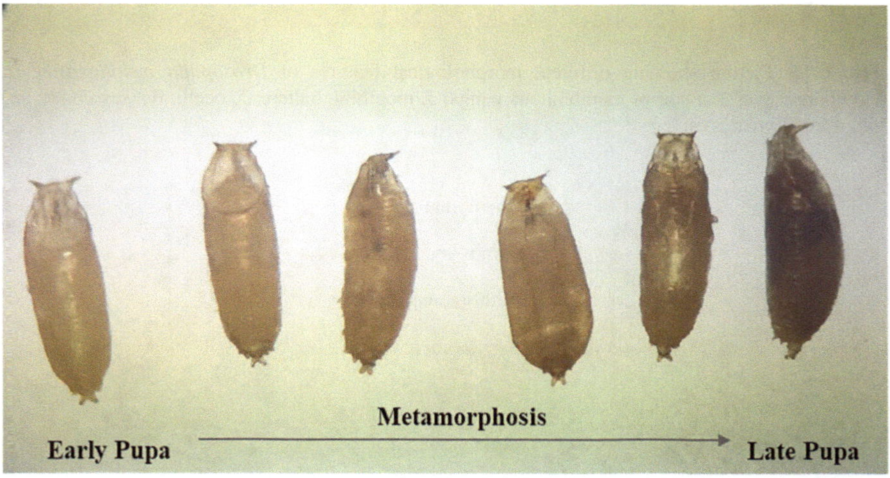

Fig. 6.14 Picture showing different stages of metamorphosis during pupation changing color from off-white as early pupa to dark brown as late pupa

- *Drosophila* has a single pair of functional fore wings, originating from the mesothorax. The wings are clear, membranous, with a distinct pattern of fine veins. These veins are important for species identification and genetic studies.
- The hindwings are modified into a pair of small, knob-like structures called halteres. They are positioned just behind the functional wings. Halters vibrate rapidly during flight and act as gyroscopes.

Fig. 6.15 Picture showing different morphological features of *Drosophila melanogaster*. *1*. Compound eye; *2*. a pair of membranous wings; *3*. mouth; *4*. haltere; *5*. ocelli; *6a*. sex comb; and *6b*. normal bristles on leg

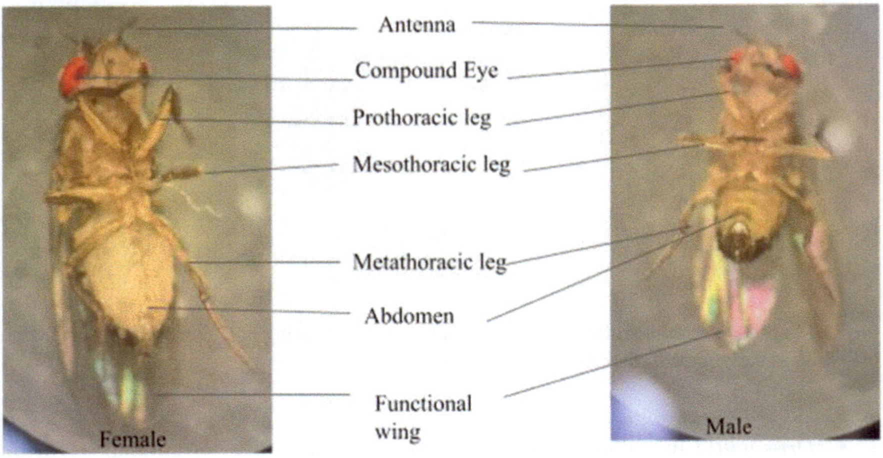

Fig. 6.16 Picture showing ventral side of female and male *Drosophila*. Female with pointed and longer abdomen and male with round and shorter abdomen

Fig. 6.17 Picture emphasizing differences between *Drosophila* male and female abdominal segments

- There are three pairs of segmented legs (prothoracic, mesothoracic, metathoracic) that extend from the ventral side of the thorax.
- The abdominal region is segmented and appears to be striped with alternating bands of light and dark pigmentation.
- The dorsal half of the abdomen is built by sclerotized cuticle plates called tergites.
- Tergites are separated from each other by membranous cuticles which impart flexibility to the abdominal region (Fig. 6.17).
- Locomotory appendages are covered with bristles in both males and females. They function for sensing.
- Males exclusively have rows of dark thickened bristles called sex comb on the tarsal segment of the first limb (Fig. 6.15).
- There is sexual dimorphism in *Drosophila* adults (Figs. 6.16 and 6.17).

Sexual Dimorphism
Drosophila exhibits sexual dimorphism; adult males and females can be distinguished based on the features tabulated in Table 6.1 (Figs. 6.16 and 6.17).

6.5 Significance/Research Applications

Drosophila was first used as a model organism for studying spontaneous mutations and their inheritance in 1905 (Carpenter 1905). For a very long time, *Drosophila melanogaster* was used extensively as a model for genetics studies like chromosomal theory, gene mapping, and analysis of genetic pathways. But subsequently,

Table 6.1 Tabulation of characteristic features of male and female adult flies

Characteristic/ feature	Male	Female
Body size	Smaller	Larger
Abdomen shape	Short and rounded	Long and pointed
Abdominal segments	Five visible (last three fused)	Seven visible out of eight
Coloration	Dark abdominal tip	Lighter with segments visible
Sex comb	Present	Absent
External genitalia	Visible at round abdominal tip (clasper) (Fig. 6.18)	Less visible (ovipositor) (Fig. 6.19)

Fig. 6.18 Picture showing male *Drosophila* with encircled clasper and sex comb

this model organism was used in various kinds of studies, and currently it is being used to study various aspects like:

- Genetics—*Drosophila* research helped establish the principles of inheritance and gene mapping and continues to be essential to research on gene regulation and functional genomics (*"Sex Limited Inheritance in Drosophila" (1910), by Thomas Hunt Morgan | Embryo Project Encyclopedia* n.d.; "The Mechanism of Mendelian Heredity" 1916).
- Developmental biology—Our knowledge of how organisms develop from a single cell has been completely transformed by discoveries made in *Drosophila*, including homeotic genes and signaling pathways like Hedgehog and Notch (Tolwinski 2017).

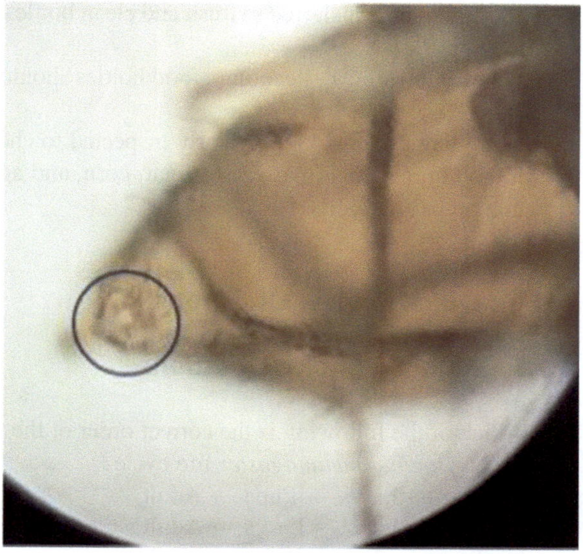

Fig. 6.19 Picture showing female *Drosophila* with encircled ovipositor

- Human disease model—*Drosophila* is used to model a number of human diseases, such as cancer, metabolic disorders, and neurodegeneration, because of its genetic resemblance to humans. It is especially effective at breaking down metabolic and carcinogenic pathways (Mohr and Perrimon 2019).
- Evolutionary biology—It is perfect for research on evolutionary biology, speciation, adaptation, and the molecular mechanisms of aging because of its quick life cycle and genetic diversity (Powell 1997a, b).
- Neuroscience and behavior—A crucial model for researching learning, memory, neural development, and the genetic foundations of behavior is the *Drosophila*. Additionally, neurodegenerative diseases like Parkinson's and Alzheimer's are modeled using it (McGurk et al. 2015; Taylor et al. 1994).

The *Drosophila* has a short life cycle and high fertility which enables large-scale genetic and developmental biology studies. Its genome is highly conserved, and most genes have orthologs in humans, thus helping in disease modeling. With well-established techniques of gene editing like CRISPR-Cas, RNAi, and transgenesis, it is possible to easily modify the fly for studies. The extensive support and free sharing of mutants also enhance the ease of scientific research with *Drosophila*. Moreover, the minimal ethical issues in using *Drosophila* for research make it a favorite model organism for research (Dähn and Wagner 2025).

6.6 Precautions

- We should try to maintain spotless laboratory environment to prevent culture contamination by mites.
- BOD should be maintained at temperature ranging from 20 to 25 °C.

- Flies should be transferred to fresh and clean bottles regularly after approximately 20–25 days.
- Excessive moisture collection in food bottles should be avoided to prevent fungal and bacterial growth.
- Culture bottles should be regularly inspected to check for any contamination.
- All food ingredients (like yeast, sugar, corn, and agar) should be properly added while preparing food.

6.7 Quick Review Questions

6.7.1 Multiple Choice Questions

1. Which of the following is the correct order of the main developmental stages in the *Drosophila melanogaster* life cycle?
   (a) Egg → Larva → Pupa → Adult
   (b) Egg → Pupa → Larva → Adult
   (c) Larva → Egg → Pupa → Adult
   (d) Egg → Adult → Larva → Pupa
   Answer: a) Egg → Larva → Pupa → Adult
2. How many larval instar stages does a *Drosophila melanogaster* pass through before entering the pupal stage?
   (a) 1
   (b) 2
   (c) 3
   (d) 4
   Answer: c) 3
3. What is the characteristic feature of the syncytial blastoderm stage in *Drosophila* embryogenesis?
   (a) Individual cells are fully separated
   (b) Nuclei are not divided into separate cells
   (c) All membranes are fully formed between cells
   (d) Adult structures begin to develop
   Answer: b) *Nuclei are not divided into separate cells*
4. Which of the following processes occurs during the pupal phase of *Drosophila* development?
   (a) Massive cell division producing new nuclei
   (b) Larva undergoes molting to new instar
   (c) Breakdown of larval tissues and formation of adult structures from imaginal discs
   (d) Laying of fertilized eggs
   Answer: c) Breakdown of larval tissues and formation of adult structures from imaginal discs
5. At what temperature does the complete life cycle of *Drosophila* (egg to adult) take around 10 days?

   (a) 4 °C
   (b) 18 °C
   (c) 25 °C
   (d) 37 °C
   Answer: c) 25 °C
6. During which stage does metamorphosis occur in *Drosophila*?
   (a) Embryo
   (b) Larva
   (c) Pupa
   (d) Adult
   Answer: c) Pupa
7. What is the main function of imaginal discs in *Drosophila* larvae?
   (a) Digest food
   (b) Develop into adult structures during metamorphosis
   (c) Protect the larvae from infection
   (d) Assist in larval movement
   Answer: b) Develop into adult structures during metamorphosis
8. In laboratory rearing, what kind of medium is commonly used for *Drosophila*?
   (a) Plain water
   (b) Nutrient agar with yeast and sugar
   (c) Soil only
   (d) Pure sugar crystals
   Answer: b) Nutrient agar with yeast and sugar
9. What should be observed to confirm the presence of *Drosophila* larvae in the rearing medium?
   (a) Changes in medium color
   (b) Small black mouth hooks and disturbance on surface
   (c) Eggs broken into fragments
   (d) Adult flies immediately after egg laying
   Answer: b) Small black mouth hooks and disturbance on surface
10. Where do *Drosophila* females typically lay their eggs in a lab setting?
   (a) Clean glass surfaces
   (b) Nutrient medium or decaying fruit medium
   (c) Soil
   (d) Water
   Answer: b) Nutrient medium or decaying fruit medium

6.7.2 Short Answer Questions

1. How can you differentiate between second and third larval instars?
2. Write a note on *Drosophila* egg and early development.
3. Why does pupation occur? Explain with respect to hormonal changes.
4. Write a short note on the advantages of using *Drosophila* as a Model organism.
5. Write a short note on food for *Drosophila*.

6. What are the different methods for immobilizing *Drosophila* for experiments?
7. Describe the life cycle of *Drosophila melanogaster*.
8. How can we identify adult males and females? Differentiate between them.
9. Elaborate on the use of *Drosophila* for Research.
10. Tabulate the identification characteristics for all stages of *Drosophila* life cycle.

Glossary

Adaptation Adaptation refers to the process by which organisms develop traits—whether anatomical, physiological, or behavioral—that increase their chances of survival and reproduction in a specific environment.

Alzheimer Alzheimer's disease is a progressive neurodegenerative disorder and the most common cause of dementia.

Anesthetize Anesthetize means to administer anesthetic drugs to a person or animal so that they become temporarily unable to feel pain, often in preparation for a medical procedure or surgery.

Antennas The antenna is a complex sensory organ located on the fly's head, and its primary function is in hearing, olfaction, mechanosensation, and balance.

Apoptosis Apoptosis is a natural, programmed process of cell death in which cells undergo an orderly sequence of events leading to their self-destruction without causing harm to surrounding cells. It is essential for maintaining healthy tissue function, development, and immune system regulation in multicellular organisms.

Behavior Behavior is the set of actions or reactions of an organism, individual, or system in response to internal or external stimuli. It includes both voluntary and involuntary actions that can be conscious or subconscious, overt or covert.

Cancer Cancer is a disease characterized by the uncontrolled growth and division of abnormal cells which can invade nearby tissues and spread to other parts of the body.

Cephalic furrow The cephalic furrow is a deep, transient epithelial fold or invagination that forms during gastrulation in Drosophila melanogaster.

Chorion The chorion is the tough, multi-layered eggshell that surrounds and protects the developing embryo inside the egg. It is formed during the late stages of oogenesis by proteins secreted from the follicle cells that surround the oocyte.

Cleavage Cleavage refers to the series of rapid mitotic cell divisions that occur immediately after fertilization, resulting in the division of the fertilized egg into many smaller cells called blastomeres.

CO_2 Carbon dioxide in the context of Drosophila is a gaseous molecule that serves as an important environmental cue influencing behavior and physiological processes.

Complete metamorphosis Complete metamorphosis is a type of insect development characterized by four distinct life stages: egg, larva, pupa, and adult.

Compound eyes Compound eyes are complex visual organs composed of approximately 700–750 repetitive units called ommatidia arranged in a crystalline hexagonal pattern.

Courtship rituals Courtship rituals in Drosophila are a series of innate, stereotyped behaviors performed primarily by males to attract and mate with females.

CRISPR-Cas CRISPR-Cas is a gene-editing technology involving two key components: a guide RNA, which directs the system to a specific DNA sequence, and the Cas9 protein, an enzyme that acts like molecular scissors to cut the DNA at the targeted site.

Cytoplasm Cytoplasm is the semi-fluid substance within a cell that is enclosed by the cell membrane but excludes the nucleus in eukaryotic cells, consisting primarily of water, salts, proteins, and various organic molecules, and it contains all the cell's organelles except the nucleus.

Ectoderm The ectoderm is the outermost of the three primary germ layers formed during early embryonic development after the process of gastrulation. It gives rise to various tissues and structures including the skin, nervous system (brain and spinal cord), eyes, ears, hair, nails, and tooth enamel.

Endoderm The endoderm is the innermost of the three primary germ layers formed during early embryonic development, alongside the ectoderm (outer layer) and mesoderm (middle layer).

Epidermis The epidermis is the outermost layer of the skin, serving as a protective barrier against environmental hazards such as pathogens, UV radiation, and water loss.

Etherizer A device used to vaporize and deliver ether to induce anesthesia.

Evolutionary biology Evolutionary biology is the branch of biology that studies the processes driving the diversity and adaptation of life on Earth over time through mechanisms such as natural selection, genetic drift, mutation, gene flow, and speciation.

Exoskeleton An exoskeleton is a hard, protective external covering that supports and shields the body of certain animals, primarily invertebrates such as insects, spiders, crustaceans, and mollusks.

Functional genomics Functional genomics is the branch of molecular biology that aims to understand the dynamic functions and interactions of genes and their products on a genome-wide scale.

Gene mapping Gene mapping is the process of determining the specific locations of genes on a chromosome and measuring the distances between them.

Gene regulation Gene regulation refers to the processes and mechanisms by which a cell controls the timing, location, and amount of gene expression.

Genetic crosses A genetic cross is the intentional mating of two selected, different individuals.

Genetic diversity Genetic diversity is the variation in genetic makeup among individuals within a population, species, or ecosystem.

Genotypes A genotype is the genetic makeup of an organism, representing the specific combination of alleles inherited from its parents.

Germ band A germ band is a specialized region of thickened cells in the blastoderm of an insect embryo from which the main body segments and structures of the organism arise.

Gonads Gonads are the primary reproductive organs in animals, responsible for producing gametes and secreting sex hormones essential for sexual development and reproduction.

Gyroscopes A gyroscope is a device used for measuring or maintaining orientation and angular velocity.

Hatching In egg-laying animals, hatching involves the escape of the embryo from its egg membranes or shell.

Hedgehog gene The Hedgehog gene is a key developmental gene that encodes secreted signaling proteins regulating cell growth, differentiation, and patterning during embryonic development.

Hemolymph Hemolymph is a body fluid found in invertebrates, especially arthropods, that functions similarly to blood in vertebrates.

Histolysis Histolysis is the biological process involving the breakdown, decay, or dissolution of organic tissues, often occurring during developmental transformations like metamorphosis.

Holometabolous Holometabolous refers to a type of insect development characterized by complete metamorphosis, which includes four distinct life stages: egg, larva, pupa, and adult.

Homeotic gene Homeotic genes are master regulatory genes that control the development and identity of anatomical structures during early embryonic development in organisms.

Hyperplasia Hyperplasia is the increase in the number of cells in an organ or tissue, leading to its enlargement.

Imaginal discs An imaginal disc is a specialized group of undifferentiated cells found in the larvae of insects that undergo complete metamorphosis, like flies and butterflies.

Instar An instar is a developmental stage of arthropods, such as insects, that occurs between each molt (ecdysis) until sexual maturity is reached.

Invagination Invagination is a fundamental morphogenetic process in which a flat sheet of cells, known as an epithelium, folds inward to form a depression or pocket.

Juvenile hormone Juvenile hormone is a hormone that regulates development, metamorphosis, and reproduction in insects. It is secreted by a pair of endocrine glands called the corpora allata, located just behind the brain.

Mesoderm The mesoderm is one of the three primary germ layers in the early embryo of most multicellular animals. It is a middle layer of embryonic cells that develops between the endoderm and ectoderm.

Metathoracic Metathoracic refers to the metathorax, which is the hindmost (last) of the three segments that make up the thorax of an insect.

Molting Molting is the process by which an animal periodically sheds its outer layer, such as an exoskeleton, skin, hair, or feathers, to allow for growth or to replace worn-out parts.

Morphogenesis Morphogenesis is the biological process that creates the shape and form of an organism.

Neurodegeneration Neurodegeneration is the progressive and irreversible loss of function and death of neurons in the central or peripheral nervous system.

Neurodegenerative diseases Neurodegenerative diseases are a group of disorders which are characterized by the progressive degeneration and death of neurons in the central and peripheral nervous systems.

Neuroscience Neuroscience is the scientific study of the nervous system.

Notch Notch refers to a transmembrane receptor protein that plays a crucial role in cell-cell communication and is vital for various developmental processes. It's part of the Notch signaling pathway.

Oviposition It is the process by which an insect or other egg-laying animal deposits its eggs.

Ovipositor The ovipositor is a specialized, retractable tube-like structure at the posterior end of a female fruit fly's abdomen. Its primary function is to deposit eggs into a suitable substrate.

Parkinson Parkinson is a progressive, neurodegenerative disorder that primarily affects the nervous system and leads to a range of motor and non-motor symptoms. It is caused by the death of nerve cells in a specific part of the brain called the substantia nigra.

Peristaltic movement Peristaltic movement, or peristalsis, is a series of involuntary, wave-like muscle contractions and relaxations.

Primordial germ cells Primordial germ cells are the embryonic precursors of the adult germ cells (sperm and eggs).

Principles of inheritance The principles of inheritance are a set of fundamental rules that explain how genetic traits are passed from parents to offspring.

Prothoracic Prothoracic refers to the prothorax, which is the foremost of the three segments that make up the thorax of an insect.

Prothoracic gland The prothoracic gland is a vital endocrine organ. It is the main source of the steroid hormone called ecdysone, which regulates molting and metamorphosis in Drosophila.

Prothoracic legs Prothoracic legs are the first pair of legs on an insect. They are attached to the prothorax, which is the foremost of the three segments that make up an insect's thorax.

Puparium A puparium is a protective case that encloses the pupa of certain insects.

Re-etherizer Use to etherize flies during the observation.

RNAi RNAi stands for RNA interference. It is a natural biological process in which small RNA molecules regulate gene expression by preventing mRNA from being translated into proteins.

Semi-solid food A semi-solid food is a type of food that is typically soft, moist, and tender, making it easy to chew, mash, and swallow without much effort.

Seminal receptacle The seminal receptacle is a coiled, tube-like organ in the female Drosophila reproductive system that functions as a primary sperm storage organ.

Sex combs A sex comb is a male-specific structure on the foreleg of many species of fruit flies. It is a row of thick, dark, and blunt bristles that are used during courtship and mating.

Sexing Sexing refers to the process of determining the biological sex of an organism.

Sexual dimorphism Sexual dimorphism is the condition in which the two sexes of the same species exhibit different characteristics beyond the differences in their sexual organs.

Speciation Speciation is the evolutionary process by which populations of a species evolve to become distinct, new species.

Spermatheca In fruit flies, such as Drosophila melanogaster, the spermathecae are a pair of mushroom-shaped organs in the female reproductive tract that serve as reservoir for long-term sperm storage.

Syncytial blastoderm A syncytial blastoderm is a developmental stage in the early embryogenesis of many insects, like the fruit fly Drosophila. It is characterized by a single, large cell that contains multiple nuclei, all within a common cytoplasm.

Transgenesis Transgenesis is the process of introducing a foreign gene, called a transgene, from one organism into the genome of another. The goal is for the recipient organism to express this new gene, giving it a novel trait.

Ventral tube The ventral tube is the structure that forms when the ventral furrow invaginates (folds inward) from the embryo's surface. This process is a key morphogenetic movement that internalizes the prospective mesoderm.

Vitelline membrane The vitelline membrane is a protective layer that surrounds the plasma membrane of an ovum (egg cell). It is a primary egg membrane, as it is secreted by the egg cell itself. It is also known as the vitelline envelope.

References

Carpenter FW (1905) The reactions of the Pomace Fly (Drosophila ampelophila Loew) to light, gravity, and mechanical stimulation. Am Nat 39(459):157–171. https://doi.org/10.1086/278502

Dähn S, Wagner AE (2025) Drosophila melanogaster as a model organism to investigate sex specific differences. Sci Rep 15(1):19648. https://doi.org/10.1038/s41598-025-04497-0

FlyBase:New to Flies—FlyBase Wiki (2013) https://wiki.flybase.org/wiki/FlyBase:New_to_Flies

Gilbert SF (2000) Developmental biology, 6th edn. Sinauer Associates, Sunderland. Available from: https://www.ncbi.nlm.nih.gov/books/NBK10081/

https://microbenotes.com/drosophila-development/

Jürgens KJ, Drechsler M, Paululat A (2024) An anatomical atlas of Drosophila melanogaster-the wild-type. Genetics 228(2):iyae129. https://doi.org/10.1093/genetics/iyae129

Kannangara JR, Mirth CK, Warr CG (2021) Regulation of ecdysone production in Drosophila by neuropeptides and peptide hormones. Open Biol 11. https://doi.org/10.1098/rsob.200373

Lakhotia SC (2021) Experiments with Drosophila for biology courses. Indian Academy of Science

Markstein M (2019) Drosophila Workers Unite!

McGurk L, Berson A, Bonini NM (2015) Drosophila as an in vivo model for human neurodegenerative disease. Genetics 201(2):377–402. https://doi.org/10.1534/genetics.115.179457

Mohr SE, Perrimon N (2019) Drosophila melanogaster: a simple system for understanding complexity. Dis Model Mech 12(10):dmm041871. https://doi.org/10.1242/dmm.041871

Petavy G, David JR, Gibert P, Moreteau B (2001) Viability and rate of development at different temperatures in Drosophila: a comparison of constant and alternating thermal regimes. J Therm Biol 26(1):29–39. https://doi.org/10.1016/s0306-4565(00)00022-x

Powell JR (1997a) Ecological genetics. In: Powell JR (ed) Progress and prospects in evolutionary biology: the Drosophila model. Oxford University Press, p 0. https://doi.org/10.1093/oso/9780195076912.003.0006

Powell JR (1997b) Progress and prospects in evolutionary biology: the Drosophila model. Oxford University Press. https://doi.org/10.1093/oso/9780195076912.001.0001

"Sex Limited Inheritance in Drosophila" (1910), by Thomas Hunt Morgan | Embryo Project Encyclopedia (n.d.) Retrieved August 14, 2025, from https://embryo.asu.edu/pages/sex-limited-inheritance-drosophila-1910-thomas-hunt-morgan

Sharma Y (2023, August 3). Drosophila development: stages, significance. https://microbenotes.com/drosophila-development/

Taylor BJ, Adriana V, Ryner LC, Baker BS, Hall JC (1994) Behavioral and neurobiological implications of sex-determining factors in *Drosophila*. Dev Genet 15(3):275–296. https://doi.org/10.1002/dvg.1020150309

The Mechanism of Mendelian Heredity (1916) Nature 97(2423):117–118. https://doi.org/10.1038/097117a0

Thomas B. Brody (2020) [No title found]. Public Library of Science (PLoS) https://doi.org/10.1371/journal.pbio.1002207.g007

Tolwinski NS (2017) Introduction: Drosophila—a model system for developmental biology. J Dev Biol 5(3):9. https://doi.org/10.3390/jdb5030009

Warren JT, Yerushalmi Y, Shimell MJ, O'Connor MB, Restifo LL, Gilbert LI (2006) Discrete pulses of molting hormone, 20-hydroxyecdysone, during late larval development of Drosophila melanogaster: correlations with changes in gene activity. Dev Dyn 235(2):315–326. https://doi.org/10.1002/dvdy.20626

Wegman LJ, Ainsley JA, Johnson WA (2010) Developmental timing of a sensory-mediated larval surfacing behavior correlates with cessation of feeding and determination of final adult size. Dev Biol 345(2):170–179. https://doi.org/10.1016/j.ydbio.2010.07.004

Wolpert L, Tickle C, Lawrence P, Arias A (2019) Principles of development. ResearchGate. https://www.researchgate.net/publication/373961893_Principles_of_Development

Study of the Developmental Stages and Life Cycle of *Caenorhabditis elegans* 7

Mukesh Behera, Gaurav Sharma, Suhani Jain, Srianshu Kumar Panda, and Narender K. Dhania

Abstract

Caenorhabditis elegans, a transparent soil-dwelling nematode, is widely studied as a model organism in different research fields. Its short life cycle, simple body design, and about 60–80% genetic similarity to humans make it useful for studying developmental biology, neurobiology, aging, and wound healing. These are maintained on high growth medium (HGM) plates seeded with *Escherichia coli* strain (OP50), which serves as a food source. The process of routine culture involves preparing nutrient agar plates, transferring worms through chunking, and storing the plates at an optimal temperature to ensure the proper development of *C. elegans* from embryo to adult. Moreover, to obtain an age-synchronized culture, homogeneous culturing techniques are employed. This involves bleaching of the culture, which removes all the larval as well as the adult worms, leaving only embryos that hatch and develop simultaneously. Age synchronization helps to study life span and stage-specific gene expression. Sub-lethal UV exposure of *C. elegans* causes tissue damage at a specific location without eradicating the entire organism. The cell or tissue rupture can be seen by using dual *acridine orange–ethidium bromide (AO–EtBr) staining*. This localized damage helps to study the repair mechanisms and stress responses of the organism. Since *C. elegans* shares significant genetic similarity with humans, such

M. Behera · G. Sharma · S. Jain · S. K. Panda · N. K. Dhania (✉)
Wound Healing Research Lab, Department of Zoology, Faculty of Sciences, University of Delhi, New Delhi, India
e-mail: narender@zoology.du.ac.in

A. C. Bharti et al. (eds.), *Textbook of Developmental Biology*,
https://doi.org/10.1007/978-981-95-7269-4_7

studies provide insights into the cellular processes underlying DNA damage, apoptosis, and recovery. Furthermore, the transparency of the worm makes it an excellent model to visually track morphological changes and evaluate the extent of UV-induced injury.

Keywords

Sub-lethal UV damage · *Caenorhabditis elegans* · *Wound healing*

Aim Study of the developmental stages and life cycle of *Caenorhabditis elegans.*

Learning Objectives
By the end of this study, students will be able to:

- Describe the morphology and life cycle of *C. elegans*
- Explain standard culturing methods and the role of *E. coli* OP50
- Perform the routine culture of *C. elegans*
- Prepare homogeneous cultures
- Perform sub-lethal UV exposure of *C. elegans*

7.1 Introduction

Caenorhabditis elegans is a minute, transparent, and non-parasitic roundworm roughly 1 mm long, commonly found in soil. Its scientific name is derived from Greek and Latin: *caeno* (recent), *rhabditis* (rod-like), and *elegans* (elegant). This worm belongs to phylum Nemathelminthes/Nematode and the genus *Caenorhabditis,* and is its most extensively researched organism (Brenner 1974).

In 1963, Sydney Brenner selected *C. elegans* as a model organism to investigate cell and neuronal development, foreseeing that molecular biology's major questions were nearing resolution. Intensive research began in 1974, leading to its status as one of the most thoroughly examined species in modern biology. It was the first multicellular organism to have its complete genome sequenced, and in 2019, researchers mapped its entire connectome (White 2018; Corsi et al. 2015).

With a straightforward body design, *C. elegans* lacks segmentation, a respiratory system, and a circulatory system. Instead, it possesses a pseudocoelom, a fluid-filled cavity serving both structural and functional purposes. The majority of individuals are hermaphrodites, equipped with both male and female reproductive organs, while males are rarely found with specialized tails adapted for mating (Brenner 1974).

Its small size, transparent body, and short life cycle allow direct microscopic observation of cells, organ development, and behavior. The life cycle involves embryonic stage, four larval stages (L1–L4), and adult stage.

Through the use of advanced microscopy and fluorescent protein tagging, scientists can track protein movement, monitor cellular changes, and study how genes regulate development. This simplicity has made *C. elegans* an indispensable tool for exploring complex biological processes, including genetics, neurobiology, aging, and responses to environmental stressors such as ultraviolet radiation. The Sun emits three types of UV rays—UVA, UVB, and UVC—distinguished by their wavelengths and intensity.

While UVA has the longest wavelength (315–400 nm) and causes minimal immediate damage, UVB (280–315 nm) is more intense and linked to DNA damage. UVC (100–280 nm) is the most harmful, but it is largely absorbed by the Earth's atmosphere. Sub-lethal UV exposure is done to investigate the cellular damage done in the epidermal layer of *C. elegans* (Silva et al. 2023).

By 2024, work involving *C. elegans* had contributed to four Nobel Prizes. Today, it is studied in more than a thousand laboratories worldwide, providing crucial insights into genetics, neuroscience, cell biology, and human disease.

7.1.1 Morphology

C. elegans is a slender, worm-like organism displaying bilateral symmetry. It is covered by a tough yet elastic cuticle secreted by the epidermis which acts like an external skeleton and is periodically shed during growth. Just under the cuticle lies the epidermis, organized into four longitudinal cords: dorsal, ventral, and two laterals (Fig. 7.1). The lateral cords contain excretory canals, while the dorsal and ventral cords are integrated with the nervous system. Along the sides, ridge-like structures known as alae provide traction for movement (Nonet 2004).

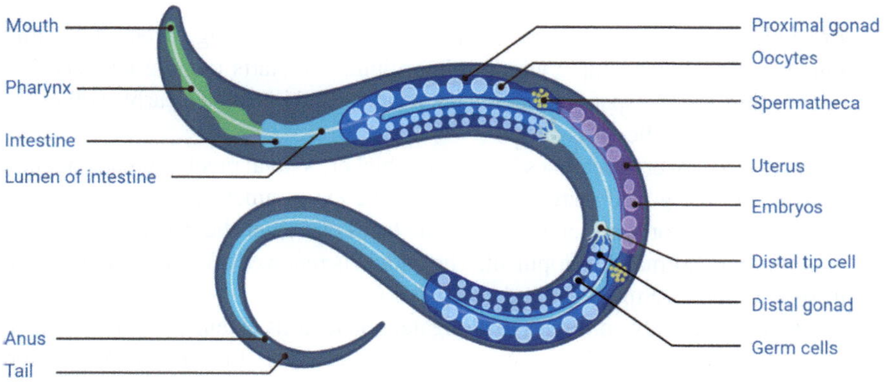

Fig. 7.1 Anatomy of adult *C. elegans*. (Created in https://BioRender.com)

The pseudocoelom is a fluid-filled cavity that acts as a hydrostatic skeleton, maintaining body shape and supporting internal organs. Movement is generated by four bands of longitudinal muscles beneath the epidermis, enabling the worm to bend dorsally and ventrally but not laterally. This results in characteristic side-to-side locomotion driven by wave-like contractions controlled by the nervous system.

Its digestive system is simple but efficient. Food enters through the mouth into the muscular pharynx, which grinds and pumps it into the intestine for digestion and absorption. Solid waste is expelled through the anus, while liquid waste is removed via the excretory canal system. Unlike mammals, *C. elegans* stores fats as neutral lipids (such as triglycerides) within the intestine, epidermis, and developing embryos, with the epidermis functioning similarly to mammalian adipose tissue.

Hermaphrodites have two gonad arms containing ovaries, oviducts, and spermathecae for storing self-produced sperm. Males possess a single-lobed gonad and a tail with spicules used in mating. The hermaphrodite nervous system contains exactly 302 neurons arranged in precise patterns, many projecting into the nerve ring, the worm's functional brain.

7.1.2 Overview of the *C. elegans* Life Cycle

C. elegans life cycle consists of an embryonic stage, four larval stages (L1–L4), and adulthood. Molting occurs between larval stages, during which a new cuticle is formed and the old one shed.

Embryogenesis lasts about 14 h, beginning with rapid cell divisions that produce a spherical embryo of 550 cells. Gastrulation follows, forming the three germ layers: ectoderm, mesoderm, and endoderm. Organogenesis shapes the embryo's structure, initiating muscle twitching, pharyngeal activity, and internal organ development. Hatching occurs at approx. 14 h, producing an L1 larva.

If food is available, the larva progresses through L2, L3, and L4 stages into adulthood. Each developmental stage signifies distinctive morphological and physiological features:

- *L1 stage*—The newly hatched larva is around 250 μm long. It begins feeding immediately, shows simple gonad primordium, and starts moving actively. The body is thin and transparent, making the pharynx and intestine clearly visible. The cuticle is soft and flexible.
- *L2 stage*—The worm becomes longer and thicker. Vulval precursor cells begin to appear, and the gonad starts extending. If the environment is stressful (lack of food, crowding, or high temperature), the larva can enter the dauer stage instead of continuing normal development. The cuticle is renewed after the first molt, and internal organs are more defined compared to L1.
- *L3 stage*—The body continues to elongate, gonad arms extend further, and many new neurons develop. The cuticle becomes stronger, the tail becomes more distinct, and the worm shows more active movement with stronger pharyngeal pumping.

- *L4 stage*—At this stage, the worm resembles a smaller adult. In hermaphrodites, the vulva is formed, while in males, the tail starts to develop copulatory structures. Worms are about 700–800 μm long and have a stouter body. Gonad arms are fully developed, and in hermaphrodites, the vulval slit is visible externally. Sexual differences between males and hermaphrodites first become clear in this stage.
- *Adult stage*—Adults are about 1 mm in length. Hermaphrodites have two gonad arms and produce both sperm (first) and oocytes (later). Males have a single gonad and a specialized tail with spicules for mating. Adults are less transparent compared to larvae, because yolk proteins accumulate in the intestine. Males have a broader fan-like tail, while hermaphrodites remain slender.

Under adverse conditions (e.g., starvation, overcrowding, temperature differences), the L2 stage may shift into a stress-resistant dauer form, capable of surviving for 10 days without feeding. Once favorable conditions return, the dauer resumes development (Artyukhin et al. 2013).

By the L4 stage, reproductive systems are fully developed. Hermaphrodites begin producing oocytes after sperm generation ceases, while males finalize tail structure (Fig. 7.2). At optimal temperatures (22–25 °C), adulthood is reached in 45–50 h post-hatching. Hermaphrodites remain fertile for 3–4 days, producing 300 offspring via self-fertilization and up to 1400 with male mating. Males remain fertile for 6 days, potentially fathering 3000 progenies (Byerly et al. 1976; Ambros 2000).

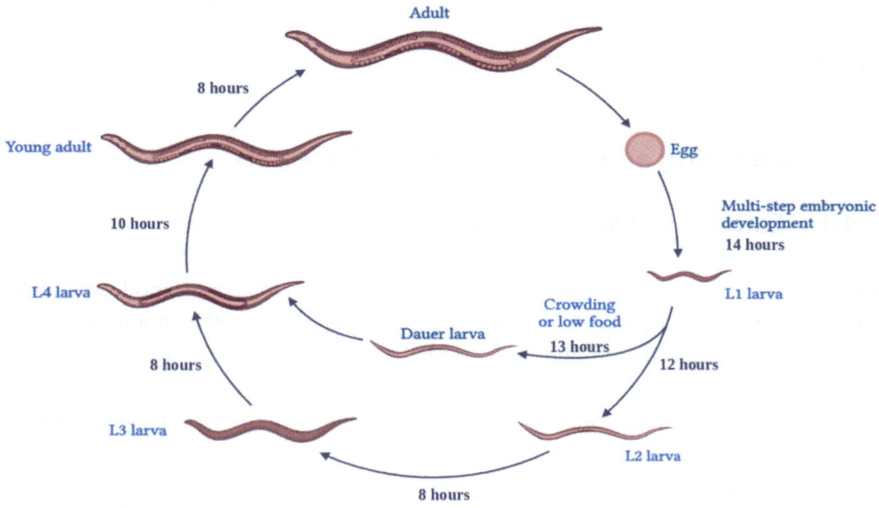

Fig. 7.2 Diagrammatic representation of developmental stages of *C. elegans*. (Created in https://BioRender.com)

7.2 Requirements

Chemicals Agar, NaCl, peptone, double-distilled water, 5 mg/ml cholesterol in ethanol, 1 M KPO_4 buffer (pH 6.0), 1 M $CaCl_2$,1 M $MgSO_4$, M9 buffer, 10 mg/ml *AO,10 mg/ml EtBr*, LB Broth, *E. coli* OP50 culture, 4% NaOCl, 5 M NaOH or KOH, double-distilled water, *C. elegans* culture plate, mineral oil.

Apparatus/Instruments Petri plates(90 mm), Schott DURAN bottle (1 L), micropipette (10 μl, 200 μl and 1000 μl), compound microscope, centrifuge, microcentrifuge tubes (2 ml), vortex, Laminar airflow, scalpel, autoclave, BOD incubator with installed UVC light, fluorescent microscope, glass slides, cover slips.

7.3 Preparation of Buffer Solutions

M9 Buffer (1 L)

(i) Weigh 3 g KH_2PO_4, 6 g Na_2HPO_4, and 5 g NaCl and dissolve in double-distilled water to prepare 1 L of buffer.
(ii) Autoclave the buffer at 15 psi, 121 °C, for 15 min.
(iii) Let the buffer cool for some time and then add 1 ml of 1 M $MgSO_4$.
(iv) Mix well and store at 4 °C.

1 M KPO_4 Buffer (1 L)

(i) Add 108.3 g of KH_2PO_4 and 35.6 g of K_2HPO_4 to distilled water to make up 1 L of solution.
(ii) Maintain pH 6.0 of the KPO4 buffer.

7.4 Culturing and Maintenance of *C. elegans*

7.4.1 Culturing of Bacteria *E. coli* Strain OP50

OP50 is an *Escherichia coli* strain mainly used as a bacterial food source for laboratory maintenance of *C. elegans* on agar plates. It is the uracil auxotroph whose growth is limited on high growth medium (HGM) plates (Arata et al. 2020).

(i) Use an inoculating loop to extract out a tiny chunk of the OP50 and dissolve in LB broth.
(ii) Leave it overnight for 16–20 h.
(iii) Check the turbidity using a spectrophotometer. The OD600 typically reaches a maximum of around 2.0 before growth slows or stops, though some studies report a maximum of around 0.6–0.8 for optimal growth.
(iv) Centrifuge at 8000 rpm for 8 min.
(v) Pellet down by aspirating the supernatant.

7.4.2 Preparation of Routine Culture

Routine culturing of *C. elegans* is carried out on HGM agar plates seeded with *Escherichia coli* OP50, which serves as the primary food source. This allows the worms to grow rapidly, reproduce, and progress through their life cycle in controlled laboratory conditions (Shaver et al. 2021; Hibshman et al. 2021).

(i) Weigh 3 g NaCl, 2.5 g peptone, and 17 g agar powder, dissolve in distilled water to make up 1 L of solution in a Schott DURAN bottle.
(ii) Autoclave the solution at 15 psi, 121 °C, for 50 min.
(iii) Let the bottle cool for 15 min.
(iv) Add 25 ml 1 M KPO_4 buffer,1 ml 5 mg/ml cholesterol in ethanol,1 ml 1 M $CaCl_2$,1 ml 1 M $MgSO_4$. Swirl to mix well.
(v) Pour it into the petri plates (90 mm).
(vi) Leave the plates to solidify in a laminar airflow (1–2 h).
(vii) Side by side prepare *E. coli* OP50 overnight LB Broth culture.
(viii) Next day, add 50–80 μl of bacteria onto each petri plate and spread it using a sterilized glass spreader.
(ix) Allow all the plates for even growth of *E. coli* on petri plates (6–8 h) at room temperature.
(x) Transferring of worms—Use previous plates for the chunking method [*Chunking*: A piece of media from stock plate with adult worms is transferred to a new plate using a sterilized scalpel] (Fig. 7.3).
(xi) After chunking, place all the plates in the incubator 22 °C. Within a day or so, the eggs will hatch, and worms emerging from the embryos will be visible.

7.4.3 Preparation of Homogeneous Culture

It is used to standardize the age of nematodes as the experiments are largely lifespan-based. This process essentially bleaches the cultures, removing all the adults and larvae, leaving only embryos, which then hatch and develop into a synchronized

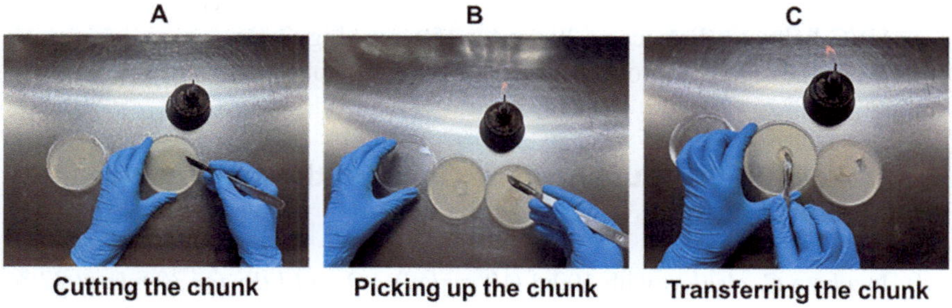

Fig. 7.3 Steps of chunk transfer from old culture media to new culture media

population. It enables precise analysis of the stage-specific traits as all the worms will be at the same developmental stage.

A. *Preparing the Bleaching Solution*
   (i) 2 mL of 4% sodium hypochlorite.
   (ii) 1 mL of 5 M KOH (or NaOH).
   (iii) 7 mL of double-distilled water.
   (iv) Mix gently so the solution is uniform.
   (v) Prepare this fresh before use.
B. *Collecting and washing the worms*
   (i) Wash the culture plates with the M9 buffer.
   (ii) Collect the wash into a 2 ml microcentrifuge tube.
   (iii) Centrifuge the tube at 2000 rpm for 2 min.
   (iv) Carefully aspirate the supernatant without disturbing the worm pellet.
   (v) Add fresh M9 buffer, mix gently, use a vortex meter if required, and centrifuge again.
   (vi) Repeat these steps 4–5 times to ensure proper washing of worms.
C. *Bleaching*
   (i) Add freshly prepared bleaching solution to the pellet.
   (ii) Gently mix by shaking for 5–7 min.
D. *Observation*
   (i) Observe the embryos using a glass slide under a microscope.
   (ii) Once the embryos are separated, transfer them into freshly prepared HGM plates.

7.4.4 Sub-lethal UV-Exposure of *C. elegans*

To scrutinize the effect of UVC light on the cuticle of *C. elegans:*

(i) Observe the culture plate under a microscope and identify the developmental stages and density of *C. elegans*.
(ii) Set the incubator temperature to 22 °C.
(iii) Once the temperature is stable, keep the culture plate open and expose it to UVC light at a distance of 30 cm for 150 min (Fig. 7.4).
(iv) Wash the culture thoroughly with M9 buffer to collect all worms.

7.4.5 Post UV-Exposure *A*cridine Orange–Ethidium Bromide (AO–EtBr) Staining

(i) Incubate *C. elegans* with 100ul (10 mg/ml) *AO and EtBr for 30 min at 22 °C.*
(ii) *Wash the worms with M9 buffer by centrifuging at 2000 rpm for 2 min.*

Fig. 7.4 Petri plate positioned 30 cm from UV light source

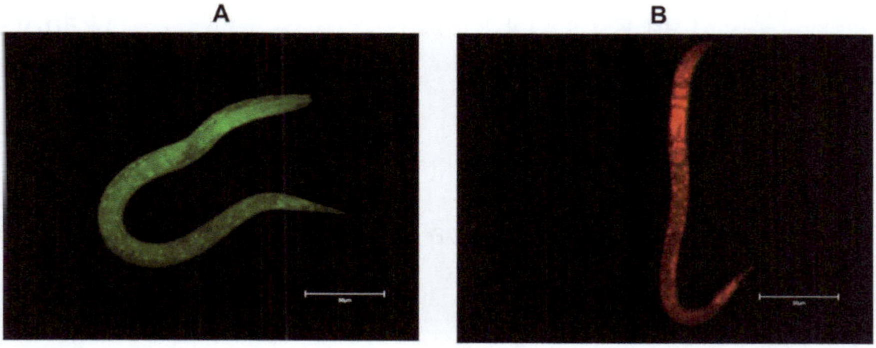

Fig. 7.5 *Acridine orange–ethidium bromide (AO–EtBr) stained C. elegans.* (**a**) Control worm showing only live cells, indicated by green fluorescence. (**b**) UV-irradiated worm exhibiting both green and red fluorescence, indicating live cells (green) and UV-damaged or apoptotic/necrotic cells (red). Red fluorescence indicating loss of membrane integrity due to UV-induced cellular damage (scale—50 nm)

(iii) *Repeat the wash step 3–4 times, until the solution appears transparent or slightly orange.*
(iv) *Visualize the stained worm under fluorescence microscope* (Fig. 7.5).

7.5 Result/Observation

7.5.1 Routine Culture

Observe the plates under a microscope after 1–2 days; different stages (L1–L4, adult) of worms will be visible, including the eggs (Fig. 7.6). Keep observing the plates for a week to analyze the exact stages of the *C. elegans*.

7.5.2 Homogeneous Culture

Synchronized stages of *C. elegans* can be seen in petri plates. These plates can be further used to study about stage-specific RNA, protein analysis, embryogenesis, larval stage transitions, and organ development.

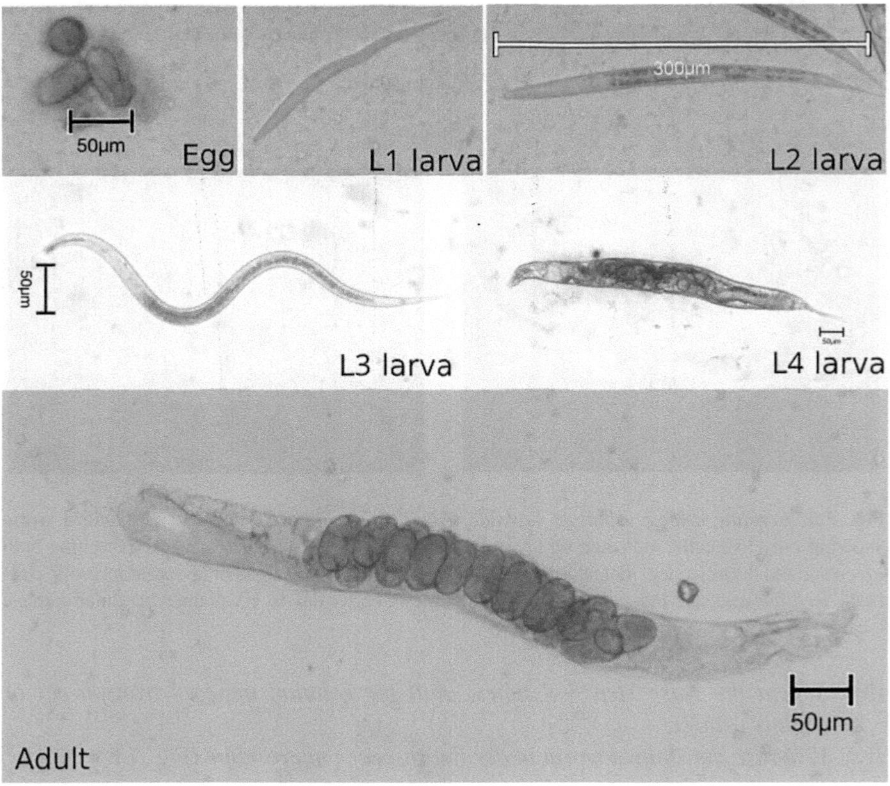

Fig. 7.6 Observed developmental stages of *C. elegans*

7.5.3 Sub-lethal UVC Exposure

Examine the fluorescently labelled sample and evaluate the extent of damage. The ruptured regions will appear blue, whereas viable cells remain colorless.

7.6 Experimental Errors

- Fungus may grow if plates are left outside for too long.
- Uneven spreading of *E. coli* can lead to irregular feeding and overcrowding of worms.
- Temperature changes may result in slow development as increase or decrease in the required temperature can cause worms to enter the dauer stage.
- Excessive bleaching can lead to the loss of viable embryos.
- Large quantity of worms is lost during washing.

7.7 Significance and Research Applications

- *C. elegans* are easy to culture in laboratory as minimal amount of equipment and chemicals are used.
- The transparent body allows the direct visualization of the structures and organ formation under a microscope.
- *C. elegans* is ideal for studying aging, genetics, and neurobiology because of its short lifespan and well-mapped genes.
- *C. elegans* is used to assess damage on the cuticle and epidermal layer caused by UV radiation, helping researchers study stress responses and repair mechanisms (Prasanth et al. 2016).

7.8 Precautions

- Freshly prepared *E. coli* OP50 strain should be used.
- It is necessary to maintain aseptic conditions throughout the experiment to prevent contamination.
- Bleaching solution must be prepared fresh before starting the experiment.
- Washing should be carefully done, as there is a chance of losing worms while aspirating the supernatant.
- Handle worms gently during chunking to avoid physical damage.
- While aspirating *AO–EtBr* solution, take care not to pipette *C. Elegans*.

7.9 Quick Review Questions

7.9.1 Multiple Choice Questions

1. What is the most commonly used food source for *C. elegans* in lab?
   (a) Chlorella
   (b) *Escherichia coli* OP50
   (c) Yeast extract
   (d) *Bacillus subtilis*
   Answer: b) Escherichia coli OP50
2. What is the primary objective of making homogeneous cultures of *C. elegans*?
   (a) To synchronize the growth stages of worms
   (b) To grow bacteria
   (c) To facilitate the growth of mixed culture
   (d) To maintain culture for long term
   Answer: a) To synchronize the growth stages of worms
3. Which tissue layer of *C. elegans* exhibits maximum damage when exposed to UVC light?
   (a) Cuticle
   (b) Syncytial epidermis
   (c) Epidermis
   (d) Dermis
   Answer: a) cuticle
4. What is the duration of lifecycle of *C. elegans*?
   (a) 14 Days
   (b) 7 Days
   (c) 3 Days
   (d) 1 Day
   Answer: c) 3 Days
5. Which mode of reproduction is mostly seen in *C. elegans*?
   (a) Hermaphrodite self-fertilization
   (b) Sexual reproduction
   (c) Parthenogenesis
   (d) Asexual reproduction
   Answer: a) Hermaphrodite self-fertilization
6. What is the ideal temperature range for *C. elegans* culturing?
   (a) 25–37 °C
   (b) 11–15 °C
   (c) 0–15 °C
   (d) d.16–25 °C
   Answer: d) 16–25 °C
7. How much time does it take to change from L1 stage to L2 stage?
   (a) 15 h
   (b) 10 h

   (c) 12 h
   (d) 24 h
   Answer: 12 h
8. Which buffer is used to wash the *C. elegans* plates?
   (a) PBS
   (b) M9 buffer
   (c) Insect ringer
   (d) Distilled water
   Answer: b) M9 buffer
9. What is the configuration of male sex chromosome in *C. elegans*?
   (a) XX
   (b) XO
   (c) XY
   (d) YY
   Answer: b) XO
10. Which of the following is not used for preparing *C. elegans* culture plates?
   (a) NaCl
   (b) Peptone
   (c) LB powder
   (d) Agar powder
   Answer: c) LB powder

7.9.2 Short Answer Questions

1. Who first described *Caenorhabditis elegans* as a model organism?
2. What is the average size of an adult hermaphrodite *C. elegans*?
3. Which larval stage can develop into the dauer stage under stress?
4. What is the reproductive mode of hermaphrodite *C. elegans*?

Glossary

Adipose tissue connective tissue that is made up of fat cells helping in energy storage and providing insulation to internal organs.
Aging biological process of growing older.
Apoptosis the process of cell death involving series of molecular steps.
Autoclave machine that uses steam (~120 °C) and pressure (~psi) to kill pathogens.
Bilateral symmetry biological symmetry where an organism can be divided into two identical halves along a single plain.
Cell biology study of cellular structure, components, and their behavior.
Connectome the wiring diagram of all neural connections.
Contamination presence of unwanted biological organism or substance having negative effect on experiment.

Copulatory structures body parts of male animals used during copulation to deposit sperm into the female reproductive tract.

Cuticle in C. elegans cuticle is an extracellular matrix that forms the barrier between the animal and its environment.

Dauer stage dormant, reversible stage of C. elegans that allows it to survive unfavorable condition.

DNA damage alteration of DNA structure by any endogenous or exogenous factors.

E. coli OP50 Escherichia coli strain used as bacterial food in maintenance of C. elegans.

Ectoderm outermost layer of germ layer in developing embryo.

Embryogenesis the process by which fertilized egg develops into embryo.

Endoderm innermost layer of germ layer in developing embryo.

Epidermis outermost layer of C. elegans that secretes cuticle.

Excretory canals *channels that connect excretory system to outside* via *excretory pore.*

External skeleton rigid outer covering that provides support and protection and prevents water loss.

Fertile ability to produce offspring.

Gastrulation stage of embryonic development where blastula transforms into three layered gastrula.

Genetics study of gene and heredity.

Germ layers three layers (ectoderm, mesoderm, and endoderm) that form during gastrulation.

Gonad primary reproductive organs that produce gametes (testes in males and ovaries in females).

Gonad primordium cells having ability to develop into testes or ovaries depending on the genetic signal.

Hatching coming out of egg.

Hermaphrodites organisms having both male and female sexual organs.

Hydrostatic skeleton *fluid-supported skeletal system seen in earthworm, octopus,* etc.

Larva juvenile form of animal that hatches from egg.

Mesoderm second and middle germ layer in developing embryo.

Molting process of shedding the old exoskeleton and growing in size.

Nematode a type of round worm.

Neurobiology study of nervous system focusing on how neurons function and interact to form the nervous network.

Neuroscience study involving neurobiology and other interdisciplinary to understand the entire nervous system.

Oocytes cell of ovary forming ovum.

Organogenesis process of formation of functional organ by three germ layers.

Pellet concentrated mass of substance post centrifugation.

Progenies offspring of an organism.

Pseudocoelom fluid-filled cavity that is not completely lined by mesoderm.
Self-fertilization process of fusion of male and female gametes of the same organism.
Sperm male reproductive cell.
Spermathecae storage organ of sperm in a female reproductive tract.
Sub-lethal UV exposure UV exposure that is non-lethal to the organism inducing only small wounds.
Supernatant liquid lying above the solid residue(pellet) after centrifugation.
Ultraviolet radiation Electromagnetic radiation having wavelengths shorter than visible light and longer than X-rays.
Vulval precursor cells cells involved in vulval development.
Wound healing process of damaged tissue repair.
Supernatant liquid lying above the solid residue(pellet) after centrifugation.

References

Ambros V (2000) Control of developmental timing in Caenorhabditis elegans. Curr Opin Genet Dev 10(4):428–433

Arata Y, Oshima T, Ikeda Y, Kimura H, Sako Y (2020) OP50, a bacterial strain conventionally used as food for laboratory maintenance of C. elegans, is a biofilm formation defective mutant. MicroPubl Biol 2020:10–17912

Artyukhin AB, Schroeder FC, Avery L (2013) Density dependence in Caenorhabditis larval starvation. Sci Rep 3(1):2777

Brenner S (1974) The genetics of Caenorhabditis elegans. Genetics 77(1):71–94

Byerly L, Cassada RC, Russell RL (1976) The life cycle of the nematode Caenorhabditis elegans: I. Wild-type growth and reproduction. Dev Biol 51(1):23–33

Corsi AK, Wightman B, Chalfie M (2015) A transparent window into biology: a primer on Caenorhabditis elegans. Genetics 200(2):387–407

Eibshman JD, Webster AK, Baugh LR (2021) Liquid-culture protocols for shavsynchronous starvation, growth, dauer formation, and dietary restriction of Caenorhabditis elegans. STAR Protoc 2(1):100276

Nonet M (2004) About the nematode Caenorhabdtis elegans

Prasanth MI, Santoshram GS, Bhaskar JP, Balamurugan K (2016) Ultraviolet-A triggers photoaging in model nematode Caenorhabditis elegans in a DAF-16 dependent pathway. Age 38(1):27

Shaver AO, Gouveia GJ, Kirby PS, Andersen EC, Edison AS (2021) Culture and assay of large-scale mixed-stage Caenorhabditis elegans populations. J Vis Exp 171:10–3791

Silva MR, Alves MF, Cunha JP, Costa JL, Silva CA, Fernandes MH et al (2023) Nanostructured transparent solutions for UV-shielding: recent developments and future challenges. Mater Today Phys 35:101131

White JG (2018) Getting into the mind of a worm—a personal view. In: WormBook: the online review of C. elegans biology [Internet]. WormBook

Study of the Developmental Stages and Life Cycle of Zebrafish

8

Radhika Gupta, Nikita Sharma, Narendra Kumar, Padmshree Mudgal, and Vikas Nishad

Abstract

The utilization of zebrafish as a model organism is crucial for understanding developmental biology, genetics, regeneration, and disease physiology in humans. Zebrafish maintenance requires low cost, and high fecundity makes it a robust model for high-throughput toxicology studies. The life cycle is short, and the larval stages are self-reliant till day 5 avoiding ethical clearances. We will be elaborating the protocols for zebrafish husbandry, natural behavior, breeding, egg development, and maintenance; moreover, the larval stages will be studied till day 5 with focus on studies to highlight the importance of the organism in high-throughput studies. The protocols will help researchers establish a zebrafish lab facility and utilize zebrafish as a model organism to understand various biological processes.

Keywords

Zebrafish · Breeding · Embryos

Aim The chapter aims to provide an insight into the developmental stages of zebrafish, their culture, maintenance, and methods to study these stages.

R. Gupta (✉) · N. Sharma · N. Kumar · P. Mudgal · V. Nishad
Department of Biochemistry, Daulat Ram College, Delhi University, New Delhi, India
e-mail: radhikagupta@dr.du.ac.in

A. C. Bharti et al. (eds.), *Textbook of Developmental Biology*,
https://doi.org/10.1007/978-981-95-7269-4_8

Learning Objectives
Upon completing the experiment, the students would gain understanding of:

- Breeding of zebrafish and spawning of embryos.
- The developmental stages from one cell embryo stage to 72 hpf (hours post fertilization) larval stage.
- The developmental stages up to 5-day-old larvae.

8.1 Introduction

Zebrafish is a relatively newer recruit as a model organism compared to the older systems such as mice, guinea pigs, primates, *C. elegans*, and *Drosophila* to study biological systems. In the year 2000, there were a total of 697 publications in the field. The publications are rising rapidly with 60,900 publications in the PubMed for the keyword zebrafish till May 2025, indicating increased importance of zebrafish as a model organism in research. Zebrafish is a vertebrate teleost primarily found in Southeast Asia and was first discovered in Assam, India. It may not seem to be an appropriate model organism to study human biology and diseases, as it lives underwater and has no arms and legs. But the increased interest in using zebrafish as a model organism to understand various biological processes as well as diseases is because zebrafish is approximately 70% genetically similar to humans (Howe et al. 2013). Additionally, about 80% of disease-associated genes of humans are present in zebrafish. Zebrafish has 25 pairs of chromosomes despite gene duplication, which is comparable to human chromosome number 23 (Postlethwait et al. 2000). The common organs between zebrafish and humans are two eyes, one mouth, spinal cord, intestines, liver, kidneys, heart, ears, brain, esophagus, cartilages, bone, and blood. Furthermore, zebrafish have very high fecundity with fast breeding and a capacity to lay hundreds of synchronized eggs in one spawning event. The fertilization event is external, allowing a large number of eggs to be produced and fertilized, and hence, post fertilization events can be studied well (Volkov et al. 2022). Zebrafish have a short generation time and develop rapidly, and the same fish can lay eggs within 4–5 days. These properties of the zebrafish are extremely important for generating reproducible results with high statistical power for high-throughput screening experiments such as genetic mutation studies and effect of chemicals on development. The zebrafish eggs in the early developmental stages lack pigments and are optically transparent, providing a high advantage of studying biological processes such as development of the organs and movement of the cell in vivo during the early stages. Further chemical methods such as treatment of 1-phenyl 2-thiourea can render the larval stages also optically transparent, hence aiding further in understanding cell division, migration, differentiation, and development in real time (Antinucci and Hindges 2016). The zebrafish egg is about 0.7 mm and is relatively large in size as compared to adult fish. The size is large enough to store egg yolk, providing proteins, lipids, and micronutrients for rapid development as the organs in the embryo start to develop within 24 h. The large embryo size is of advantage as it is easier to perform microinjections in the egg for injecting the morpholinos (Sant and Timme-Laragy 2018). Zebrafish system requires less space and cost and is relatively easy to maintain. It is required to preserve the water quality and food for the fish husbandry. In this chapter, we will be exploring the process of zebrafish development from an one-cell egg to a 5-day-old larva that can be used for experimental purposes.

8.2 Requirements

Chemicals: sea salt.

Apparatus/Instruments: Incubator, refrigerator, fish breeding tanks (made of transparent plastic such as polycarbonate), separators, fish nets, sieves, wash bottles, 90 mm petri plates, 12-well and 6-well plates, microscope.

8.3 Zebrafish Husbandry

8.3.1 System Maintenance

Zebrafish are maintained in a laboratory with 28 ± 2 ° C ambient temperature. They can be maintained in individual fish tanks (RO water with 0.06 g/liter sea salt) where the water is changed daily or in a circulatory system with aquatic environment ambient for the fish. There are filters in the system to remove excreta and food particles. The laboratory also maintains a photoperiod of 10:14 dark: light for the healthy growth of fish. The circulatory system has 2.5-liter and 5-liter tanks to store the fishes. Male and female fishes are kept separately. RO water is used in the circulatory system. The pH of the system water is maintained between 6.8 and 7.5. Fish tanks are cleaned once in a week to remove any dirt and algae on the side or bottom of tank. The tanks are well labeled with the sex, the date of birth of each fish also mentioning when the fishes are breed last (Fig. 8.1). The filters in the circulatory system are regularly checked and maintained.

8.3.2 Feeding

Zebrafish are given two kinds of food. The dry flak fish feed (Tetramin, Germany) is given twice daily by grinding the feed to powder form in a mortar and pestle (Fig. 8.2c). The Brine shrimp (*Artemia*) is live feed hatched in the laboratory and given once daily. 1.5-liter RO water and 5 g *Artemia* dry are taken in a 2-liter bottle cut from below and 5 g sea salt is added. Air pump is used to provide air in the water, and the temperature is kept at 28 °C using a lamp. The system is kept for 24 h to

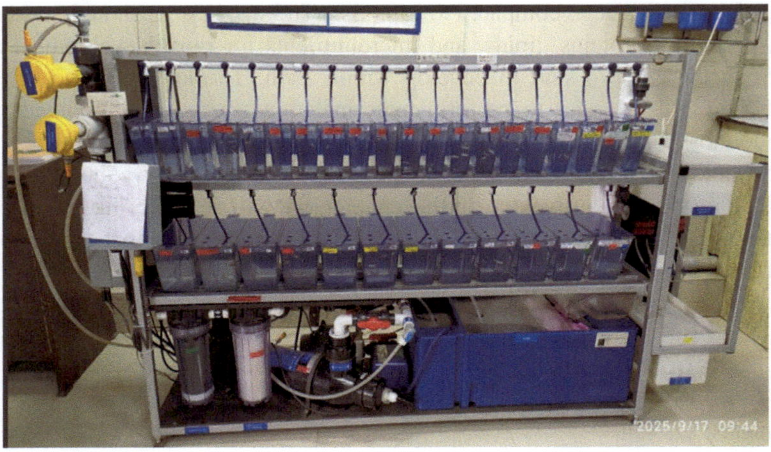

Fig. 8.1 Water recirculatory system for husbandry of fishes. This is a Pentair water system to keep the fishes in ambient environment. There are 19, 1.5-liter tanks and 12, 3-liter tanks to store the fishes made of polycarbonate. There is a UV, light, and temperature control system. The pumps feed water into the system, and the wastewater is filtered and recirculated. The water feed into the system is RO. The temperature is maintained at 28 °C, and there are 10 h of dark and 14 h of light cycle. Chlorine in the water is removed by carbon filters present in the system

hatch the shrimps. The culture is settled in the bottle and the solution is removed and water is changed. The Brine shrimps are collected in squeeze bottle and food is dispensed in adult fish tanks according to the number of fishes in each tank (Fig. 8.2d–f). *Paramecium* culture is given as feed to 5-day-old larvae up to 1-month-old baby fishes are developed. The culturing is done by taking 250 ml RO water and adding 5 g wheat germ in a beaker and heating for 60 s in microwave; 1 g yeast powder is added to the solution and allowed to cool down. 1 ml of old culture is added to the solution and kept at room temperature for a day for the *Paramecium* culture to grow. 10 drops of culture is added to 100 larvae in a 90 mm petri plate (Fig. 8.2a, b).

8.3.3 Breeding

Male and female fishes are procured from local pet shop. Male and female zebrafishes can be distinguished easily. Males are thin, streamlined, and gold yellow in color, and female fishes have a protruding belly with white/silver blue strips (Fig. 8.3).

The breeding in the zebrafish is initiated at the onset of light and hence maintenance of light:dark photoperiod of 14:10 is extremely essential for laboratory breeding. The breeding tanks are set up one day before in the evening. They are filled upto one-third with systems water. In the breeding tank, a plastic separator is inserted in the center, and male and female fishes are kept in the two separate sides in the tank. The breeding tank has breeding traps where there are two containers that fit

Fig. 8.2 Feed for larvae, juvenile, and adult zebrafish. (**a**) The larvae after 5 dpf are fed with *Paramecium* culture. The preparation of culture is explained in the text. (**b**) *Paramecium* under 4X magnification. (**c**) The juvenile and adult fishes are given a combination of dry and live feed. The dry feed is powdered using motor and pestle as shown in the figure. (**d**) The live feed is prepared from dried artemia cysts. (**e**) Setup for *Artemia* preparation. (**f**) Figure showing the live *Artemia* after hatching

into each other. The inner container holds the fish and has grids from where the eggs fall into the bottom container (Fig. 8.4a–c). The fishes are first fed in their tanks and then an hour later are transferred to the breeding tanks and are left undisturbed overnight (Fig. 8.4d). The next morning, the separators are removed at the onset of

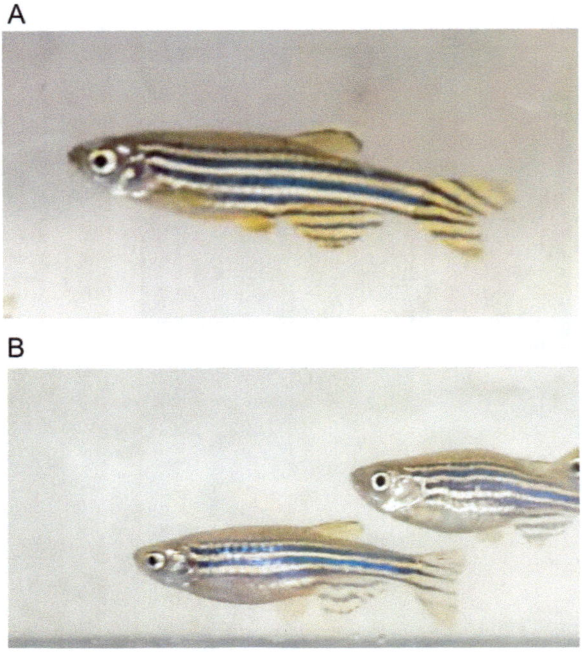

Fig. 8.3 Figure depicting adult male and adult female zebrafish. (**a**) Male fish has slender and streamlined body with golden hue. (**b**) Female fish has protruded belly where the eggs are stored. The body has silverfish blue hue

light, and fishes are allowed to mate for around 20–30 min undisturbed. The fertilized eggs are collected at the bottom of the breeding tanks. The eggs are collected in embryo water containing RO water with 0.06 g/liter concentration of ocean sea salt and are grown in the incubator at 28 ± 2 °C. Fertilized eggs and unfertilized eggs are separated from each other using a pipette carefully.

8.3.4 Maintaining and Observing Developmental Growth Stages of Larvae

After the eggs are fertilized, they can be immediately visualized to observe different developmental stages. The first stage is the single-cell stage and usually all the eggs from the same clutch are synchronized. Stereo microscope at 4X–20X can be used to observe the different developmental stages. Every 20–30 min, the egg divides and the development proceeds. Some of the defective eggs will either freeze at a particular stage or will be cloudy in appearance. Each stage can be documented by taking pictures and collating together. Each stage is given a name for effective identification as described in the figure. Hatching of the eggs/embryos takes place

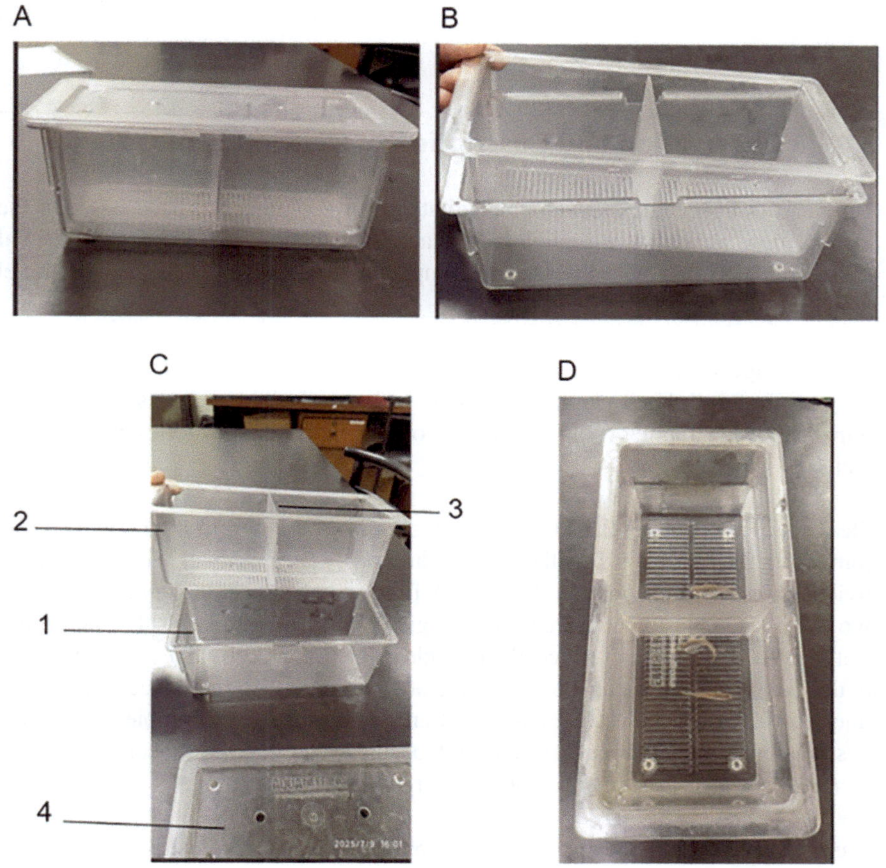

Fig. 8.4 Breeding setup for breeding of zebrafish male and female fishes to obtain embryos. (**a**) Breeding tank setup. (**b**) There are four components in the breeding tank setup. (**c**) (1) The bottom container where the eggs are collected; (2) the inner container with grids; (3) the separator to separate male and female fishes; (4) the lip to prevent fishes from jumping. (**d**) The breeding setup with male and female fishes separated with the help of the separator. The separator is removed in the morning for the spawning to start. The fishes are periodically bred every 7 days

by 72 h and the development of the larvae can be observed on each day till day 5 as till that time they are not free feeding and animal ethics are not required.

8.3.5 Raising of Larvae

The fertilized eggs are kept in 28 °C incubator. Till day 5, the larvae feed on the egg yolk, and after that *Paramecium* is given to them as a feed. The larvae are kept in 90 mm petri dishes till day 14 and are then transferred to fish tanks and maintained manually. The water is changed every second day, and dead and diseased larvae are regularly removed using pipettes. It takes 3 months for the eggs to develop into adult fishes.

8.4 Observation and Description of Developmental Stages

Time and morphological developmental stages of zebrafish fertilized egg: On the basis of time as well as morphological features of the embryo, the developmental stages are determined visually and are compiled with details of each stage (Kimmel et al. 1995).

0 hour stage or zygote period one cell stage Formation of blastodisc due to movement of yolk free cytoplasm at the animal pole. Hence, a clear segregation of animal and vegetal poles (yolk rich) can be observed. The first zygotic cell cycle will occur (Fig. 8.5a).

Cleavage period (0.45–2.15 h) The cell goes from a 2-cell to a 64-cell stage by going from 2-cell cycle to 7-cell cycle synchronously (Fig. 8.5b–e). The blastomeres divide every 15 min by partial division. A furrow develops in the animal pole, and two equal-sized blastomeres are formed to give rise to two-cell stage. The blastodisc is ellipsoidal in shape, and hence, the next cleavage to four-cell stage is at right angle to the plane of first cleavage occurring after 15 min. At the 16-cell stage, the innermost 4 cells divide completely and are surrounded by incompletely divided blastomeres. The subsequent divisions will occur. At the end of sixth cell cycle, there are three layers of blastomeres that are overlapping.

Blastula period (2.15–4.3 h) This is a rapid period with 8- and 14-cell cycles till the gastrulation period. Various important processes occur during this period leading to epiboly (Fig. 8.5f–i). At the 128-cell stage, the planes cannot be defined and there are about 5 irregular tiers of cells. The ninth cell cycle is the last synchronous cellular division stage with 7-tier blastomere formation. At the tenth cell cycle stage, the mid-blastula transition occurs. The marginal blastomeres start to lose their border, and during division cytokinesis does not occur; this is the beginning of yolk syncytial layer (YSL) formation. The length of the cell division also increases after this cycle. At the 11th cell cycle, there is the formation of irregular YSL with single row of nuclei and 11 tiers of blastomeres. Next is the high stage, where the blastodisc sits on top of the yolk layer and starts to flatten out. Here there are two rows of YSL nuclei and the division is asynchronous. In the next division cycle, the cell gets an oblong shape and YSL nuclei increase in number there is more flattening of the blastodisc. At 4 h post fertilization, the cell is spherical in shape with flat borders between blastodisc and yolk. At the spherical shape stage, the egg remains spherical. Yolk cell bulges (doming) toward animal pole as epiboly begins. The formation of blastoderm starts with thinning of the blastodisc, and in this cycle, the margin reaches 30% distance between vegetal and animal poles (Fig. 8.5i).

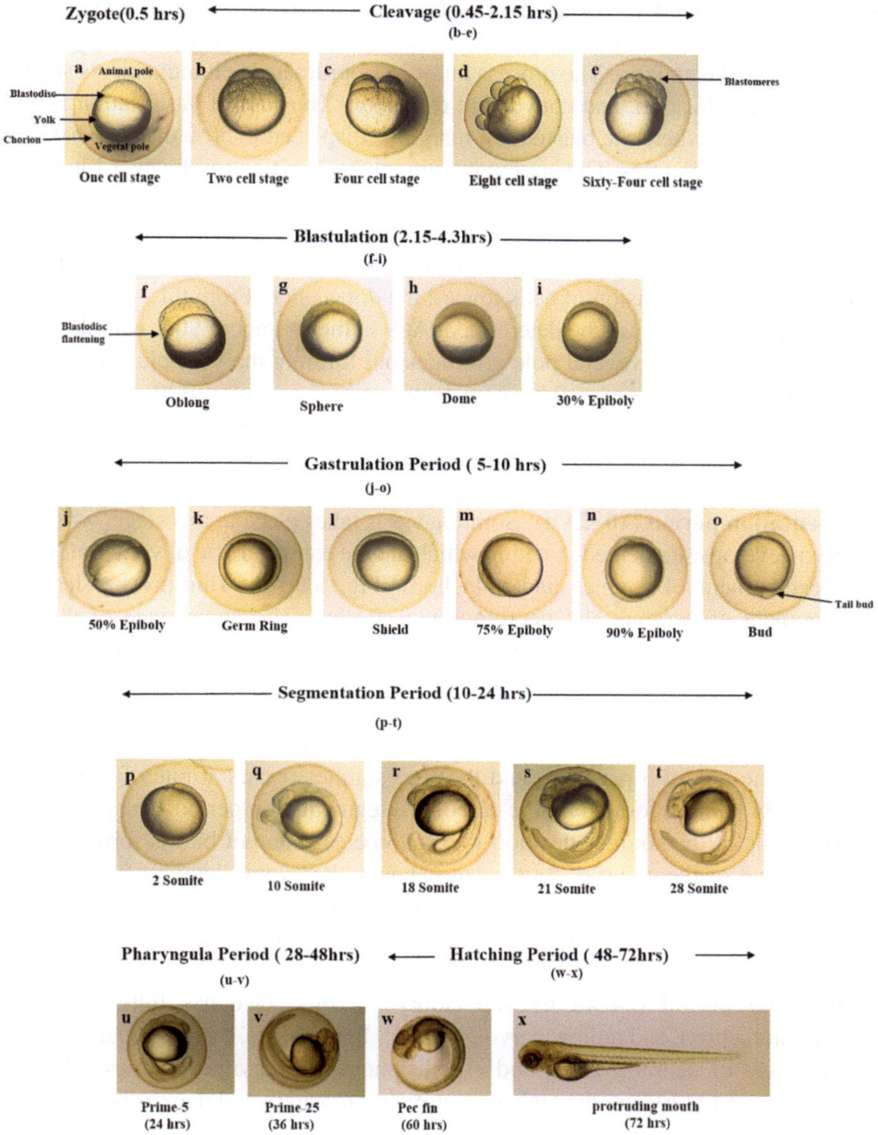

Fig. 8.5 Depiction of developmental stages of zebrafish embryo from 0 hpf to 72 hpf. (**a**) Zygote period, where single cell can be visualized. (**b–e**) Cleavage period, where embryo transitions from single-cell stage to multicellular stage, from cycle 2–7 of cell division. (**f–i**) Blastulation period: beginning of epiboly indicating movement of cells and formation of various layers. (**j–o**) Gastrulation period: basic body plan of the larva begins with processes like convergence and extension. (**p–t**) Segmentation period: brain and spinal cord start to develop with somite development and the beginning of organogenesis. (**u–v**) Pharyngula period: pigmentation starts and circulatory system and fins start to develop. (**w–x**) Hatching period: rapid progression of organogenesis; embryo hatches from the chorion

Gastrulation period (5–10 h) This is an important period as the single-layered blastula transforms into multi-layered gastrula with three germ layers—endoderm, mesoderm, and ectoderm. The process involves coordinated cell movements such as epiboly, internalization, and convergent extension. The start of gastrulation is with 50% (Fig. 8.5j), where the blastoderm margin is displaced to 50% distance between vegetal and animal poles. The margin remains at this position for the next 1 h. At the blastoderm margin, the germ ring is formed as thickened annulus. Within the germ ring, there are two layers: 2–3-cell thick epiblast and 1–2-cell thick hypoblast adjacent to the YSL. The next stage is embryonic shield formation at about 6 h by convergence and extension of cells toward the dorsal side. From this stage onward, the axis of the embryo can be determined. At 9 h, there is about 90% epiboly where the remaining yolk of vegetal pole protrudes out forming the yolk plug (Fig. 8.5n). The next is the bud stage at 10 h where 100% epiboly is achieved by closing of the yolk plug. Within 10–15 min of yolk plug closure at the dorsal side of the closure, there is formation of tail bud at the caudal end of the embryonic axis (Fig. 8.5o).

Segmentation period (10–24 h) Somitogenesis takes place in this stage, where the embryo develops segmented structure which are the characteristic of vertebrates with the development of somites and notochord (Fig. 8.5p–t). This process leads to development of nervous and neuromuscular system. The appearance of the first somite occurs from the mesoderm, leading to the development of muscles and vertebrae. The anteroposterior and the dorso-ventral axis are well established.

The somites appear in both the tail and the trunk region with the anterior somites developing first and posterior somites developing in the end. The early somites appear at the rate of 3 per hour, and the later somites occurs at rate of 2 per hour. Total 30–34 pair are formed by the end of the segmentation period. The majority of interior cells of the somites develop into myomeres. The tail bud is formed in the posterior end of the embryo. As the tail extends, the overall length of the embryo increases. The development of the rudimentary nervous and organ systems starts to takes place in this period. Along the tail, rudimentary fin folds start to appear.

Pharyngula period (24–48 h) The embryo at this stage has fully developed somites, notochord, and hollow nervous system. In this stage, there is organogenesis, development of neural system, and establishment of pharyngeal arches. The first stage is prim 5 (24 h) here the embryo is growing in length and there is beginning of pigmentation in retina and skin. At the 32 h stage, the heart is visible and begins to beat. The tail pigmentation starts by 36 h (prim 25), and the blood circulation becomes stronger. At 42 h stage (prim 25), the heart beats at the rate of 180 beats a minute. The body axis straightens from its earlier curvature, and the pectoral fins start developing (Fig. 8.5u–v). Behavioral changes start with the appearance of tactile sensitivity leading to uncoordinated movements.

Hatching period (48–72 h) After 72 h, the embryos are called larvae as they are released from the chorion. During this period, the embryo continues to grow. The development of many organs is complete but is not visible in the embryo as they are

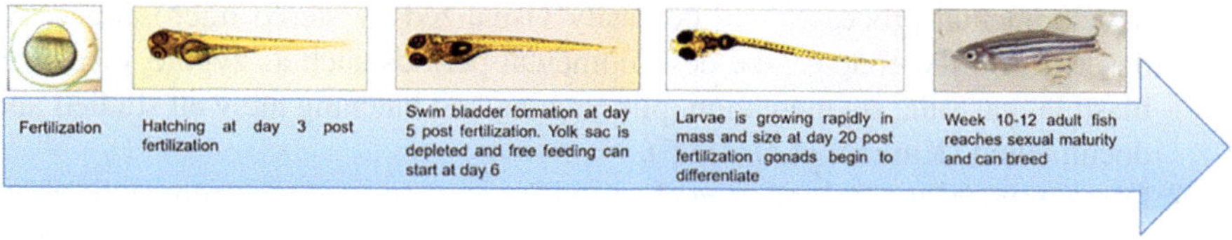

Fig. 8.6 Life cycle of zebrafish depicting different stages with respect to time of development

deeply positioned (Fig. 8.5w, x). The head straightens as the cartilage develops hatching occurs asynchronously. The yolk sac starts to diminish as it is providing nourishment to the growing embryo. The pectoral fin continues to develop. There are prominent changes in the pharyngeal region such as initially the mouth is present between the eyes region but towards the end of hatching due to jaw morphogenesis (cartilage development) the mouth positions anterior wards beyond the eyes and opens up. The melanophores are well defined and densely populated from midline anterior trunk to tail. Pigmentation of the retina is very dense, and the iridophores form a ring around the lens of the eyes. The heart is beating strongly and the blood is circulating. Environmental factors can affect the rate of hatching of the embryo.

Early larval development After hatching by day 3, the larva continues to grow rapidly and has completed its morphogenesis. By day 4, the protrusion of the mouth continues and the inflation of the swim bladder begins. During hatching, the embryo does not move much, but early larva starts to swim actively and moves its jaws, pectoral fins, and eyes. These developments produce swift escape responses and herald respiration, the seeking of prey, and feeding. The external feeding of the larvae starts within 24–48 h after the inflation of swim bladder (Fig. 8.6).

Juvenile to adult development There is a rapid development of the larva after day 5. It develops food-seeking and risk-avoidance behavior. There is a rapid development of jaws, fins, pigmentation, and gills. By 3–4 weeks post fertilization, the larvae grow rapidly in size and mass and develop into juveniles reaching a size of about 11 mm. By this time, the larvae start to resemble adult fishes, and male-female differentiation also starts. This phase lasts from 4 to 12 weeks approximately. By 10–12 weeks, fishes become sexually mature. At sexual maturity, the fish produces viable gametes and develops secondary sexual characteristics which include yellow tinge in male and abdominal protrusion in females (Ribas and Piferrer 2014) (Fig. 8.6).

8.5 Significance and Research Application

(i) Developmental biology: Developmental processes and disease modeling are extremely difficult to determine in humans. Zebrafish is a valuable model to study developmental biology its embryos are transparent and the developmental processes can be easily visualized by stereo microscopes. As described in the chapter, the developmental periods such as zygote, cleavage, blastula, gastrula, segmentation, pharyngula, and hatching are well studied and documented (Kimmel et al. 1995).

(ii) Disease modeling: Zebrafish can be used for modeling various human diseases, including cancer, neurological disorders, cardiovascular diseases, and infectious diseases. The cardiovascular system of zebrafish and humans is very similar at the genetic level. High-resolution microscopy can be performed to monitor the development of the heart and heart anomalies in zebrafish embryo with the development of tissue-specific transgenic lines of zebrafish and advanced microscopy (Santoro et al. 2019). Both forward and reverse genetic techniques are applied to successfully understand the cardiovascular system. Due to the development of innovative genetic tools, both physiological and pathological aspects of cardiovascular disease are studied. Recently, the role of Sox 18 in vertebrates has been elucidated in the development of lymphatic vessels (François et al. 2008; Cha et al. 2012).

(iii) Toxicology and drug target discovery: Zebrafish larvae are effective models in assessing the toxic effects of chemicals, environmental pollutants, and drugs. The process of screening compounds can be mechanized, and screening can be done at large scale. The chemical screening can also be used to assess the drugs on larval disease models. The chemicals can be dissolved in fish water and absorbed by the skin. The toxicological assays can be performed in a shorter time frame and in a high-throughput manner to provide statistical significance. The results can be extrapolated to humans due to high genetic and organ similarity with zebrafish. The effects of chemicals on larval growth and hatching are important parameters and can be measured easily. Physiological and morphological deformities can be observed under a low-magnification microscope. These observations help in calculating the lethal dose 50 for any chemical. A notable example is use of zebrafish larvae in studying the effect of tartrazine, a common food additive. It was shown that tartrazine induces early hatching in the embryos and induces oxidative stress in the larvae along with morphological and developmental defects (Gupta et al. 2019). The disease models in zebrafish larvae can be studied to identify novel genes. Eight small molecules discovered in zebrafish are currently under trial for various diseases (Mandelbaum et al. 2018).

(iv) Neuroscience: The zebrafish brain develops and is fully functional by day 3 post fertilization. It has structural, functional, genetic, and anatomical similarity to mammalian brain, and various genes responsible for neurodegenerative diseases, such as SNCA, PINK1, PSEN1, PSEN2, SOD1, and HD, are present (Pickrell and Youle 2015). The Parkinson disease model of zebrafish is well established, and knockdowns of disease-associated human genes such as PARKIN1 and PONK1 have replicated the symptoms of the disease such as locomotion, behavioral effects, and neuro-inflammation (Doyle and Croll 2022).

(a) Zebrafish larvae are also used as an anxiety, behavioral, and stress model. It is known to release cortisol when exposed to stress, as the neuroendocrine axis is the same in humans (HPA) and zebrafish (HPI) (Basnet et al. 2019). Changes in swimming behavior and locomotion are indicators of anxiety, stress, learning, and memory and can be modeled for such studies. An anxiety response of zebrafish larvae is thigmotaxis or wall-hugging behavior. In thigmotaxis there is tendency of larvae to avoid the center and move toward the vertical surfaces or edges when they are anxious (Lamprea et al. 2008). This is observed when the larvae are under stress (Mudgal et al. 2023). This model can be explored for high throughput screen of neuroactive compounds.

(v) Regenerative medicine: Zebrafish has high regenerative potential, and with genetic manipulation tools, it is an effective model to study regeneration. Unlike other animal models, zebrafish can regenerate fins, heart, kidneys, pancreas, liver, and spinal cord. It can also be used as a model for cardiovascular and neurological injuries. Zebrafish has conserved pathways such as wnt, hedgehog, bmp, and notch, which are reported to play an important role in regeneration (Stoick-Cooper et al. 2007). Additionally, newer pathways are identified in the larval as well as adult fish models of regeneration (Pei et al. 2008). The comparison of these pathways in humans can provide insights into the diminished regenerative capacity of humans during the evolution of vertebrates. For example, the larval model for spinal cord regeneration has high potential for the discovery of roles of new genes, molecules, and cell types, identifying novel molecular and cellular components of regeneration (Alper and Dorsky 2022; Keatinge et al. 2021). Larval model can be used to identify drugs in the treatment of spinal cord injuries (Chapela et al. 2019). The larval fin regeneration model is another attractive system to study regeneration. The fin regenerates within 3 days post injury, and hence, the assays can easily be performed in a short time span. A recent study has shown that zebrafish larvae with tail fin amputation shows regeneration within 4 h instead to 3 days on treatment with berberine. This accelerated regeneration has therapeutic benefits (Gupta et al. 2025).

(vi) Cancer research: Many factors such as cell cycle genes, tumor suppressors, oncogenes, and epigenetic markers involved in cancer development and progression are highly conserved in humans and zebrafish (Stell and Sieger 2020). By expressing human oncogenes that are responsible for inducing cancer, models for the disease can be generated in zebrafish. It has been shown in preneoplastic cells that akt1 expression induces oncogenic cell proliferation (Chia et al. 2018). The zebrafish larval system is an excellent model for xenoplastic studies. As the adaptive immune system is not fully developed and is immature, the human cancer cells survive, proliferate, and metastasize in the larva. Fluorescent labeling can be used to visualize tumor progression of the transplanted cells (Ignatius and Langenau 2011). In addition to these studies, large-scale chemical screening can be performed to identify novel molecules for therapeutics. Fluorescent microscopy is used to identify phenotypes in transgenic zebrafish. Compounds are recently identified using Tg(lck:EGFP), a T-cell reporter line for identification of molecules against leukemia (Ridges et al. 2012). Hence, various methods such as cell transplantation, high-throughput screening, live imaging, and fluorescent microscopy have enhanced our understanding of cancer biology using zebrafish as a model organism.

8.6 Precautions

- For standard growth of the larvae, the temperature has to be maintained at 28° C ± 2.
- The dark and light cycle of 10:14 should be maintained.
- Four to six fishes per liter should be the storage capacity to avoid stress to the fish. 100 embryo/35 ml is recommended in 90 mm petri dish, and 5–10 dpf larvae should be stored in 250/l density.
- Zebrafish should be kept in groups as they are shoaling species.
- Zebrafish lack stomach, so the feeding is done 2–3 times a day with a combination of live and dry feed.
- Inventory should be maintained for the fishes. This includes date of fertilization also genetic features should be mentioned for each fish.
- The water should be filtered by active charcoal to avoid presence of chlorine as it is toxic to the fish.
- Zebrafish can jump and hence it is important to always keep the tanks covered.
- It is important to remove unfertilized eggs from the fertilized eggs during egg collection. The unfertilized eggs can provide nutrition to fungi and bacteria leading to contamination of water.

8.7 Quick Review Questions

8.7.1 Multiple Choice Questions

1. When does hatching occur in the zebrafish embryo?
   (a) 24–48 h
   (b) 48–72 h
   (c) 72–96 h
   (d) 96–120 h
   Answer: b) 48–72 h
2. What is the dark and light cycle for the zebrafish?
   (a) 10:14
   (b) 12:12
   (c) 10:10
   (d) 14:10
   Answer: a) 10:14

3. Which molecule can be used to render larvae optically transparent?
   (a) 1-phenyl-2-thio urea
   (b) Poly urea
   (c) 2-poly-4-thio urea
   (d) Thio urea
   Answer: a) 1-phenyl-2-thio urea
4. What is the mode of cleavage in the zebrafish embryo?
   (a) The entire egg cleaves
   (b) Only in the vegetal pole
   (c) Only in the blastodisc atop the yolk
   (d) Equally in both blastodisc and yolk
   Answer: c) Only in the blastodisc atop the yolk
5. When does Zebrafish reach sexual maturity?
   (a) 1 year
   (b) 6 months
   (c) 3 months
   (d) 1 month
   Answer: c) 3 months
6. YSL stands for.
   (a) Yellow superior layer
   (b) Yolk syncytial lump
   (c) Yellow syncytial lump
   (d) Yolk syncytial layer
   Answer: d) Yolk syncytial layer
7. When does the first somite develop?
   (a) 10 h
   (b) 10.5 h
   (c) 9 h
   (d) 9.5 h
   Answer: b) 10.5 h
8. Tail bud formation occurs at the end of gastrulation stage.
   (a) True
   (b) False
   Answer: a) True.

8.7.2 Short Answer Questions

1. How many cells does the embryo have after fertilization?
2. Till which stage are the cell divisions synchronous?
3. What is the time interval between early cell division?
4. What is the feed given to 3 dpf larvae, 10 dpf larvae, juvenile, and adult fish?
5. How do you differentiate male and female zebrafish?

Experimental Errors

1. Water is the medium in which fishes are surviving; hence, water quality maintenance is very important in the absence of a conducive environment, survival of fishes will be difficult and experimental errors will increase. The pH, salt concentration, and temperature of the water should be maintained.
2. Genetic pool of the fish has to be maintained. Same population of fishes has to be used for all experiments. The ideal will be to use sequenced organism for experimental work to avoid experimental errors.
3. The light and dark cycle for the fishes has to be maintained; otherwise, behavioral changes can be observed leading to errors.
4. Feeding cycle and feed quality have to be maintained to avoid effect on growth and reproductive efficiency of fishes.
5. The housing density of the fishes needs to be appropriate as both high and low density affects behavior and reproduction efficiency.

Glossary

Animal ethics The philosophy of moral consideration of non-human animal treatment

Animal pole Region of the egg where nucleus is present and rapid cell division takes place

Antero-posterior axis The principal body axis that runs from head (anterior) to the tail (posterior) region

Artemia These are aquatic crustaceans also called brine shrimps and can be used as a feed for zebrafish.

Blastoderm A single layer of embryonic epithelial tissue that makes up the blastula

Blastodisc A region at the surface of a developing embryo germinal disc on the yolk of the egg

Blastomeres Cells generated by cell division of zygote

Bmp Bone morphogenetic protein

Cartilage Strong flexible connective tissues

Cleavage Cell division in the early embryo development

Clutch Batch of eggs laid by a female fish at one time

Cortisol Natural steroid stress hormone

Dorso-ventral axis Imaginary line that define front (ventral) and back (dorso) of an organism

Ectoderm The outermost layer of the three primary germ layers in the embryo

Embryonic axis Development of the embryo axis that delineates the head and tail region

Endoderm The innermost layer of the three primary germ layers in the embryo

Epiblast The embryonic ectoderm

Epiboly Process of development during gastrulation were cells move to cover the yolk cells completely and is required to form ectoderm, mesoderm, and endoderm.

Epigenetic markers Effect on gene expression without modification on DNA

Fecundity Fertility and capability to produce healthy and abundant offspring.

Gametes Reproductive haploid cell that fertilizes with another haploid cell

Generation time The development of the egg from fertilization to sexual active adulty

Germ layers Primary layer of cells that are formed during embryonic development

Hatching Emergence of larva from the egg

Hedgehog Secretory chemical signals that induce normal growth, differentiation, and patterning in the embryo

Herald respiration Early respiratory changes

Hpa Hours post amputation

Hpi Hypothalamic-pituitary-interrenal axis

In vivo Experiments conducted within a living organism

Iridophores Specialized pigment cells in zebrafish that give a shiny appearance and reflect light

Juvenile It is the developmental stage, which comes before sexual maturity and after larva

Larva It is the free-swimming, immature developmental stage after hatching

Live feed These are the natural live food organisms (e.g., Rotifer, Paramecia) given to fish larvae and juveniles

Melanophores These are the pigment cells present in zebrafish which contain melanin

Mesoderm It is the middle layer of the three germ layers or mass of cell that lies between the ectoderm and endoderm.

Metastasize It is usually referred to as spread of cancer cells from primary tumor to other parts of the body

Microinjections It is a method of gene transfer used to inject genetic material or drug directly into the cells or embryos

Micronutrients These are the minerals and vitamins that are required in trace amount for proper growth and development of the body

Morpholinos These are the synthetic antisense molecules used to block gene expression in zebrafish embryos

Myomeres These are the sequenced blocks of skeleton muscle which are commonly found in aquatic chordates.

Neuroendocrine axis The term neuroendocrine axis can be simply defined as the axis in which neurological and endocrine systems work together to control hormone release and cause significant physiological changes

Neuro-inflammation These are the inflammatory response within the nervous system which are commonly associated with neurodegenerative disorders

Neurological disorders It is defined as any disorder that affect brain, nerves, or spinal cord

Notch It is a cell surface receptor which follows juxtracrine signaling by interacting with delta protein

Organogenesis It is defined as the stage of embryonic development which involves migration and differentiation of cells to form organs and tissues from three germ layers (endoderm, mesoderm, and ectoderm)

Parkin1 It is a gene encoding an E3 ubiquitin ligase, involved in PINK-1/PARKIN signaling, plays a key role in mitophagy, and its mutation is linked to Parkinson's disease

Parkinson It is a neurodegenerative disorder which is mainly associated with central nervous system. It affects movement, tremor, slowness, and stiffness

Pectoral fins These are the paired fins which are located on each side of fish, behind the operculum

Pentair water system These systems are designed to provide temperature regulation, optimal water quality, and filtration to ensure the proper development of zebra fish in a healthy environment

Pharyngeal arches In zebrafish, pharyngeal arches are tissue structures with seven pairs of segmented, embryonic, transient, and soft tissues that are essential for cranofacial development of adult

Phenotypes These are the set of observable traits of an organism which are determined by environment, genes, or random variation

Photoperiod This is the duration of daily light and dark cycles, important for zebrafish breeding and physiology

Pink1 It is a gene which encodes a serine threonine protein kinase (PINK-1 kinase); its mutation is associated with Parkinson's disease

Pluripotent These are the embryonic stem cell which are all capable of dividing and differentiating into all cell types of the body

Preneoplastic cells These are defined as the abnormal cells which may progress to cancer

Psen1 Psen-1 gene is involved in γ-secretase activity, which is associated with Delta notch pathway (Juxtracrine signaling); mutation in this gene is linked to Alzheimer's disease

Psen2 This gene is similar to Psen-1 gene. It is also associated with Alzheimer's disease or neurodegeneration

Regeneration It is the ability of an organism to regrow damaged tissue like fins. Zebrafish are also capable of regeneration

Segmentation It is defined as the division of the embryo in somites, which are repeated structural units

Sexually mature This is the stage when zebrafish can reproduce successfully

Snca (α-synuclein) It is gene which is associated with synaptic function; its mutation can lead to familial Parkinson's disease

Sod1 (superoxide dismutase-1) It is a gene that encodes for Cu/Zn superoxide dismutase enzyme, which plays a role in antioxidant defense

Somites These are the group of bilaterally paired paraxial mesoderm blocks that arise during the embryonic stage of somitogenesis

Somitogenesis It is defined as the process of somite formation during development

Sox 18 It is a transcription factor which is involved in vascular and lymphatic development

Spawning It is a form of external fertilization where male and female fishes release eggs

Swim bladder These are the gas-filled organs used to regulate buoyancy in fish

Teleost These are the group of bony fishes that includes zebrafish; this group has distinct features such as movable upper jaw and more equal lobbed tail fish

Tg(Lck:Egfp) It is the transgenic model of zebrafish, where tg stands for transgenic and lck stands for lymphocyte-specific kinase.

Thigmotaxis Also known as "wall hugging." It is the tendency of fishes to stay near the periphery or tank walls, commonly used to measure anxiety-like behavior.

Transgenic A foreign genetic material is introduced into an organism by recombinant DNA technology.

Vegetal poles It is the yolk-rich region of the egg which is opposite to animal pole

VEGF-A and VEGF-B These are the vascular endothelial growth factor genes that are important for blood vessel development in zebrafish

Wnt It is a glycoprotein, which is associated with Wnt signaling pathways; this pathway is essential for regulating cell fate, embryonic axis formation, and organ development

Xenoplastic It is a type of transplantation of cells or tissue between two different species

Yolk plug It is a temporary, small, circular patch of yolk cells that remains exposed to vegetal pole during gastrulation. It is the last uncovered part of yolk before epiboly

Yolk sac It is the extra-embryonic structure which provides nutrition to the developing embryo

Yolk syncytial layer It is the multinucleated layer, formed at the yolk-blastoderm interface which is critical for signaling and morphogenesis

Zebrafish husbandry It is defined as the feeding, breeding, maintenance, and care of zebrafish in laboratory conditions

Zygote It is the single-cell stage of the embryo formed immediately after fertilization

References

Alper SR, Dorsky RI (2022) Unique advantages of zebrafish larvae as a model for spinal cord regeneration. Front Mol Neurosci 15:983336. https://doi.org/10.3389/fnmol.2022.983336. PMID: 36157068; PMCID: PMC9489991

Antinucci P, Hindges R (2016) A crystal-clear zebrafish for in vivo imaging. Sci Rep 6:29490. https://doi.org/10.1038/srep29490. PMID: 27381182; PMCID: PMC4933947

Basnet RM, Zizioli D, Taweedet S, Finazzi D, Memo M (2019) Zebrafish larvae as a behavioral model in neuropharmacology. Biomedicine 7(1):23. https://doi.org/10.3390/biomedicines7010023. PMID: 30917585; PMCID: PMC6465999

Cha YR, Fujita M, Butler M, Isogai S, Kochhan E, Siekmann AF, Weinstein BM (2012) Chemokine signaling directs trunk lymphatic network formation along the preexisting blood vasculature. Dev Cell 22(4):824–836. https://doi.org/10.1016/j.devcel.2012.01.011. PMID: 22516200; PMCID: PMC4182014

Chapela D, Sousa S, Martins I, Cristóvão AM, Pinto P, Corte-Real S, Saúde L (2019) A zebrafish drug screening platform boosts the discovery of novel therapeutics for spinal cord injury in mammals. Sci Rep. 9(1):10475. https://doi.org/10.1038/s41598-019-47006-w. Erratum in: Sci Rep 2020 Jan 15;10(1):749. doi:10.1038/s41598-020-57443-7. PMID: 31324865; PMCID: PMC6642202

Chia K, Mazzolini J, Mione M, Sieger D (2018) Tumor initiating cells induce Cxcr4-mediated infiltration of pro-tumoral macrophages into the brain. elife 7:e31918. https://doi.org/10.7554/eLife.31918. PMID: 29465400; PMCID: PMC5821457

Doyle JM, Croll RP (2022) A critical review of zebrafish models of Parkinson's disease. Front Pharmacol 13:835827. https://doi.org/10.3389/fphar.2022.835827. PMID: 35370740; PMCID: PMC8965100

François M, Caprini A, Hosking B, Orsenigo F, Wilhelm D, Browne C, Paavonen K, Karnezis T, Shayan R, Downes M, Davidson T, Tutt D, Cheah KS, Stacker SA, Muscat GE, Achen MG, Dejana E, Koopman P (2008) Sox18 induces development of the lymphatic vasculature in mice. Nature 456(7222):643–647. https://doi.org/10.1038/nature07391. Epub 2008 Oct 19

Gupta R, Ranjan S, Yadav A, Verma B, Malhotra K, Madan M, Chopra O, Jain S, Gupta S, Joshi A, Bhasin C, Mudgal P (2019) Toxic effects of food colorants erythrosine and Tartrazine on zebrafish embryo development. Curr Res Nutr Food Sci 7(3). https://doi.org/10.12944/CRNFSJ.7.3.26

Gupta R, Bhasin C, Joshi A, Agarwal N, Aggarwal A, Mudgal P (2025) Transcriptome analysis of Berberine induced accelerated tail fin regeneration in zebrafish larvae. Gene Expr Patterns 55: 119390. https://doi.org/10.1016/j.gep.2025.119390. Epub 2025 Feb 9

Howe K et al (2013) The zebrafish reference genome sequence and its relationship to the human genome. Nature 496(7446):498–503. https://doi.org/10.1038/nature12111. Epub 2013 Apr 17. Erratum in: Nature. 2014;505(7482):248. Cooper, James [corrected to Cooper, James D]; Eliott, David [corrected to Elliot, David]; Mortimer, Beverly [corrected to Mortimore, Beverley]; Begum, Sharmin [added]; Lloyd, Christine [added]; Lanz, Christa [added]; Raddatz, Günter [added]; Schuster, Ste. PMID: 23594743; PMCID: PMC3703927

Ignatius MS, Langenau DM (2011) Fluorescent imaging of cancer in zebrafish. Methods Cell Biol 105:437–459. https://doi.org/10.1016/B978-0-12-381320-6.00019-9. PMID: 21951542; PMCID: PMC3959906

Keatinge M, Tsarouchas TM, Munir T, Porter NJ, Larraz J, Gianni D, Tsai HH, Becker CG, Lyons DA, Becker T (2021) CRISPR gRNA phenotypic screening in zebrafish reveals pro-regenerative genes in spinal cord injury. PLoS Genet 17(4):e1009515. https://doi.org/10.1371/journal.pgen.1009515. PMID: 33914736; PMCID: PMC8084196

Kimmel CB, Ballard WW, Kimmel SR, Ullmann B, Schilling TF (1995) Stages of embryonic development of the zebrafish. Dev Dyn 203(3):253–310. https://doi.org/10.1002/aja.1002030302

Lamprea MR, Cardenas FP, Setem J, Morato S (2008) Thigmotactic responses in an open-field. Braz J Med Biol Res 41(2):135–140. https://doi.org/10.1590/s0100-879x2008000200010

Mandelbaum J, Shestopalov IA, Henderson RE, Chau NG, Knoechel B, Wick MJ, Zon LI (2018) Zebrafish blastomere screen identifies retinoic acid suppression of MYB in adenoid cystic carcinoma. J Exp Med. 215(10):2673–2685. https://doi.org/10.1084/jem.20180939. Epub 2018 Sep 12. PMID: 30209067; PMCID: PMC6170170

Mudgal P, Gupta R, Joshi A, Prakash C, Gupta K, Sachdeva R, Joshi N (2023. Jun 13 [cited 2025 Jul. 30]) Assessment of Anxiolytic Activity of Brahmi (*Bacopa monnieri*) in Zebrafish Model System. JNR [Internet] 23(2):661–670

Pei DS, Sun YH, Chen CH et al (2008) Identification and characterization of a novel gene differentially expressed in zebrafish cross-subfamily cloned embryos. BMC Dev Biol 8:29. https://doi.org/10.1186/1471-213X-8-29

Pickrell AM, Youle RJ (2015) The roles of PINK1, parkin, and mitochondrial fidelity in Parkinson's disease. Neuron 85(2):257–273. https://doi.org/10.1016/j.neuron.2014.12.007. PMID: 25611507; PMCID: PMC4764997

Postlethwait JH, Woods IG, Ngo-Hazelett P, Yan YL, Kelly PD, Chu F, Huang H, Hill-Force A, Talbot WS (2000) Zebrafish comparative genomics and the origins of vertebrate chromosomes. Genome Res 10(12):1890–1902. https://doi.org/10.1101/gr.164800

Ribas L, Piferrer F (2014) The zebrafish (D anio rerio) as a model organism, with emphasis on applications for finfish aquaculture research. Rev Aquac 6(4):209–240

Ridges S et al (2012) Zebrafish screen identifies novel compound with selective toxicity against leukemia. Blood 119(24):5621–5631. https://doi.org/10.1182/blood-2011-12-398818

Sant KE, Timme-Laragy AR (2018) Zebrafish as a model for toxicological perturbation of yolk and nutrition in the early embryo. Curr Environ Health Rep 5(1):125–133. https://doi.org/10.1007/s40572-018-0183-2. PMID: 29417450; PMCID: PMC5876134

Santoro MM, Beltrame M, Panáková D, Siekmann AF, Tiso N, Venero Galanternik M, Jung HM, Weinstein BM (2019) Advantages and challenges of cardiovascular and lymphatic studies in zebrafish research. Front Cell Dev Biol 7:89. https://doi.org/10.3389/fcell.2019.00089. PMID: 31192207; PMCID: PMC6546721

Stell KR, Sieger D (2020) Zebrafish in vivo models of cancer and metastasis. Cold Spring Harb Perspect Med 10(8):a037077. https://doi.org/10.1101/cshperspect.a037077. PMID: 31615862; PMCID: PMC7397842

Stoick-Cooper CL, Weidinger G, Riehle KJ, Hubbert C, Major MB, Fausto N, Moon RT (2007) Distinct Wnt signaling pathways have opposing roles in appendage regeneration. Development 134(3):479–489. https://doi.org/10.1242/dev.001123. Epub 2006 Dec 21

Volkov M, Machikhin A, Bukova V et al (2022) Optical transparency and label-free vessel imaging of zebrafish larvae in shortwave infrared range as a tool for prolonged studying of cardiovascular system development. Sci Rep 12:20884. https://doi.org/10.1038/s41598-022-25386-w

Examination of Histological Sections of Gametes from Photomicrographs/Permanent Slides of Rat/Human: Testis and Ovary

9

Anna Senrung and Alok Chandra Bharti

Abstract

This practical chapter focuses on the histological examination of gametes, spermatozoa, and ova through the analysis of photomicrographs or permanent slides of rat/human testis and ovary. Gametes are the fundamental reproductive cells required for sexual reproduction, and their development within the gonads is tightly regulated by hormonal and structural mechanisms. The testis functions as both a gametogenic and an endocrine organ. Histological sections reveal seminiferous tubules (ST) lined with stratified germinal epithelium housing spermatogenic cells, spermatogonia, spermatocytes, spermatids, spermatozoa, supporting Sertoli cells, and testosterone-secreting interstitial Leydig cells. The spermatozoon, a motile cell with distinct head, midpiece, and tail regions, is structurally specialized for fertilization. The ovary, by contrast, contains follicles at successive stages of development. These structures reflect the cyclical nature of oogenesis and are influenced by hormonal signals, notably Gonadotropin-releasing hormone (GnRH), follicle-stimulating hormone (FSH), Luteinizing hormone (LH), estrogen, and progesterone. This exercise trains students to identify and interpret the histoarchitecture of gonadal tissues, recognize stages of gametogenesis, and correlate microscopic features with functional and hormonal processes. Through careful observation, labelled diagramming, and structured recording, learners reinforce theoretical concepts and acquire practical skills in tissue analysis. This foundational understanding is critical not only for academic learning but also for applications in reproductive biology, endocrinology, developmental science, infertility assessment, and reproductive toxicology. By integrating

A. Senrung (✉)
Department of Zoology, Daulat Ram College, University of Delhi, New Delhi, India

A. C. Bharti (✉)
Molecular Oncology Lab., Department of Zoology, University of Delhi, New Delhi, India

A. C. Bharti et al. (eds.), *Textbook of Developmental Biology*,
https://doi.org/10.1007/978-981-95-7269-4_9

histology with physiological context, the exercise builds competence in evaluating both normal and pathological aspects of reproductive health.

Keywords

Ovary · Testis · Ovum · Sperm · Ovarian follicle

Aim To examine and identify histological sections of gametes from photomicrographs or permanent slides of rat/human testis and ovary, and to understand their structural organization and functional significance in gametogenesis.

Learning Outcomes

By the end of this practical exercise, students will be able to:

- Identify and describe the microscopic structure of the testis and ovary.
- Identify key histological structures of the testis, seminiferous tubules (ST) with various stages of spermatogenesis, the lumen in the testis, and Leydig cells.
- Distinguish between various ovarian follicles with developing oocytes and other structures, including corpus luteum, corpus albicans, and medullary tissue.
- Correlate the histological features observed with the functional aspects of male and female gametogenesis.

9.1 Introduction

The study of reproductive histology is a fundamental factor in understanding human and mammalian biology. Gametes are the specialized reproductive cells responsible for sexual reproduction. In mammals, these are spermatozoa (male gametes) and ova (female gametes), produced within specialized gonads—the testis and ovary, respectively. The microscopic structure of these reproductive organs offers crucial insights into gametogenesis, the process through which functional gametes are formed. The testis comprises seminiferous tubules (ST) where spermatogenesis occurs, whereas the ovary contains follicles at different stages of oogenesis.

In males, the testis is responsible for spermatozoa and testosterone production. Histological sections reveal the organization of the ST, the spermatogenic lineage, and supporting cells (Sertoli and Leydig) (Gurung et al. 2023). In females, the ovary produces ova and estrogen and progesterone hormones. Histological section of the ovary demonstrates various follicular development stages, corpus luteum, and stromal tissues (Richards and Pangas 2010).

Studying histological sections of these organs using photomicrographs or permanent slides helps students understand the organization, cellular composition, and dynamic processes that occur during gamete formation. Through this practical exercise, learners will identify and differentiate key cellular stages of spermatogenesis and oogenesis, recognize structural components of the testis and ovary, and correlate microscopic features with physiological functions.

This exercise lays the foundation for understanding broader concepts in reproductive biology, endocrinology, fertility studies, and developmental biology.

9.2 Requirements

- Prepared permanent histological slides/photomicrographs of testis and ovary (rat/human)
- Compound microscope
- Observation record sheet
- Labelled diagrams or reference images
- Pencil and lab notebook

9.3 Methodology

- Place the slide under a compound microscope at appropriate magnification.
- Make labelled diagrams of testis and ovary sections under high magnification.
- Note structural differences between spermatogenesis and oogenesis.
- Record developmental stages, cell types, and functional structures observed.

9.3.1 Observation of Testis Section

- Observe and identify the ST, which appear as circular or oval structures, and examine the various stages of spermatogenic cells within ST:
  - *Spermatogonia:* Located near the basement membrane.
  - *Primary spermatocytes:* Larger cells moving inward from the basal lamina.
  - *Spermatids:* Small, round cells closer to the lumen.
  - *Spermatozoa:* Slender, elongated cells with tails in the lumen.
- Identify interstitial (Leydig) cells between the tubules that secrete testosterone.

9.3.2 Observation of Ovary Section

- Focus on the cortex region of the ovary containing follicles at different stages:
  - *Primordial follicles:* Small, with a single layer of follicular cells.
  - *Primary follicles:* Enlarged oocytes with cuboidal follicular cells.

  - *Secondary follicles:* Presence of fluid-filled antrum and multiple granulosa layers.
  - *Graafian follicle (mature follicle):* Secondary oocyte, large antrum, cumulus oophorus.
  - Identify *corpus luteum or corpus albicans* in the post-ovulatory phase if present.
- Observe the medulla region containing blood vessels and connective tissue.

9.4 Description of Histology and Function of the Gonads

9.4.1 Testis

The testis (Fig. 9.1a, b) is the male primary reproductive organ producing sperm and male sex hormones. Histologically, it comprises a dense connective tissue capsule (tunica albuginea) surrounding the testis, which thickens posteriorly to form the mediastinum testis. From the mediastinum, a septum arises and spreads into the parenchyma, dividing the testis into approximately 370 testicular lobules (Suede et al. 2023). Each lobule contains one to four highly coiled ST, where spermatogenesis occurs. Between the ST lies interstitial connective tissue that contains testosterone-secreting Leydig cells. The ST drain into a network of straight tubules and rete testis, which ultimately lead to the efferent ductules and epididymis. Blood vessels, lymphatics, nerves, and macrophages are also present within the interstitial tissue. The overall histological organization of the testis reflects its dual role in gametogenesis and hormone production.

9.4.1.1 Seminiferous Tubules (ST)

ST (Fig. 9.1c) are lined by germinal epithelium and comprises Sertoli cells and spermatogenic cells at different developmental stages involved in spermatogenesis (Meneses et al. 2015).

9.4.1.1.1 Spermatogenic Cells

Spermatogenic cells (Fig. 9.1d) include spermatogonia, spermatocytes (primary and secondary), spermatids, and spermatozoa (Suede et al. 2023). These are arranged in a stratified and orderly manner within the seminiferous epithelium of the ST. The earliest cells, type A spermatogonia, are located near the ST basement membrane, and progressively more differentiated stages are found nearer to the lumen. This unidirectional organization, beginning from the basal compartment toward the lumen, mirrors the sequential process of spermatogenesis, ensuring the continuous and efficient production of spermatozoa that are eventually released to the tubular lumen.

Spermatogonia These are the immature male germ cells near the basement membrane of the ST and represent the initial stage of spermatogenesis (Fig. 9.1d). They can be broadly categorized into three kinds (Gilbert 2000a): type A, intermediate,

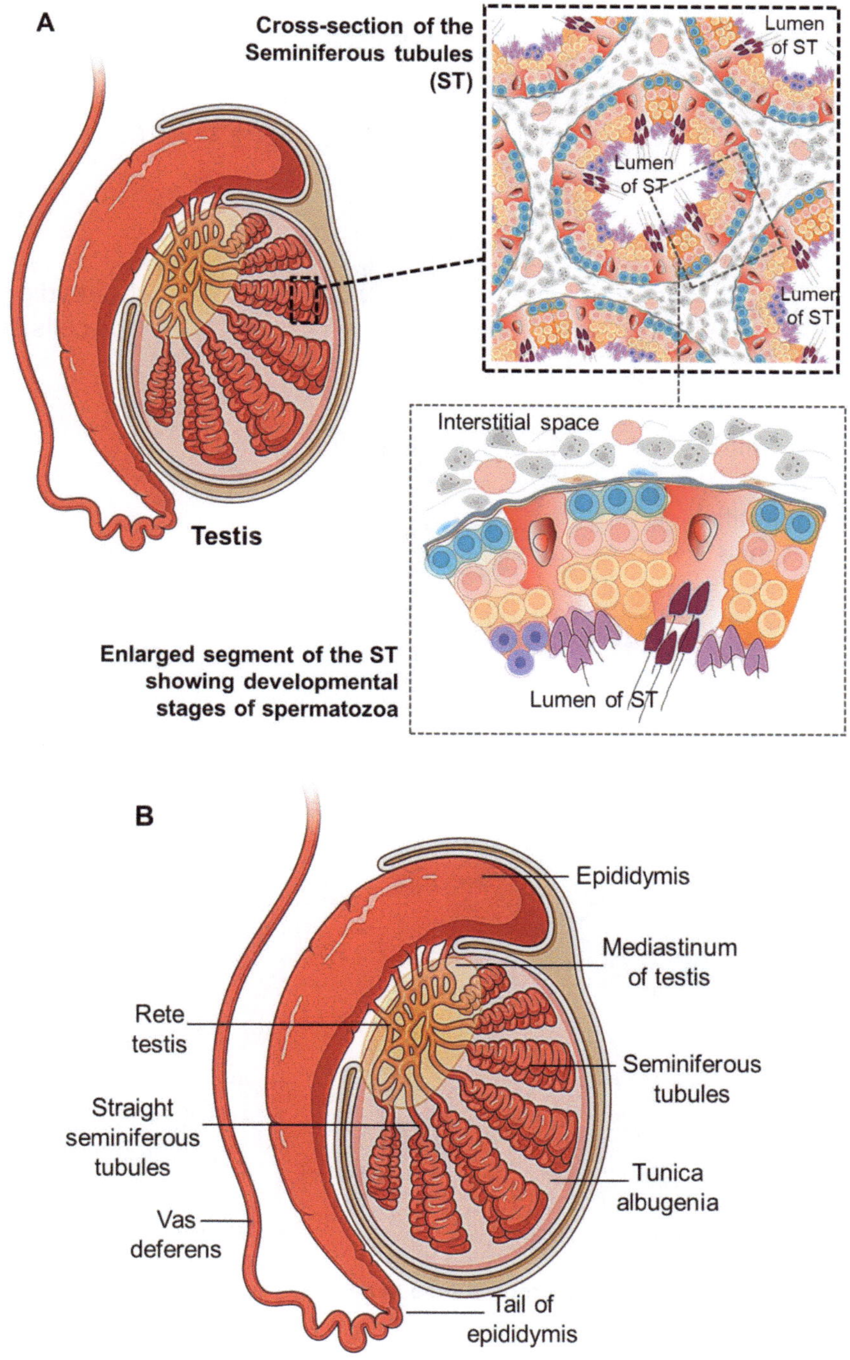

Fig. 9.1 Schematic diagram illustrating human testis and seminiferous tubules (ST) structure. (**a**) Overview of the testis and seminiferous tubules. (**b**) Testis. (**c**) Section of the ST. (**d**) Enlarged segment of the ST showing developmental stages of spermatozoa. (The diagrams have been prepared using information derived from (Aladamat and Tadi 2022; Balinsky 1960; Durairajanayagam et al. 2015; Eroschenko 2008; Obukohwo et al. 2021). Created in https://BioRender.com)

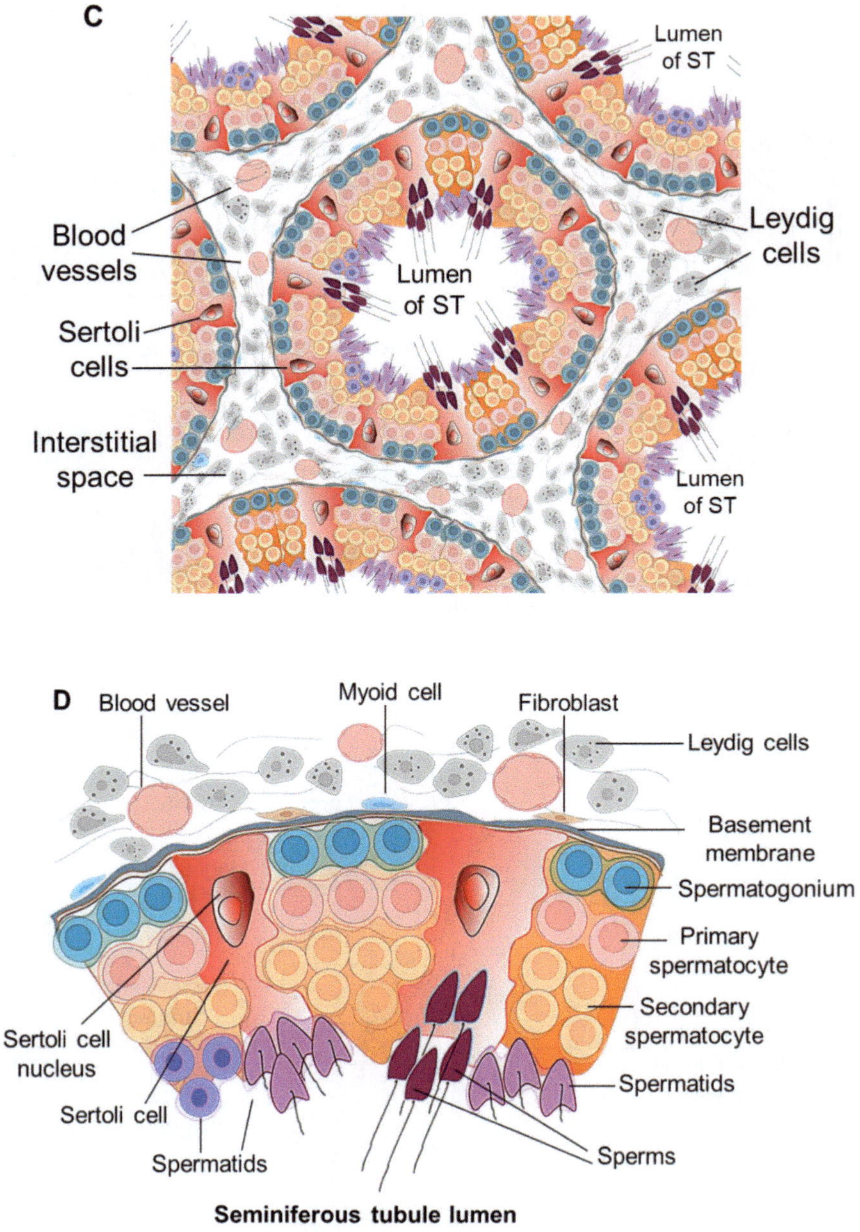

Fig. 9.1 (continued)

and type B spermatogonia. Type A serve as stem cells and are further subclassified into A1, A2, A3, and A4 based on their stage of differentiation. Type A1 spermatogonia, situated closest to the basement membrane, function as true stem cells. Upon mitotic division, they either self-renew (producing more A1 cells) or give rise to A2 spermatogonia. This progression continues through A3 to A4 spermatogonia. Type A4 spermatogonia represent a critical decision point; they can either undergo self-renewal, initiate apoptosis, or differentiate into intermediate forms, which are committed to the pathway of sperm production. Intermediate spermatogonia then undergo mitotic divisions and produce type B, which differentiate into primary spermatocytes, marking the beginning of meiosis. This hierarchical and regulated progression ensures the maintenance of the germ cell pool while enabling continuous sperm production. During this stage, nuclear gene transcription is actively initiated, whose products are essential for the formation of the acrosome and axoneme.

Spermatocytes Spermatocytes are of two types: primary and secondary spermatocytes (Fig. 9.1d). Each spermatocyte (primary) undergoes the first meiotic division to form two haploid spermatocytes (secondary), which then rapidly undergo the second meiotic division and produce two haploid spermatids each, ultimately yielding four spermatids from a single primary spermatocyte. Secondary spermatocytes are smaller than primary spermatocytes and have a very brief existence. Notably, cytokinesis is incomplete during the divisions of spermatogonia and spermatocytes, resulting in interconnected cells via cytoplasmic bridges (Alberts et al. 2002b). These bridges allow the exchange of ions, molecules, or gene products, leading to cells within the same cohort progressing through spermatogenesis synchronously (Gilbert 2000a).

Spermatids These, at first, are rounded, unflagellated cells that appear markedly dissimilar from spermatozoa (Fig. 9.1d). They are situated near the lumen of the ST. Early spermatids possess a small, spherical nucleus with a distinct nucleolus. In some species, two generations of spermatids may be observed simultaneously. Like their precursors, the spermatocytes, early spermatids are often interconnected by narrow cytoplasmic bridges, facilitating synchronous development. These round spermatids undergo a transformation known as *spermiogenesis* (or *spermateliosis*), during which they lose their cytoplasmic connections and complete differentiation into mature sperm cells. Through this process, they are structurally and functionally modified into elongated, motile spermatozoa.

Spermatozoon Spermatozoon (Fig. 9.1d) is the fully matured gamete formed at the end of spermiogenesis. It is a highly specialized, motile cell designed for fertilization and is generally found toward the ST lumen. Structurally, it consists of a head and a flagellum or tail consisting of a midpiece, principal, and an end piece. The head contains a condensed nucleus capped by the acrosome, which houses enzymes vital for penetrating the ovum during fertilization. The midpiece is rich in mitochondria that generate energy for motility, while the tail facilitates active propulsion of the sperm. After being released into the ST lumen, spermatozoa travel through the male reproductive tract, maturing further in the epididymis before becoming capable of fertilization. Generally, four spermatozoa are formed from a single primary spermatocyte.

9.4.1.1.2 Sertoli Cells

Sertoli cells (Fig. 9.1d), though not spermatogenic, are a characteristic and essential component of the seminiferous epithelium. Named after its discoverer Enrico Sertoli in 1865 (Franca et al. 2016). They are derived from the ST epithelium and are distinguished by their tall, columnar shape. These cells spread from the basement membrane to the lumen, spanning the full thickness of the germinal epithelium. Their nuclei are typically located near the basal lamina and are identifiable by their tripartite, indented outline and prominent nucleolus (Nistal and Paniagua 2020). In adults, Sertoli cells often assume highly irregular shapes due to the constant remodelling caused by developing sperm cells embedded within their cytoplasm (Nistal and Paniagua 2020). Sertoli cells rarely divide postnatally. Functionally, they are essential for spermatogenesis, providing structural and nutritional support to developing germ cells and facilitating their proliferation and maturation (Shah et al. 2021). Their activity is regulated by follicle-stimulating hormone (FSH). Additionally, Sertoli cells secrete several proteins, including the hormones activin and inhibin. Activin enhances FSH secretion, while inhibin regulates negative feedback to suppress FSH production, thus contributing to the fine hormonal control of spermatogenesis (Shah et al. 2021).

9.4.1.2 Leydig Cells

Interstitial cells of the testis, the Leydig cells, are located in the interstitial connective tissue between ST (Fig. 9.1d). These cells are polygonal with eosinophilic cytoplasm and a prominent, round nucleus. Ultrastructurally, Leydig cells show steroid-secreting characteristics with abundant smooth endoplasmic reticulum, large lipid droplets, and mitochondria containing distinctive tubulovesicular cristae (Aladamat and Tadi 2020). Additional cytoplasmic inclusions include lipofuscin granules, which appear as irregular, rounded bodies, and rod-like Reinke crystalloids. Leydig cells primarily synthesize and secrete androgens, chiefly testosterone, the principal male sex hormone. These hormones regulate spermatogenesis and are vital for developing and maintaining male secondary sexual features. Specifically, they are crucial for the proper development and functioning of the accessory reproductive organs, the seminal vesicles, prostate, and epididymis. They also induce male secondary sexual characteristics, including voice deepening, facial hair growth, and body hair distribution in the axillary and pubic regions. Leydig cell activity is regulated by luteinizing hormone (LH) produced from the anterior pituitary gland essential for sustaining spermatogenesis and maintaining male fertility (Lei et al. 2025).

9.4.1.3 Spermatogenesis

All the differential developmental stages of the male gamete, beginning with the spermatogenic germ cell, spermatogonia through spermatocyte, spermatids, and ultimately, the mature sperm cell or spermatozoa located within the ST are subsequent products of the stages formed via a complex process of cell division and differentiation process known as *spermatogenesis*. Spermatogenesis can be broadly divided into the following events: the development and multiplication (mitotic division) of spermatogonia and further development to spermatocytes (*spermatocytogenesis*), reductional differentiation of spermatocytes to spermatids, and maturation of spermatids into spermatozoa (*spermiogenesis*).

Spermatogenesis (Fig. 9.2) begins with the mitotic division of spermatogonia (2n), giving rise to two primary diploid spermatocytes. Each primary spermatocyte undergoes meiosis I to produce two haploid secondary spermatocytes. And each secondary spermatocytes produce two spermatids. The spermatids, without further division, transformed into mature motile sperm or spermatozoa. In rats, spermatogenesis takes approximately 34.5 days, while in humans, it takes 65 days (Gilbert 2000a).

Type A1 spermatogonia function as stem cells, enabling the continuous process of spermatogenesis. Their capacity for self-renewal ensures a persistent supply of differentiating germ cells. As a result, sperm production continues uninterrupted throughout the male's reproductive lifespan. In humans, approximately 100 million sperms are produced daily in each testis. Over a lifetime, this adds up to an estimated

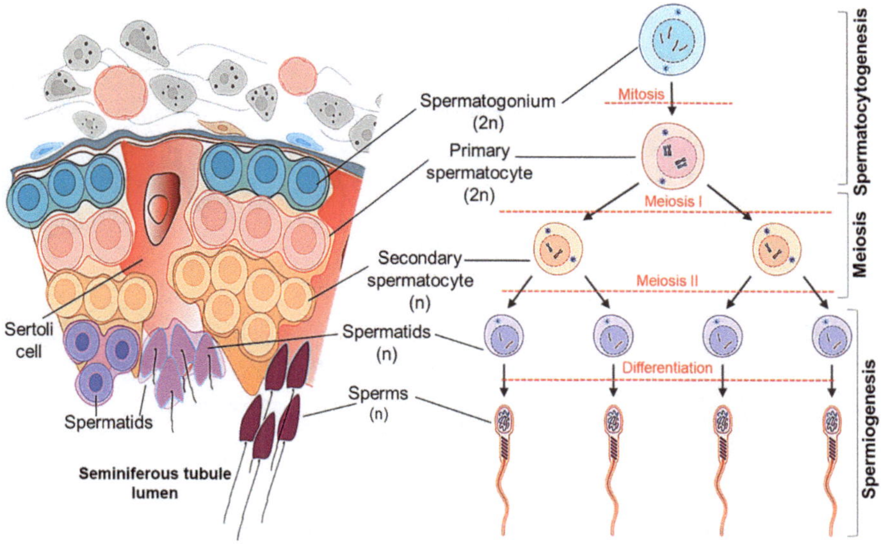

Fig. 9.2 Diagrammatic illustration of spermatogenesis in human male. (The diagrams have been prepared using information derived from (Aladamat and Tadi 2022; Balinsky 1960; Durairajanayagam et al. 2015; Eroschenko 2008; Obukohwo et al. 2021). Created in https://BioRender.com)

total of 10^{12} to 10^{13} sperm, highlighting the remarkable efficiency and sustained productivity of the male reproductive system.

9.4.1.3.1 Spermiogenesis

Spermiogenesis, or spermateleosis, is the final phase of spermatogenesis where haploid spermatids undergo complex morphological and structural transformations to become mature, motile spermatozoa. This process does not involve any further cell division but comprises distinct stages: the Golgi, cap, acrosome, and maturation phases (Creasy and Chapin 2013; Huang and Ho 2006).

In the *Golgi and cap phase* (Fig. 9.3a–d), the Golgi apparatus gives rise to an acrosomal vesicle that begins to position over the nucleus. This is followed by the *cap phase* in which the acrosomal granule spreads over the nucleus's anterior surface, forming a cap-like structure. The nucleus then rotates, positioning the acrosomal cap toward the basement membrane, while the developing flagellum, originating from the distal centriole, orients toward the lumen.

The *acrosome phase* (Fig. 9.3e, f) involves elongation of both the acrosomal granule and condensing and reshaping of the nucleus. The acrosomal cap extends posteriorly, cytoplasm diminishes, and excess cytoplasmic material is shed as residual bodies, subsequently phagocytosed by Sertoli cells.

In the final *maturation phase* (Fig. 9.3g), mitochondria coil helically around the proximal flagellum, forming the midpiece that generates energy for sperm motility.

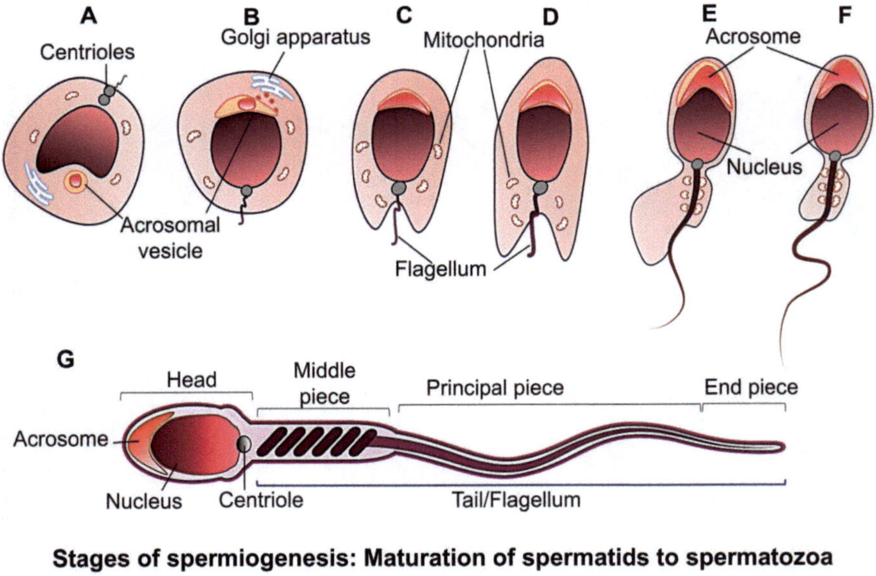

Fig. 9.3 Schematic diagram illustrating spermiogenesis (maturation of spermatids into spermatozoa) in the human male. (The diagrams have been prepared using information derived from (Durairajanayagam et al. 2015; Eroschenko 2008; Gilbert 2000a, b; Obukohwo et al. 2021). Created in https://BioRender.com)

One of the key nuclear events during this phase is the replacement of histones by protamines, whose transcription occurred earlier in the haploid spermatid, resulting in a highly compacted nucleus (Rathke et al. 2014). These orchestrated changes render the spermatozoon structurally specialized, streamlined, and capable of fertilizing the ovum.

9.4.1.4 Morphology and Functional Parts of the Spermatozoon

A spermatozoon (Fig. 9.4a) is a specialized male gamete structurally adapted for efficient motility and fertilization. The majority of its volume is occupied by nuclear material, reflecting its primary function, to deliver the paternal genetic material to the ovum and trigger its activation. Unlike the female gamete (ovum), the sperm lacks cytoplasmic reserves and protective coverings. It is essentially a motile nucleus equipped with a propulsive apparatus powered by mitochondria.

A typical mammalian spermatozoon is about 60 μm long and comprises two main regions: the head and tail (flagellum)

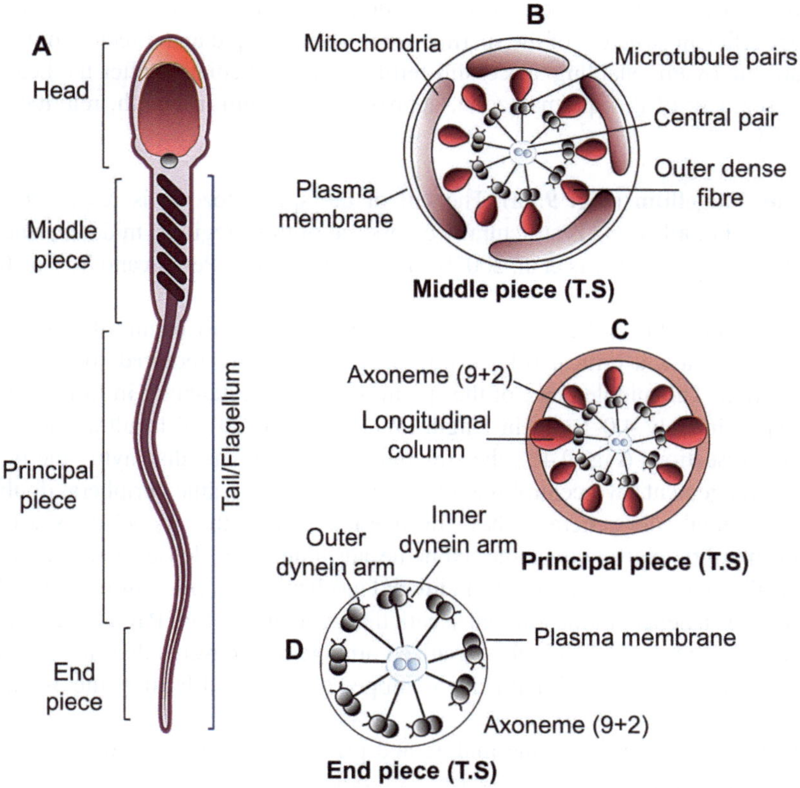

Fig. 9.4 Diagrammatic representation of the structure and functional components of the human spermatozoon. TS - Transverse section. (The diagrams have been prepared using information derived from (Balinsky 1960; Eroschenko 2008; Gilbert 2000a, b). Created in https://BioRender.com)

Head (Fig. 9.4a): The head is typically flattened and oval or spoon-shaped of about 5 μm length and 3–3.5 μm width (Wilding et al. 2011). It comprises two main components, the acrosome and the nucleus (Toshimori and Eddy 2015). The acrosome, a cap-like structure derived from the Golgi apparatus, contains hydrolytic enzymes like hyaluronidase and proteases that enable sperm to penetrate the egg's outer layers during fertilization (Alberts et al. 2002b). It plays a critical role in both the physical penetration of the egg membranes and the activation of the oocyte. The nucleus occupies most of the head and contains densely packed chromatin, representing the paternal genetic material. A shallow depression on the posterior side of the nucleus houses two centrioles, proximal and distal, positioned at right angles to each other. The proximal centriole is essential during fertilization as it initiates the formation of the mitotic spindle required for the zygote's first cleavage division. The distal centriole, aligned with the sperm's longitudinal axis, forms the axoneme, the flagellum's core structure, and functions as the basal body for flagellar assembly. Surrounding the posterior region of the nucleus during spermiogenesis is the manchette, a transient microtubular structure that assists in nuclear shaping and intracellular transport. Though it disappears in mature spermatozoa, it plays an important role in their structural organization during development. This compact, energy-efficient structure of the spermatozoon ensures rapid and directed movement toward the ovum, enabling successful fertilization. Immediately after the head is a small segment of the spermatozoon, approximately 1 μm in length, referred to as the *neck*.

Tail or Flagellum (Fig. 9.4a) The tail of the spermatozoon is responsible for energy and motility and is structurally composed of three regions: middle, principal, and the end piece (Alberts et al. 2002b; Guseva et al. 2024; Pereira and Sousa 2023).

Middle Piece This is the segment of the sperm located just behind the head-neck region and plays a critical role in powering the motility required for successful fertilization. The middle piece of the sperm is roughly rectangular in form, measuring approximately 4.0–5 μm in length (Wilding et al. 2011). In ultrastructural or transverse sections (Fig. 9.4b), the structure is the axoneme, displaying the typical 9 + 2 arrangement, two central microtubules encircled by nine peripheral doublets. This axonemal core generates the whip-like movements that propel the sperm tail during motility. Surrounding the axoneme are nine outer dense fibers, unique to mammalian sperm, which provide additional mechanical support. Surrounding these structures is a helical mitochondrial sheath that generates the ATP required to power flagellar movement. This highly organized arrangement ensures that the spermatozoon remains motile and functionally equipped to reach and fertilize the ovum.

At the junction between the middle and principal tail pieces, a dense ring-like structure known as the annulus or ring centriole is often observed. While it is not structurally identical to a typical centriole, it serves as a morphological boundary between the middle piece and the tail.

Additionally, during the later stages of spermiogenesis, a transient microtubular structure called the manchette forms around the posterior part of the nucleus and extends into the middle piece. It plays a critical role in shaping the nucleus and transporting proteins during sperm maturation. All these internal components are enclosed within the plasma membrane, completing the compact and efficient architecture of the spermatozoon.

Principal and End Piece Principal piece (~45 μm) and the end piece (~5 μm) regions (Fig. 9.4c, d) share the axonemal complex as their core, featuring two central microtubules encircled by nine peripheral doublets, the characteristic 9 + 2 arrangement. In the principal piece, this axial structure is further reinforced by a fibrous sheath that runs longitudinally, providing structural support and flexibility during flagellar movement. The end piece represents the terminal region of the tail and is the most simplified; it contains only the central axoneme surrounded by a thin cytoplasmic layer and plasma membrane. Lacking supportive fibers or sheaths, this tapering end ensures the streamlined shape necessary for efficient propulsion through the female reproductive tract.

The electron microscopic structure of the sperm head, neck, and tail is shown in Figs. 9.5 and 9.6. Additionally, microscopic cross-sectional images of the mammalian testis and ST are provided in Fig. 9.7a–d.

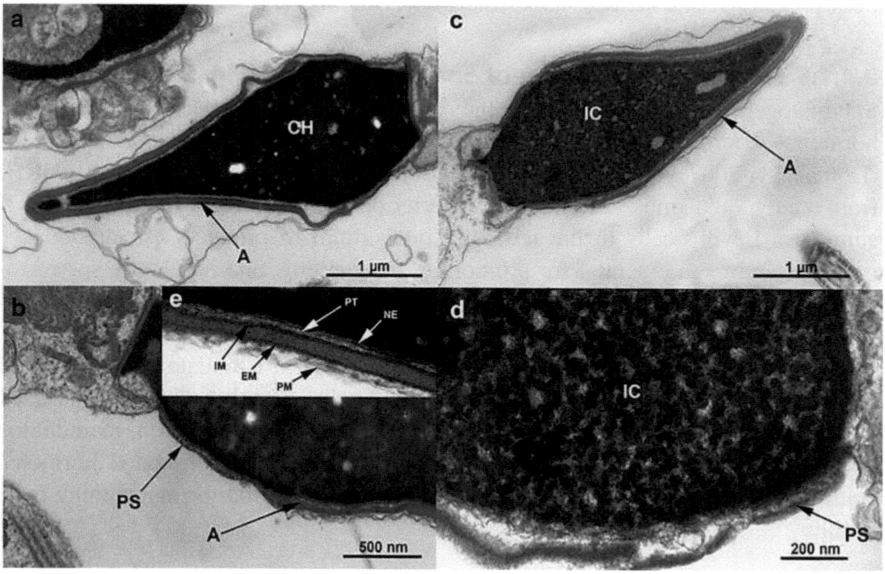

Fig. 9.5 Electron microscopic structure of the sperm head. Spermatozoon with (**a**) condensed chromatin (CH) and (**c**, **d**) immature chromatin (IC). (**b**) Postacrosomal segment of perinuclear theca (PS). (**e**) Sperm acrosome fragment. NE - nuclear envelope, PT - perinuclear theca, IM - inner acrosome membrane, EM - extra acrosome membrane, PM - plasma membrane. (The image is adapted from (Bragina and Bocharova 2018) and is licensed under Creative Commons)

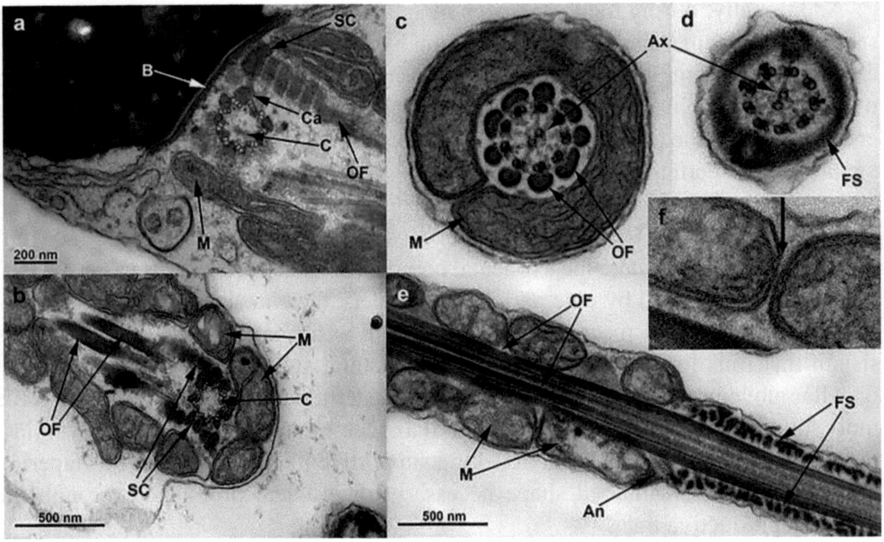

Fig. 9.6 Electron microscopic structure of the sperm neck and tail. Spermatozoon (**a**) neck, (**b**) decapitated, (**c**) tail mid-piece (transverse section), (**d**) tail principal piece (transverse section), (**e**) tail middle and principal piece (longitudinal section), (**f**) mitochondria contact site (arrow). B - basal plate, C - centriole, Ca - capitulum, SC - striated column, OF - outer dense fibers, M - mitochondria, Ax - axoneme, FS - fibrous sheath, An - annulus. (The image is adapted from (Bragina and Bocharova 2018) and is licensed under Creative Commons)

9.4.1.5 Hormonal Regulation of Spermatogenesis (Fig. 9.8)

Spermatogenesis, occurring within the ST of the testis, is precisely regulated by hormonal coordination among the hypothalamus, pituitary gland, and testis (Domanskyi 2007). Pulsatile secretion of gonadotropin-releasing hormone (GnRH) from the hypothalamus stimulates the anterior pituitary to release LH and FSH. LH stimulates Leydig cells in the testicular interstitium to produce testosterone, the principal androgen essential for promoting, maintaining, and regulating spermatogenesis. FSH acts on Sertoli cells in the ST, enhancing germ cell development by stimulating androgen-binding protein (ABP) synthesis and other supportive factors. Testosterone, accumulated by ABP, works with FSH to stimulate spermatogonial maturation into functional spermatozoa. Additionally, Sertoli cells secrete inhibin, which negatively regulates FSH secretion through feedback inhibition, maintaining hormonal balance and modulating spermatogenesis. This coordinated hormonal control ensures constant production of viable sperm essential for maintaining male fertility.

9.4.1.6 Sperm Defects (Figs. 9.9 and 9.10)

Sperm defects, also known as teratozoospermia, refers to abnormalities in the shape, structure, or function of sperm cells that can impair male fertility. These defects may arise during spermatogenesis due to genetic factors, infections, exposure to toxins, hormonal imbalances, or lifestyle-related causes such as smoking and excessive

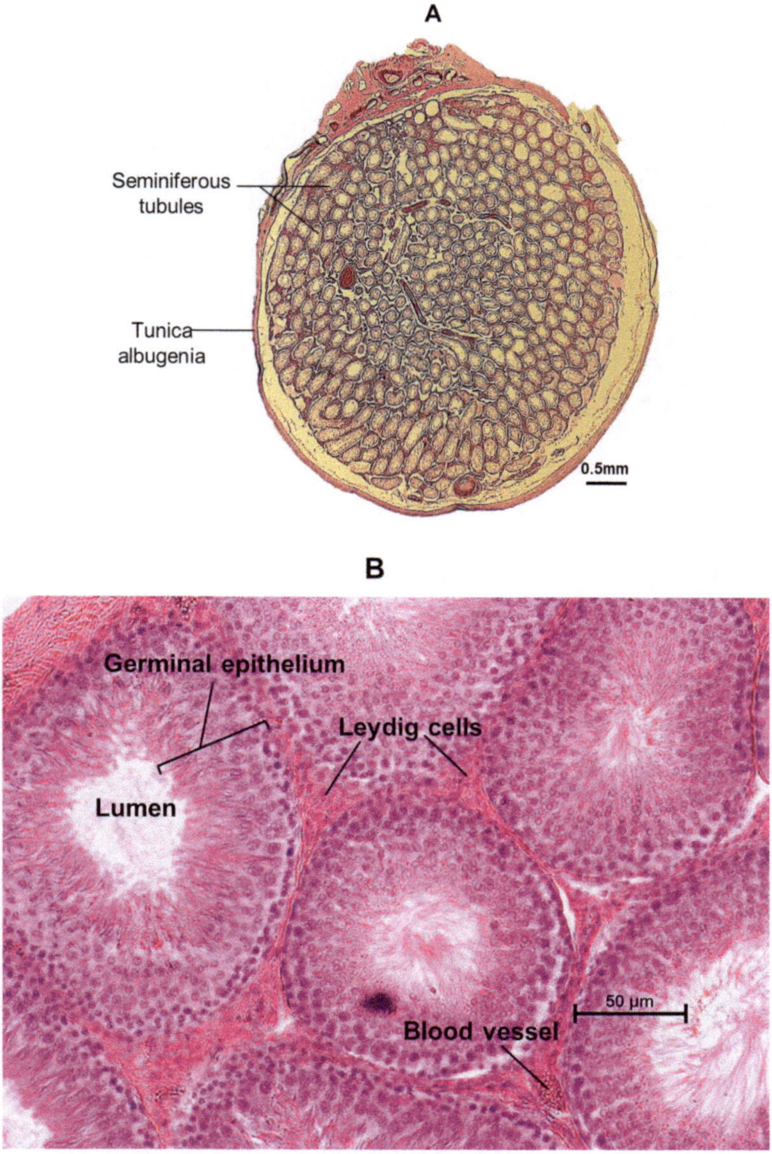

Fig. 9.7 Microscopic cross-sectional images of the mammalian testis and seminiferous tubules (ST). (**a**) Cross-section of the whole testis tissue, (**b–d**) Cross-section of ST. (Source: Zoology Department, Daulat Ram College, University of Delhi, INDIA)

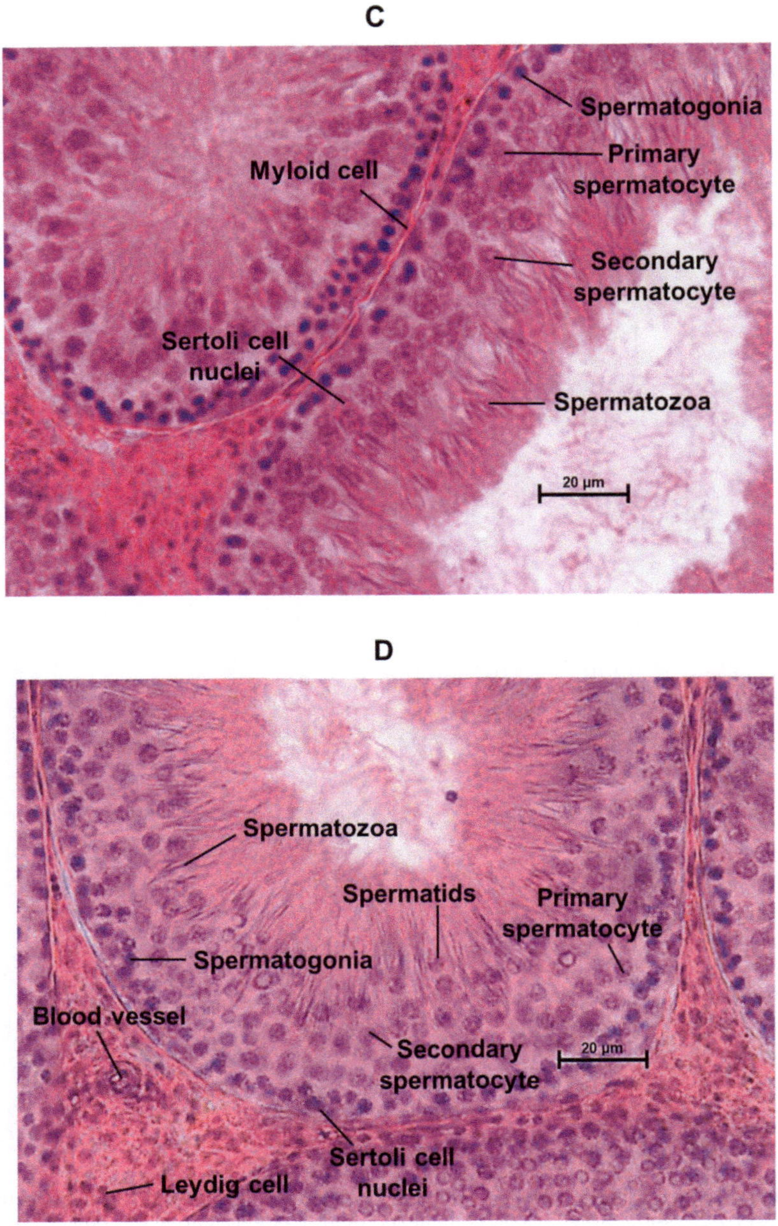

Fig. 9.7 (continued)

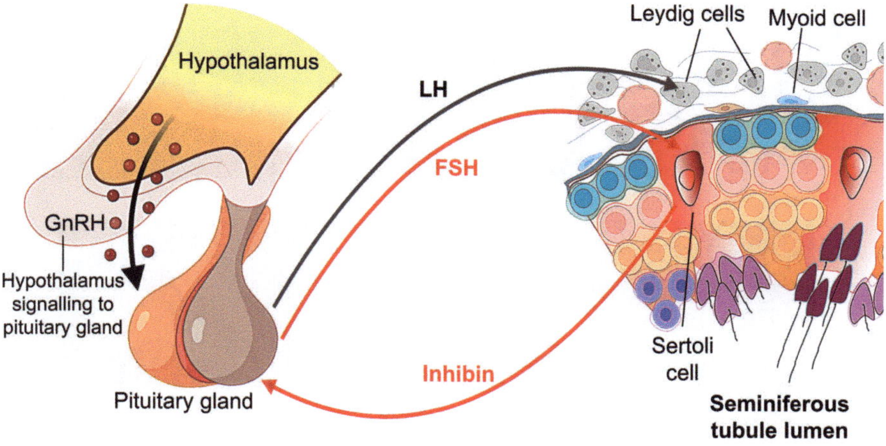

Fig. 9.8 Schematic representation of the hormonal regulation of male gamete formation (spermatogenesis). FSH - Follicle-stimulating hormone, GnRH - Gonadotropin-releasing hormone, LH - Luteinizing hormone. (The diagrams have been prepared using information derived from (Balinsky 1960; Eroschenko 2008; Gilbert 2000a, b). Created in https://BioRender.com)

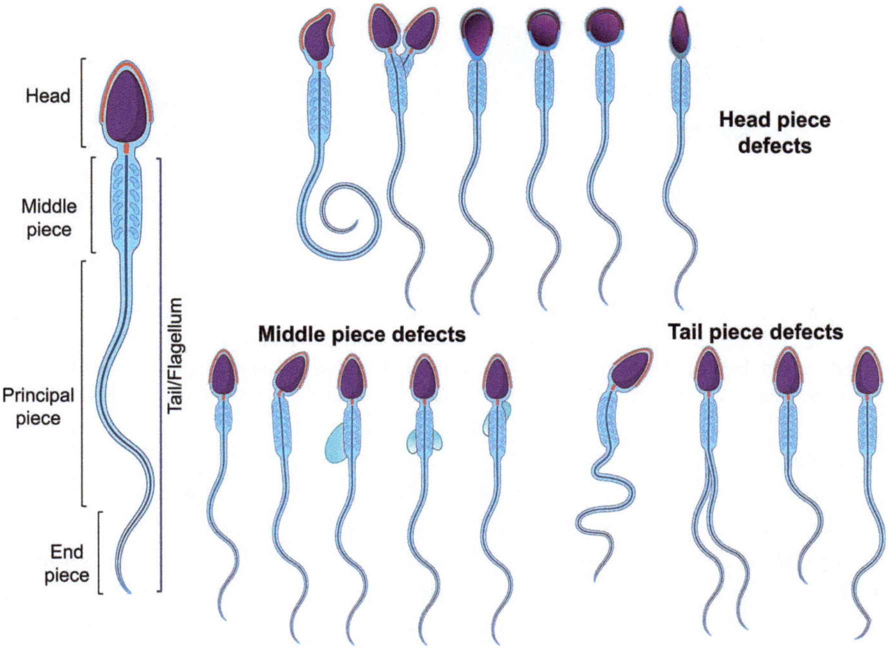

Fig. 9.9 Diagram representing types of defects that can be seen in male gametes or spermatozoa. (The diagrams have been prepared using information derived from (Itoi et al. 2022; Nassir et al. 2022; Rengan et al. 2012). Created in https://BioRender.com)

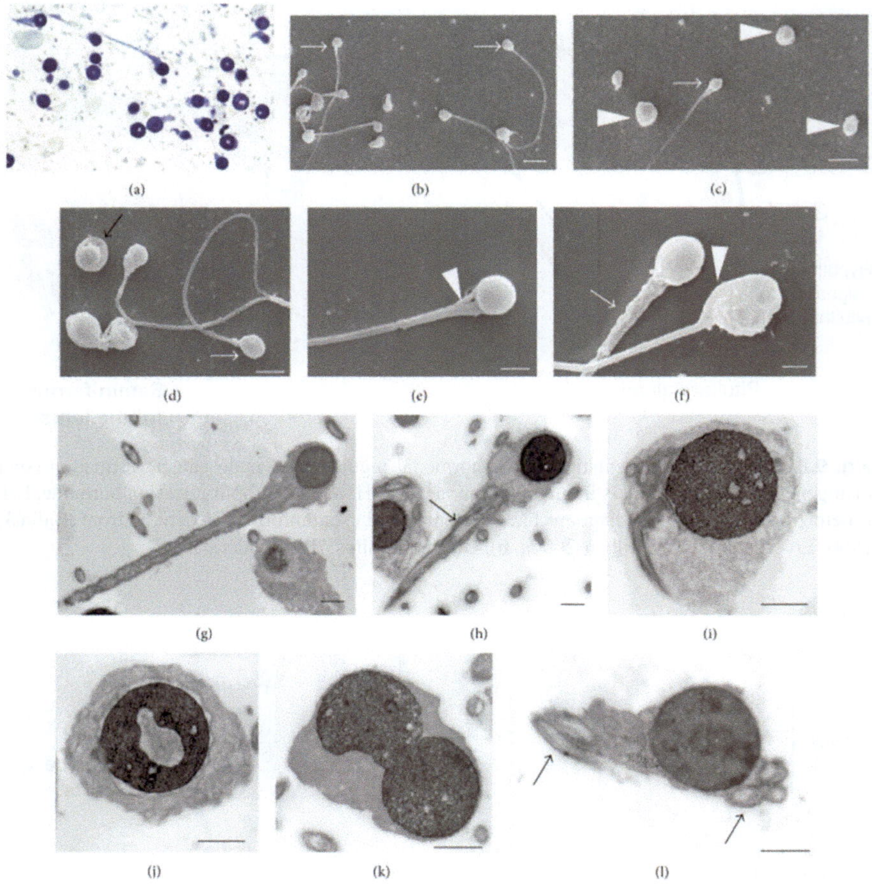

Fig. 9.10 Optical microscopy and ultrastructural images of human spermatozoa with round heads. (**a–d**) With and without tail; (**e**) conical neck; (**f**) thick midpiece with cord-like structure near the head; (**g**) rounded head and nuclei; (**h**) anomalous axoneme; (**i**) bent tail; (**j**) detached nuclear envelope with lacunar chromatin defect; (**k**) binucleated head; (**l**) axoneme sections enclosed within the same head. (Adapted from (Ricci et al. 2015). Image is licensed under Creative Commons)

heat. Sperm defects are typically categorized into three main types: head defects, midpiece defects, and tail defects. *Head defects* include large, small, tapered, double, or amorphous heads, often associated with impaired DNA integrity or fertilization ability. *Midpiece defects* involve irregularities such as bent or swollen midpieces, which can disrupt the energy supply needed for motility. *Principal and end piece defects* may present as coiled, short, multiple, or absent tails, leading to poor sperm motility. In some cases, multiple abnormalities may occur simultaneously, a condition referred to as multiple morphological abnormalities of the sperm flagella (MMAF) (Wang et al. 2020). The presence of a high percentage of defective sperm can significantly reduce the chances of natural conception, making morphological analysis an important part of fertility assessments.

9.4.2 Ovary

The ovary (Fig. 9.11), the primary female reproductive gland, is oval-shaped and situated on the pelvic wall near the anterior aspect of the uterus. Each ovary is attached posteriorly by a mesentery known as the *mesovarium*, a fold of the peritoneum through which blood vessels, lymphatics, and nerves enter the ovary via a stalk-like structure called the hilus. Functionally, the ovary serves two main roles: it is the site of oogenesis (the formation of ova) and hormone secretion (estrogen and progesterone).

Structurally, the ovary consists of two main regions: an inner medulla and an outer cortex. The ovarian cortex, covered by the germinal epithelium, contains follicles at various developmental stages and is responsible for oocyte growth and maturation. Additionally, the cortex also contains corpora lutea and their degenerating forms known as *corpus albicans/atretic follicles*. On the other hand, the medulla, the central region of the ovary, consists mainly of loose fibroelastic connective tissue, smooth muscle fibers, nerves, blood vessels, and lymphatics that support its metabolic and endocrine functions. The boundary between the cortex and medulla is indistinct, with the cortical region interrupted at the ovarian hilus. Externally, the ovary is enclosed by the tunica albuginea, a dense fibrous connective tissue capsule that provides structural support and protection.

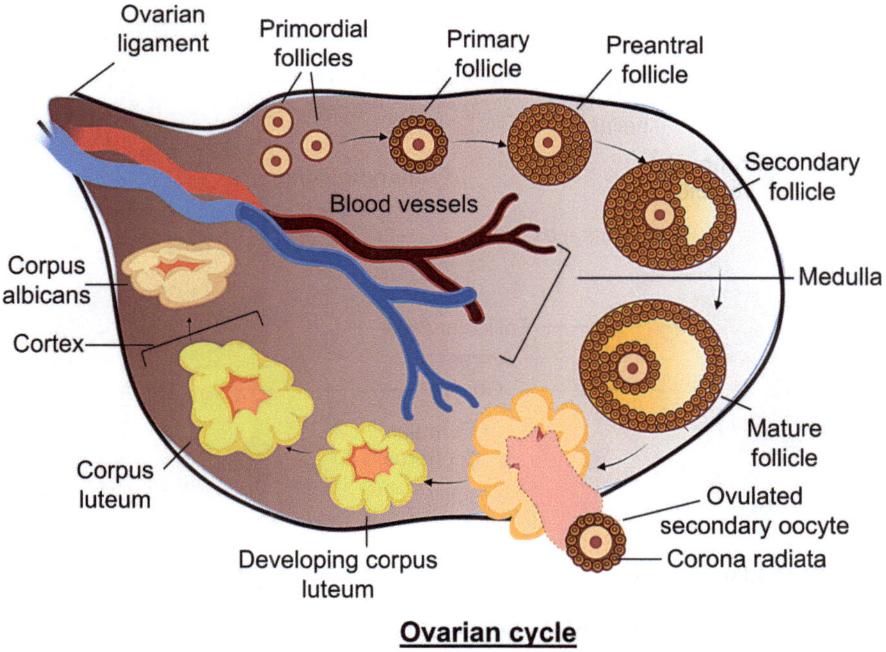

Fig. 9.11 Schematic representation of the ovary, ovarian cycle, and follicles. (The diagrams have been prepared using information derived from (Balinsky 1960; Eroschenko 2008; Gilbert 2000a, b; Jinno 2025; Williams and Erickson 2012). Created in https://BioRender.com)

9.4.2.1 Ovarian Cycle (Menstrual Cycle)

The ovarian cycle, a vital component of female reproduction, comprises three phases: follicular (first day of menstruation to ovulation, lasting about 14 days—5 days of menstruation and 9 days of follicular maturation), ovulatory (release of egg by ovary at about day 14), and luteal (ovulation to beginning of menstruation, lasting about 14 days) (Kirkendoll and Bacha 2023). During this cycle, ovarian follicles develop, and one mature oocyte is typically released into the uterine tube for possible fertilization. The cycle generally spans about 28 days on an average.

Development of Ovarian Follicles (Follicular Stage) (Fig. 9.12) In the earliest stage, the oocyte is surrounded by a single layer of flattened follicular (epithelial) cells, forming the primary follicle. As the follicle grows, these cells proliferate, forming multiple layers around the ovum. As development progresses, secretory activity within the follicular cells leads to the formation of small vacuoles (fluid-filled spaces) among them. These vacuoles gradually merge to form a large central cavity called the antrum, marking the transition to the antral follicle stage, which includes the secondary and mature (Graafian) follicles. The Graafian follicle is a vascularized structure approximately 5 mm in diameter. Its wall is composed of two connective tissue layers: an inner theca interna (rich in blood vessels and hormone-

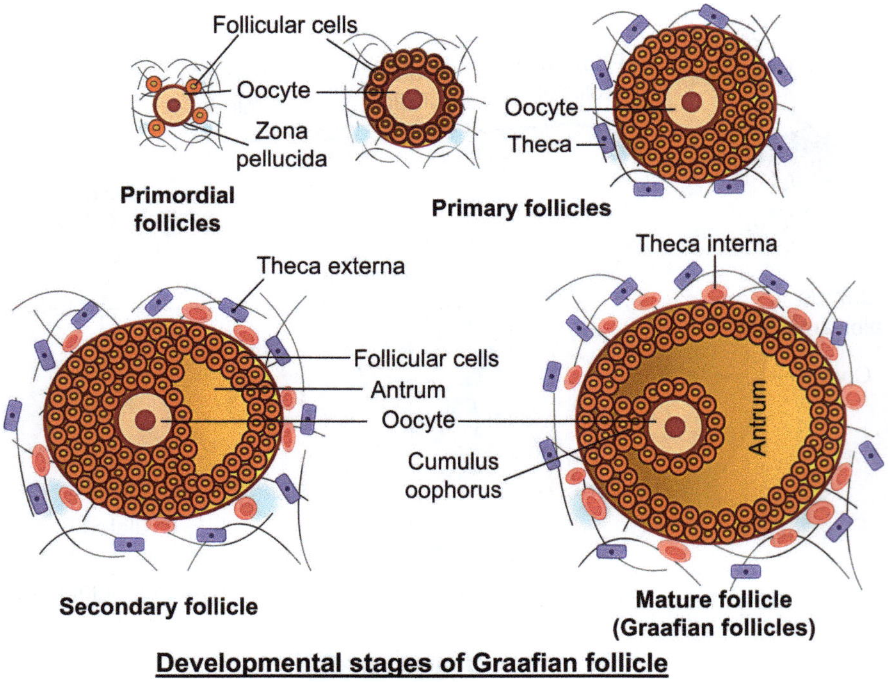

Fig. 9.12 Diagrammatic representation of the development of ovarian follicles. (The diagrams have been prepared using information derived from (Balinsky 1960; Eroschenko 2008; Gilbert 2000a, b; Jinno 2025; Williams and Erickson 2012). Created in https://BioRender.com)

secreting cells) and an outer theca externa (fibrous support tissue). These connective tissue layers are parted from the inner follicular (granulosa) cells by a distinct basement membrane. The antrum is lined by the membrana granulosa, a multilayered epithelium of granulosa cells. A specialized cluster of these cells, the *cumulus oophorus*, extends into the antrum and encloses the ovum. The ovum itself is surrounded by the zona pellucida, a thick glycoprotein-rich layer, which is further encircled by a single or multilayered ring of granulosa cells, the *corona radiata*.

Ovulation Ovulation is the physiological process through which a mature ovum is released from a fully developed Graafian follicle. As the follicle enlarges, it protrudes outward, creating a noticeable bulge on the ovarian surface. The ovarian wall at this point becomes stretched and thin. With continued secretion of follicular fluid, intra-follicular pressure rises, eventually causing the follicle to rupture. Following follicular rupture, the ovum is released and guided into the fallopian tube's opening (ostium) by the coordinated movement of cilia lining its inner surface. It then travels through the tube, where fertilization by sperm may occur.

In lower mammals, ovulation typically occurs spontaneously during the period of sexual excitement, known as the *estrous cycle* or "*heat*," during which the female is sexually receptive. These animals are referred to as *spontaneous ovulators*. In contrast, in species like the rabbit, ovulation is induced and occurs only following copulation or sexual stimulation. In humans and monkeys, ovulation occurs periodically, approximately every 28 days, and is spontaneous, not associated with a distinct period of sexual receptivity or heat. In human females, ovulation begins at puberty and continues until menopause.

Corpus Luteum (Body Yellow) Following ovulation, the ruptured Graafian follicle collapses and differentiates into a yellow, hormone-producing structure, the *corpus luteum*. This temporary endocrine gland regulates the reproductive cycle and, if pregnancy ensues, enlarges and remains active to sustain hormone secretion. The corpus luteum consists of granulosa and thecal cells supported by blood vessels. Granulosa cells release large quantities of progesterone and a moderate amount of estradiol; both the hormones are crucial for sustaining the uterine lining and ensuring early pregnancy maintenance (Jones and Lopez 2014; Kirkendoll and Bacha 2023). The thecal cells synthesize androstenedione (Kirkendoll and Bacha 2023). If fertilization does not take place, the corpus luteum regresses after a brief functional phase, forming the corpus luteum of menstruation (or ovulation) of about 2 cm diameter (Jones and Lopez 2014). Its degeneration reduces hormone secretion, ultimately triggering menstruation. Histologically, a newly formed corpus luteum appears as a compact nodule composed predominantly of dense connective tissue within the ovarian cortex (Kirkendoll and Bacha 2023). Ultimately, the regressed corpus luteum is replaced by a small, whitish, fibrous scar called the *corpus albicans*, which persists as a residual structure within the ovarian cortex.

Corpus Albicans The corpus albicans represents the terminal stage in the ovarian follicle's life cycle, forming after the degeneration of the corpus luteum in the absence of fertilization. During this process, lutein cells lose their structure and function, and the glandular tissue is progressively replaced by dense fibrous connective tissue. This results in the formation of a small, white, scar-like structure, the corpus albicans (a term meaning "white body" in Latin) (Jones and Lopez 2014). It remains embedded within the ovarian cortex as a histological remnant of the ovulatory cycle. Over time, multiple corpus albicantia may accumulate in the ovary, especially in post-reproductive age, serving as markers of past ovulations (Hariri et al. 2009).

In species such as humans, cows, monkeys, and elephants, typically one Graafian follicle matures during each reproductive cycle. However, in other animals such as rats, bats, and cats, multiple follicles may mature and ovulate simultaneously.

9.4.2.2 Oogenesis

Oogenesis (Fig. 9.13) is the biological process responsible for the formation of female gametes, or ova (oocyte), within the ovaries (Krajnik et al. 2023). This process starts during fetal development and persists until menopause. In the

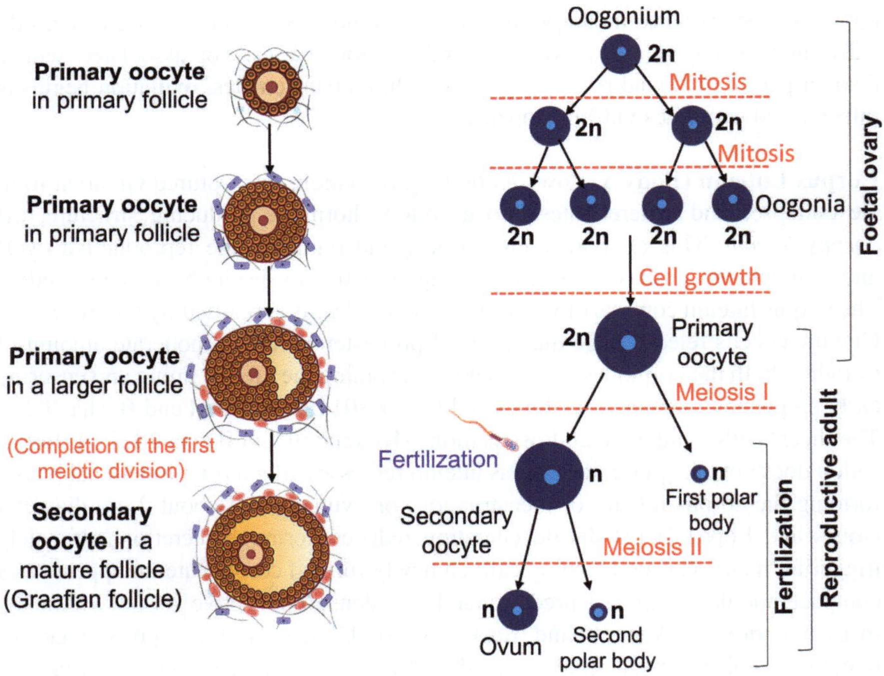

Fig. 9.13 Diagram showing the process of oogenesis. (The diagrams have been prepared using information derived from (Balinsky 1960; Eroschenko 2008; Gilbert 2000a, b; Jinno 2025; Williams and Erickson 2012). Created in https://BioRender.com)

developing fetal ovary, primordial germ cells transform into diploid oogonia (2n). These oogonia proliferate through mitotic divisions and subsequently initiate meiosis I, transforming into primary oocytes. However, these cells pause their development at the diplotene stage of prophase I and remain in this dormant state until puberty. Each primary oocyte is enclosed within a primordial follicle, and no additional oogonia are formed after birth. Studies indicate that a female is born with approximately one to two million oocytes, but by the onset of puberty, this number declines to around 400,000 (Yatsenko and Rajkovic 2019).

During each menstrual cycle, FSH stimulates a few primary oocytes to resume meiosis I, with one follicle eventually becoming dominant and releasing an ovum during the ovulatory phase (Krajnik et al. 2023). The meiosis I division results in the formation of a secondary oocyte and a smaller first polar body, which typically degenerates. The secondary oocyte then initiates meiosis II but pauses at the metaphase II stage. It is released during ovulation and completes meiosis II only if fertilization takes place, yielding a mature ovum and a second polar body (Alberts et al. 2002a).

Oogenesis produces a single, large ovum rich in nutrients, along with small polar bodies that degenerate. This asymmetrical division preserves cytoplasmic contents such as enzymes, mRNAs, metabolic substrates, and organelles within the ovum, providing essential materials to sustain early stages of embryonic growth and development (Gilbert 2000a, b).

9.4.2.3 Egg/Ovum

The ovum (Fig. 9.14), or egg cell, represents the mature female gamete in humans, uniquely adapted for the process of sexual reproduction. It is among the largest human cells, visible without magnification, and is haploid (n), possessing 23 chromosomes. The ovum has a spherical form and is enclosed by several protective layers. Surrounding it is the zona pellucida, a dense glycoprotein coat that facilitates sperm binding and recognition. Encircling the zona pellucida is the corona radiata, a sheath of specialized granulosa cells (follicular cells) that nourishes

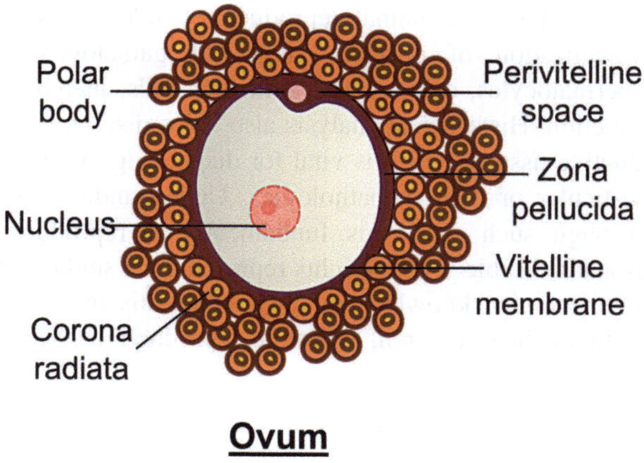

Fig. 9.14 Diagram of Ovum. (Created in https://BioRender.com)

and protects the ovum (Williams and Erickson 2012). The ooplasm (egg cytoplasm) is abundant in nutrients, organelles, and maternal mRNA vital for supporting early embryonic growth and development. The centrally placed nucleus holds the female genetic material and completes the second meiotic division only upon fertilization. Following ovulation, the ovum remains viable for approximately up to 24 h (Oliver and Basit 2019); if fertilization does not occur, it degenerates and is subsequently shed during menstruation.

The electron microscopic structure of the human ovarian tissue is shown in Fig. 9.15. Additionally, microscopic sectional images of the mammalian ovary and follicles are provided in Fig. 9.16.

9.4.2.4 Hormonal Regulation of Ovarian Cycle

Oogenesis in humans is controlled by a coordinated network of hormones originating from the hypothalamus, pituitary, and ovaries. The hypothalamus releases GnRH in rhythmic pulses, stimulating the anterior pituitary to produce FSH and LH (Fig. 9.17). FSH facilitates the development and maturation of ovarian follicles that enclose the primary oocytes. As the follicles mature, granulosa cells begin producing estrogen, which stimulates the proliferation of the endometrial lining while providing negative feedback to the hypothalamus and pituitary, regulating the secretion of FSH and LH (Thiyagarajan et al. 2024). A sharp rise in LH levels, stimulated by the peak in estrogen secretion, triggers ovulation, leading to the release of the secondary oocyte from the mature Graafian follicle (Kirkendoll and Bacha 2023). Following ovulation, the remaining follicular structure transforms into the corpus luteum, which produces progesterone and small amounts of estrogen to sustain the uterine lining for potential implantation (Thiyagarajan et al. 2024). In the absence of fertilization, the corpus luteum regresses, hormone levels decline, and menstruation begins, signaling the conclusion of one cycle and the onset of another.

9.5 Significance/Research Applications

The study of histological sections of the testis and ovary provides critical insights into human and animal reproductive biology. Examining these sections allows visualization of different stages of gametogenesis, such as spermatogonia, spermatocytes, oocytes, and follicular cells, thereby clarifying their structure and function. Histological analysis also helps distinguish normal from abnormal reproductive tissues, which is vital for diagnosing infertility, hormonal imbalances, and testicular or ovarian pathologies. Understanding spermatogenesis and oogenesis through such studies is fundamental to reproductive biology and biomedical research. Table 9.1 highlights representative studies demonstrating the applications of histological knowledge of the human testis, ovary, and gametes (sperm and ovum) in biomedical research, as well as reproductive toxicology.

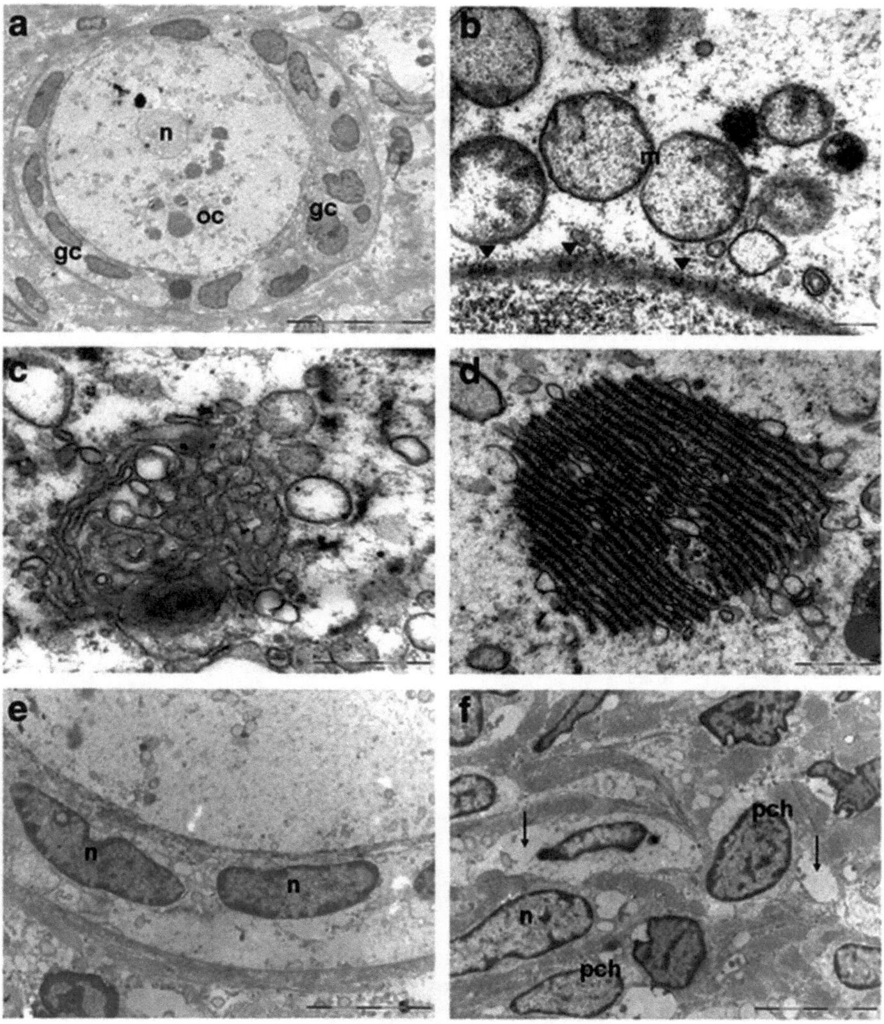

Fig. 9.15 Electron microscopic structure of human ovarian tissue. (**a**) Primordial/primary follicle oocyte showing nucleus (n) with chromatin and cytoplasm (oc) containing organelles; (**b**) perinuclear assemblies of round mitochondria (m) and moderate electron dense matrices (arrow head) nuclear pores; (**c**) Golgi apparatus; (**d**) annulate lamellae; (**e**) granulosa cells with oval nuclei (n); (**f**) stromal cells with oval nuclei (n). gc - granulosa cells, m - mitochondria (m), n - nuclei, pch - peripherally condensed chromatin, *down arrow* empty interstitial areas. (The image is adapted from (Fabbri et al. 2016) and licensed under Creative Commons)

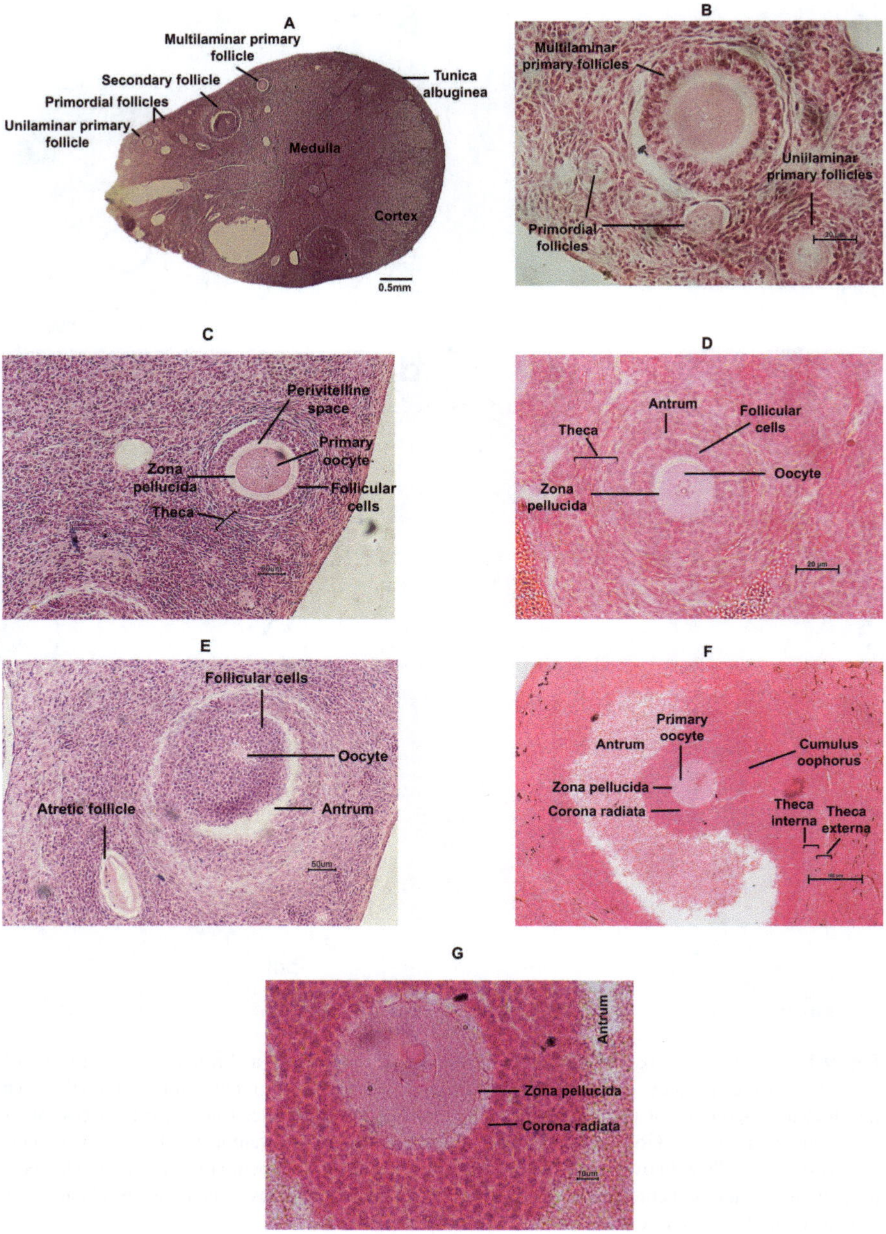

Fig. 9.16 Microscopic images of the mammalian ovary and follicles. (**a**) Ovary with follicles. (**b**) Ovarian follicles. (**c**) Late primary follicle. (**d**) Secondary follicle. (**e**) Secondary follicle. (**f**) Mature follicle. (**g**) Oocyte. (Source: Zoology Department, Daulat Ram College, University of Delhi, INDIA)

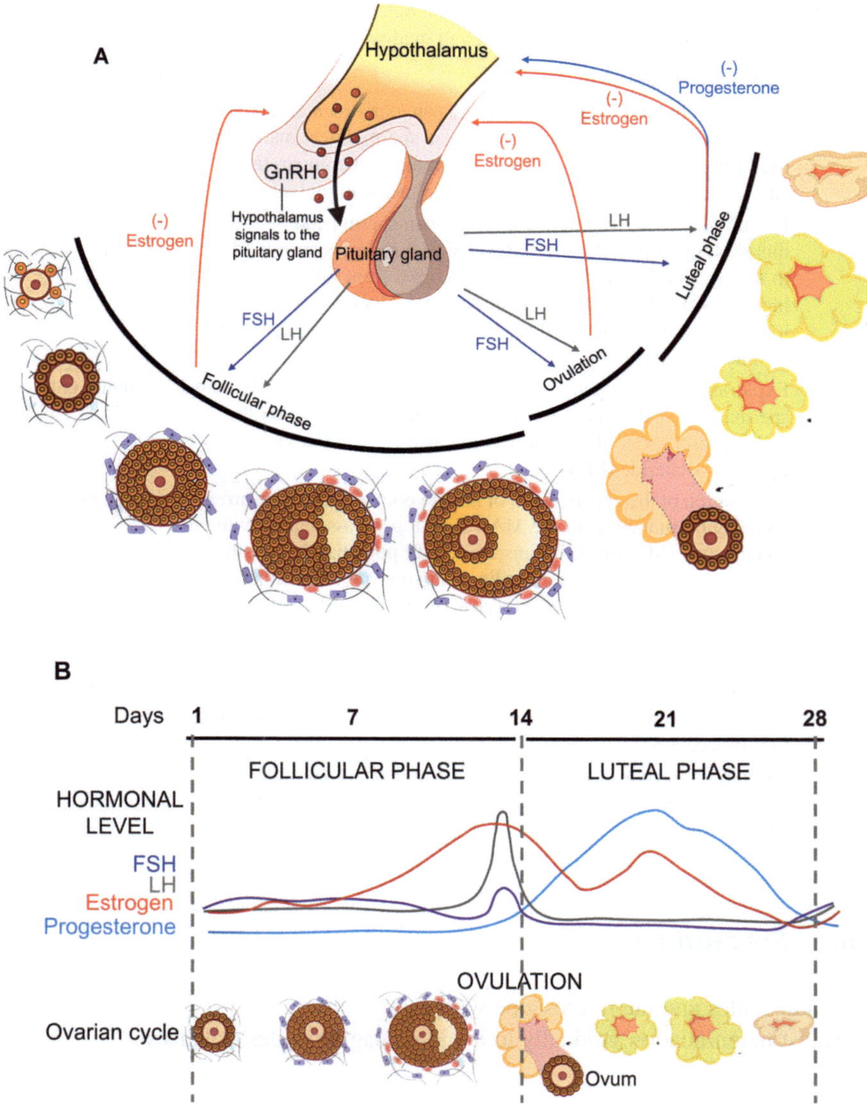

Fig. 9.17 Diagrammatic representation of hormonal regulation of ovarian cycle. (The diagrams have been prepared using information derived from (Balinsky 1960; Eroschenko 2008; Gilbert 2000a, b; Jinno 2025; Williams and Erickson 2012). Created in https://BioRender.com)

Table 9.1 Representative studies highlighting the outcomes of gamete research and its contribution to improving human reproductive health and quality

S. no.	Type of study/research	Reference
1.	Evaluation of GOPC knockout mice as an experimental model for investigating human globozoospermia, a sperm defect marked by round-headed, acrosome-lacking sperm, through comparative analysis of semen samples from affected men and mutant mice	Ricci et al. (2015)
2.	Examining the association between human sperm head morphology, especially the acrosomal region, and sperm binding to the zona pellucida. Using semen samples from infertile men, it highlights the diagnostic value of analyzing sperm–zona interactions, which are vital for successful fertilization and male fertility assessment	Garrett et al. (1997)
3.	Exploration of the function of toll-like receptor 4 (TLR4) in human spermatozoa and its involvement in infertility, focusing on the cellular response to the gram-negative bacterial endotoxin lipopolysaccharide (LPS)	Sahnoun et al. (2017)
4.	Exploration of the genetic basis of human oocyte maturation arrest, focusing on mutations in the MAD2L1BP gene associated with metaphase I (MI) arrest. It aims to identify potential therapeutic targets for primary female infertility using oocytes from both fertile donors and infertile women experiencing recurrent IVF/ICSI failure, obtained with informed consent	Huang et al. (2023)
5.	Investigates abnormal cortical contractions in mammalian oocytes as potential indicators of oocyte quality and female fertility outcomes, validated using surplus human oocytes obtained from IVF procedures	Nikalayevich et al. (2024)
6.	Examining the correlation between sperm PAH-DNA adduct levels and semen quality in Han Chinese men seeking evaluation for infertility	Ji et al. (2013)

9.6 Precautions

- Handle slides and the microscope with care.
- Adjust magnification gradually to avoid damaging slides or lenses.

9.7 Quick Review Questions

9.7.1 Multiple Choice Questions

1. Which of the cells in the testis support and nourish the developing spermatogenic cells?
   (a) Sertoli cells
   (b) Leydig cells
   (c) Spermatogonia
   (d) Spermatids
   Answer: a) Sertoli cells

2. What is ovulation?
   (a) Fertilization of egg by a sperm cell
   (b) Shedding of the uterine lining during menstruation
   (c) Release of a mature oocyte from the ovary
   (d) Development of follicles in the ovary
   Answer: c) Release of a mature oocyte from the ovary
3. What is a Graafian follicle?
   (a) A degenerated follicle after ovulation
   (b) An immature follicle containing a primary oocyte
   (c) A fully matured ovarian follicle ready to release an ovum
   (d) A structure formed after fertilization of the ovum
   Answer: c) A fully matured ovarian follicle ready to release an ovum
4. Leydig cells are primarily responsible for:
   (a) Formation of the blood-testis barrier
   (b) Secretion of testosterone
   (c) Phagocytosis of residual bodies
   (d) Formation of spermatozoa
   Answer: B) Secretion of testosterone
5. Which of the following correctly represents the sequence of spermatogenic cell development?
   (a) Spermatid → Spermatocyte → Spermatogonia → Spermatozoa
   (b) Spermatogonia → Spermatocyte → Spermatid → Spermatozoa
   (c) Spermatocyte → Spermatogonia → Spermatid → Spermatozoa
   (d) Spermatozoa → Spermatid → Spermatocyte → Spermatogonia.
   Answer: B) Spermatogonia → Spermatocyte → Spermatid → Spermatozoa
6. Which cells in the ovary secretes estrogen?
   (a) Granulosa
   (b) Theca interna
   (c) Oocytes
   (d) Corpus luteum
   Answer: A) Granulosa cells
7. The primary oocyte in humans is arrested at which stage of meiosis at birth?
   (a) Prophase I
   (b) Metaphase I
   (c) Anaphase I
   (d) Metaphase II
   Answer: A) Prophase I
8. The corpus luteum is formed from:
   (a) Primordial follicles
   (b) Graafian follicle after ovulation
   (c) Primary oocyte
   (d) Secondary oocyte
   Answer: B) Graafian follicle after ovulation

9. In histological sections of testis, the lumen of seminiferous tubules contains:
   (a) Spermatogonia
   (b) Spermatocytes
   (c) Spermatids and spermatozoa
   (d) Sertoli cell nuclei
   Answer: C) Spermatids and spermatozoa
10. Which of the following features distinguishes primary follicles from secondary follicles in the ovary?
   (a) Presence of zona pellucida
   (b) Presence of antrum
   (c) Multiple layers of granulosa cells
   (d) Presence of oocyte
   Answer: c) Multiple layers of granulosa cells

9.7.2 Short Answer Questions

11. Define the term "gamete."
12. Name the cell types found in the seminiferous tubules of the testis.
13. What is the function of Sertoli cells in the testis?
14. Identify the site of testosterone production in the testis.
15. Explain the role of Leydig cells in male reproductive physiology.
16. Define oogenesis and its primary stages in the human ovary.
17. How does a primary ovarian follicle differ from a secondary ovarian follicle?
18. Describe the histological features of a Graafian follicle.
19. What is the corpus luteum, and what is its function in the ovary?
20. What structures in the ovary are responsible for hormone secretion?

Glossary

Acrosomal vesicle A Golgi-derived vesicle in developing sperm that forms the acrosome, containing enzymes essential for egg penetration during fertilization.

Acrosome A cap-like structure on the sperm head containing enzymes that enable penetration of the ovum's outer layers during fertilization.

Androgen Steroid hormones like testosterone that regulate male reproductive functions and development of secondary sexual characteristics.

Antrum Fluid-filled cavity in a growing ovarian follicle supplying nutrients and hormonal support to the oocyte.

Atretic follicles Degenerating ovarian follicles that fail to mature and undergo regression.

Axoneme It is the central core of a cilium or flagellum, composed of microtubules arranged typically in a "9 + 2" pattern, which provides structural support and enables movement.

Centriole A cylindrical microtubule-based organelle that helps in spindle formation during cell division and in the development of cilia and flagella.

Corpus luteum A temporary endocrine gland formed post-ovulation from the ruptured follicle, secreting progesterone and estrogen to sustain the uterine lining for pregnancy.

Corpus albicans A whitish, fibrous structure in the ovary formed after corpus luteum degeneration when pregnancy fails.

Corana radiata The innermost granulosa cell layer surrounding the oocyte, providing nourishment and protection after ovulation.

Cumulus oophorus A cluster of granulosa cells anchoring the oocyte to the follicle wall, offering structural support, nutrients, and signaling cues for oocyte maturation and release.

Dynein arm A motor protein complex on the outer microtubule doublets of cilia and flagella that hydrolyzes ATP to generate bending motion for motility.

Epididymis A highly coiled duct on the testis where sperm mature, are stored, and transported to the vas deferens.

Estrogen A primary female sex hormone produced by the ovaries that regulates reproductive cycles, secondary sexual traits, and uterine lining development.

Fertilization The process where a male gamete (sperm) fuses with a female gamete (ovum) to form a zygote, initiating embryonic development.

Fibroblast A connective tissue cell that produces collagen and extracellular matrix, aiding in tissue structure and repair.

Flagella *Long, whip-like structures found on certain cells* (e.g., *sperm) that provide motility to cells.*

Follicular cells Epithelial cells encasing the oocyte within an ovarian follicle, supplying nutrients, hormonal regulation, and structural support during oocyte growth and maturation.

Follicular phase The initial stage of the ovarian cycle during which FSH stimulates follicular growth, leading to estrogen production, and oocyte maturation, preparing the ovum for eventual ovulation.

Gonadotropin-releasing hormone (GnRH) *Hypothalamic hormone that stimulates the anterior pituitary gland to release reproduction-regulating hormones,* i.e., *luteinizing hormone and follicle-stimulating hormone.*

GOPC knockout mice Male mice lacking the Golgi-associated PDZ- and coiled-coil motif-containing protein (GOPC) gene, displaying infertility and globozoospermia with round-headed sperm and tails coiled around the nucleus.

Graafian follicles A mature ovarian follicle characterized by a large antrum, a fully developed oocyte, and surrounding layers including the granulosa cells, theca cells, and cumulus oophorus.

Hypothalamus A small region of brain below the thalamus that regulates vital physiological processes such as hormone secretion, autonomic functions, and homeostasis through the pituitary gland.

Intracytoplasmic sperm injection (ICSI) A fertility technique involving the injection of single sperm into the cytoplasm of a mature egg to achieve fertilization.

Idiosome Cell body of a spermatozoon, containing the nucleus and cytoplasm, excluding the tail (flagellum).

Inhibin A gonadal hormone that suppresses FSH production.

In vitro *fertilization (IVF)* External fertilization of egg and sperm in a laboratory dish, followed by embryo transfer into the uterus.

Leydig cells Testicular interstitial cells that secrete testosterone, the principal male hormone essential for spermatogenesis and secondary sexual characteristics development.

Luteinizing hormone Secreted by the anterior pituitary, it induces ovulation and corpus luteum formation in females and stimulates testosterone production in male Leydig cells.

Luteal phase Post-ovulation phase where the corpus luteum forms and secretes progesterone, preparing the endometrium for possible implantation of a fertilized egg.

Manchette A temporary microtubular structure surrounding the spermatid nucleus, crucial for shaping the sperm head during the process of spermiogenesis.

Mediastinum of testis A fibrous central region anchoring septa that divide testicular lobules, housing the rete testis, vessels, and ducts aiding sperm transport.

Mitotic division (mitosis) Cell division process whereby a single cell divides to produce two genetically identical diploid daughter cells.

Meiotic division (meiosis) A specialized cell division whereby the chromosome number is reduced by half.

Myoid cell Specialized contractile cells surrounding seminiferous tubules that offer structural support and, through rhythmic contractions, propel sperm and testicular fluid toward the rete testis.

Ovulation Release of mature egg from the ovary.

Oogonia Primordial diploid (2n) germ cells, which served as precursors to oocytes.

Perivitelline space Space between the egg membrane and zona pellucida. Plays a key role in preventing polyspermy during fertilization.

Pituitary gland (master gland) A pea-sized endocrine gland secreting growth, metabolism, and reproduction-regulating hormones.

Polar body A small, nonfunctional haploid cell produced during oogenesis alongside the ovum, which typically degenerates, serving to discard the extra set of chromosomes during meiosis.

Primordium germ cell Earliest embryonic precursors of gametes that migrate to the gonadal ridges during development and differentiate into spermatogonia or oogonia.

Primary follicle An early ovarian follicle stage where the oocyte is surrounded by cuboidal granulosa cells, marking the onset of follicular growth from the primordial stage.

Primary oocyte An immature egg cell derived from oogonia arrested in prophase I within an ovarian follicle.

Primary spermatocyte A diploid male germ cell derived from spermatogonia undergoing the first meiotic division.

Primordial follicles Reserve or dormant ovarian follicles consisting of a single layer of granulosa cells and a primary oocyte.

Progesterone A steroid hormone produced primarily by the corpus luteum that regulates the menstrual cycle, prepares the endometrium for implantation, and supports pregnancy maintenance.

Rete testis Interconnected tubulae network in the testis mediastinum that channels sperm from seminiferous tubules to efferent ducts.

Secondary follicle An ovarian follicle with multiple granulosa cell layers, a developing antrum, and differentiated theca cells.

Secondary oocyte A haploid egg cell arrested in metaphase II. Upon fertilization, it completes meiosis II to form the ovum and a second polar body.

Secondary spermatocyte A haploid cell formed from a primary spermatocyte that divides to produce spermatids.

Seminiferous tubules Tightly coiled testicular tubes where sperm are produced and mature.

Sertoli cells Specialized somatic cells within the testes (seminiferous tubules) that support and nourish developing sperm cells. They secrete hormones and factors essential for spermatogenesis.

Sperm PAH-DNA adducts Chemical complexes formed when polycyclic aromatic hydrocarbons (PAHs), environmental pollutants, bind covalently to sperm DNA. Linked to male infertility factors such as reduced sperm quality and impaired fertilization potential.

Spermatids Immature haploid male germ cells that develop into sperm.

Spermatocytogenesis The formation of primary spermatocytes from spermatogonia.

Spermatogonia The male undifferentiated germ stem cells that produce primary spermatocytes.

Spermatozoa Mature male gametes specialized for fertilization.

Spermiogenesis The transformation of spermatids into mature, motile spermatozoa.

Theca interna Inner layer of cells surrounding an ovarian follicle lying just outside the granulosa cells. They are highly vascularized, producing androgens for estrogen synthesis.

Theca externa Outer layer of connective tissue surrounding an ovarian follicle, lying outside the theca interna, providing structural support to the growing follicle.

Tunica albuginea A dense, fibrous protective layer around the testis or ovary.

Vas deferens Muscular duct carrying sperm from epididymis to ejaculatory ducts.

Vitelline membrane Protective layer surrounding the yolk, maintaining egg structural integrity and regulating sperm entry during fertilization.

Zona pellucida Glycoprotein-rich layer around an oocyte that protects it and facilitates sperm binding and prevents polyspermy.

References

Aladamat N, Tadi P (2020) Histology, leydig cells. StatPearls Publishing

Aladamat, N., & Tadi, P. (2022). Histology, leydig cells. StatPearls Publishing. https://www.ncbi.nlm.nih.gov/books/NBK556007/

Alberts B, Johnson A, Lewis J, Raff M, Roberts K, Walter P (2002a) Eggs. In: Molecular biology of the cell. Garland Science, New York. https://www.ncbi.nlm.nih.gov/books/NBK26842/

Alberts B, Johnson A, Lewis J, Raff M, Roberts K, Walter P (2002b) Sperm. In: Molecular biology of the cell. Garland Science, New York. https://www.ncbi.nlm.nih.gov/books/NBK26914/

Balinsky BI (1960) An introduction to embryology. W. B. Saunders Company. https://ia801907.us.archive.org/7/items/in.ernet.dli.2015.145984/2015.145984.An-Introduction-To-Embryology_text.pdf

Bragina EE, Bocharova EN (2018) Ultrastructure of spermatozoa from infertility patients. In: Meccariello R, Chianese R (eds) Spermatozoa–facts and perspectives. IntechOpen, pp 71–99

Creasy DM, Chapin RE (2013) Male reproductive system. In: Haschek and Rousseaux's handbook of toxicologic pathology, pp 2493–2598

Domanskyi A (2007) Androgen receptor-dependent and independent functions of ARIP4. Helsingin yliopisto Helsingfors universitet University of Helsinki

Durairajanayagam D, Rengan AK, Sharma RK, Agarwal A (2015) Sperm biology from production to ejaculation. In: Unexplained infertility: pathophysiology, evaluation and treatment. Springer, pp 29–42

Eroschenko VP (2008) Difiore's atlas of histology with functional correlations, vol 213, 11th edn. Lippincott Williams & Wilkins. https://doi.org/10.1111/j.1469-7580.2008.00956.x

Fabbri R, Macciocca M, Vicenti R, Pasquinelli G, Caprara G, Valente S, Seracchioli R, Paradisi R (2016) Long-term storage does not impact the quality of cryopreserved human ovarian tissue. J Ovarian Res 9(1):50. https://doi.org/10.1186/s13048-016-0261-8

Franca LR, Hess RA, Dufour JM, Hofmann MC, Griswold MD (2016) The Sertoli cell: one hundred fifty years of beauty and plasticity. Andrology 4(2):189–212. https://doi.org/10.1111/andr.12165

Garrett C, Liu DY, Baker HW (1997) Selectivity of the human sperm-zona pellucida binding process to sperm head morphometry. Fertil Steril 67(2):362–371. https://doi.org/10.1016/S0015-0282(97)81924-3

Gilbert SF (2000a) Developmental biology, 6th edn. Sinauer Associates, Sunderland

Gilbert SF (2000b) An introduction to early developmental processes. In: Developmental biology, 6th edn. Sinauer Associates

Gurung P, Yetiskul E, Jialal I (2023) Physiology, male reproductive system. In: StatPearls [Internet]. Statpearls publishing

Guseva EA, Buev VS, Mirzaeva SE, Pletnev PI, Dontsova OA, Sergiev PV (2024) Structure and composition of spermatozoa fibrous sheath in diverse groups of metazoa. Int J Mol Sci 25(14):7663

Hariri LP, Bonnema GT, Schmidt K, Winkler AM, Korde V, Hatch KD, Davis JR, Brewer MA, Barton JK (2009) Laparoscopic optical coherence tomography imaging of human ovarian cancer. Gynecol Oncol 114(2):188–194. https://doi.org/10.1016/j.ygyno.2009.05.014

Huang W-P, Ho H-C (2006) Role of microtubule-dependent membrane trafficking in acrosomal biogenesis. Cell Tissue Res 323(3):495–503

Huang L, Li W, Dai X, Zhao S, Xu B, Wang F, Jin RT, Luo L, Wu L, Jiang X, Cheng Y, Zou J, Xu C, Tong X, Fan HY, Zhao H, Bao J (2023) Biallelic variants in MAD2L1BP (p31(comet)) cause female infertility characterized by oocyte maturation arrest. elife 12. https://doi.org/10.7554/eLife.85649

Itoi F, Miyamoto T, Himaki T, Honnma H, Sano M, Ueda J (2022) Importance of real-time measurement of sperm head morphology in intracytoplasmic sperm injection. Zygote 30(1):9–16. https://doi.org/10.1017/S0967199421000307

Ji G, Yan L, Wu S, Liu J, Wang L, Zhang S, Shi L, Gu A (2013) Bulky DNA adducts in human sperm associated with semen parameters and sperm DNA fragmentation in infertile men: a cross-sectional study. Environ Health 12(1):82. https://doi.org/10.1186/1476-069X-12-82

Jinno M (2025) Ovarian stimulation by promoting basal follicular growth. Reprod Biol Endocrinol 23(1):35

Jones RE, Lopez KH (2014) The female reproductive system. In: Jones RE, Lopez KH (eds) Human reproductive biology. Academic, pp 23–50. https://doi.org/10.1016/B978-0-12-382184-3.00002-7 (Reprinted from Fourth Edition)

Kirkendoll SD, Bacha D (2023) Histology, Corpus Albicans. In: StatPearls [Internet]. StatPearls Publishing

Krajnik K, Mietkiewska K, Skowronska A, Kordowitzki P, Skowronski MT (2023) Oogenesis in women: from molecular regulatory pathways and maternal age to stem cells. Int J Mol Sci 24(7):6837

Lei T, Yang Y, Yang W-X (2025) Luteinizing hormone regulates testosterone production, leydig cell proliferation, differentiation, and circadian rhythm during spermatogenesis. Int J Mol Sci 26(8):3548

Meneses M, Bernardino R, Sousa M, Silva B, Sá R, Oliveira P, Alves M (2015) Regulation of testicular glucose metabolism in (pre) diabetes and its implication to male reproductive health. In: Advances in medicine and biology. Nova Science Publishers, New York, pp 29–56

Nassir M, Levi M, Dardikman-Yoffe G, Mirsky SK, Shaked NT (2022) Prediction of sperm progression in three dimensions using rapid optical imaging and dynamic mechanical modeling. Cells 11(8). https://doi.org/10.3390/cells11081319

Nikalayevich E, Letort G, de Labbey G, Todisco E, Shihabi A, Turlier H, Voituriez R, Yahiatene M, Pollet-Villard X, Innocenti M, Schuh M, Terret ME, Verlhac MH (2024) Aberrant cortex contractions impact mammalian oocyte quality. Dev Cell 59(7):841–852. e847. https://doi.org/10.1016/j.devcel.2024.01.027

Nistal M, Paniagua R (2020) Non-neoplastic diseases of the testis. In: Urologic surgical pathology. Elsevier, Inc., Philadelphia, p 614

Obukohwo OM, Kingsley NE, Rume RA, Victor E (2021) The concept of male reproductive anatomy

Oliver R, Basit H (2019) Embryology, fertilization

Pereira R, Sousa M (2023) Morphological and molecular bases of male infertility: a closer look at sperm flagellum. Genes 14(2):383

Rathke C, Baarends WM, Awe S, Renkawitz-Pohl R (2014) Chromatin dynamics during spermiogenesis. Biochimica et Biophysica Acta Gene Regul Mech 1839(3):155–168

Rengan AK, Agarwal A, van der Linde M, du Plessis SS (2012) An investigation of excess residual cytoplasm in human spermatozoa and its distinction from the cytoplasmic droplet. Reprod Biol Endocrinol 10:92. https://doi.org/10.1186/1477-7827-10-92

Ricci G, Andolfi L, Zabucchi G, Luppi S, Boscolo R, Martinelli M, Zweyer M, Trevisan E (2015) Ultrastructural morphology of sperm from human Globozoospermia. Biomed Res Int 2015:798754. https://doi.org/10.1155/2015/798754

Richards JS, Pangas SA (2010) The ovary: basic biology and clinical implications. J Clin Invest 120(4):963–972

Sahnoun S, Sellami A, Chakroun N, Mseddi M, Attia H, Rebai T, Lassoued S (2017) Human sperm toll-like receptor 4 (TLR4) mediates acrosome reaction, oxidative stress markers, and sperm parameters in response to bacterial lipopolysaccharide in infertile men. J Assist Reprod Genet 34(8):1067–1077. https://doi.org/10.1007/s10815-017-0957-8

Shah W, Khan R, Shah B, Khan A, Dil S, Liu W, Wen J, Jiang X (2021) The molecular mechanism of sex hormones on Sertoli cell development and proliferation. Front Endocrinol 12:648141

Suede SH, Malik A, Sapra A (2023) Histology, spermatogenesis. In: StatPearls [Internet]. StatPearls Publishing

Thiyagarajan DK, Basit H, Jeanmonod R (2024) Physiology, menstrual cycle. In: StatPearls [Internet]. StatPearls Publishing

Toshimori K, Eddy EM (2015) The spermatozoon. In: Knobil and Neill's physiology of reproduction. Elsevier, London, pp 99–148
Wang WL, Tu CF, Tan YQ (2020) Insight on multiple morphological abnormalities of sperm flagella in male infertility: what is new? Asian J Androl 22(3):236–245. https://doi.org/10.4103/aja.aja_53_19
Wilding M, Coppola G, Di Matteo L, Palagiano A, Fusco E, Dale B (2011) Intracytoplasmic injection of morphologically selected spermatozoa (IMSI) improves outcome after assisted reproduction by deselecting physiologically poor quality spermatozoa. J Assist Reprod Genet 28(3):253–262
Williams CJ, Erickson GF (2012) Morphology and physiology of the ovary. Endotext [Internet]
Yatsenko SA, Rajkovic A (2019) Genetics of human female infertility. Biol Reprod 101(3):549–566

Study of Different Types of Placenta (Photomicrographs/Slides)

10

Sunita K. Yadav and Anna Senrung

Abstract

The placenta is a vital, temporary organ in eutherian mammals that facilitates the exchange of gases, nutrients, and waste products between the mother and developing fetus. This chapter explores the detailed structure and classification of the placenta, highlighting the diversity of placental types among mammalian species. Two primary criteria form the basis for classification: the gross morphological distribution of fetal-maternal contact sites and the number of tissue layers separating maternal and fetal blood. Placental types are described based on shape as well as histological intimacy. Comparative examples across species provide insight into evolutionary adaptations.

Keywords

Placenta · Placentation · Chorionic villi · Epitheliochorial · Hemochorial · Maternal-fetal interface

S. K. Yadav (✉)
Department of Zoology, Daulat Ram College, University of Delhi, New Delhi, India
e-mail: sunitayadav@dr.du.ac.in

A. Senrung
Department of Zoology, Daulat Ram College, University of Delhi, New Delhi, India

Neuropharmacology, Drug delivery, and Cancer Research Laboratory (NDCRL), Daulat Ram College, University of Delhi, New Delhi, India

A. C. Bharti et al. (eds.), *Textbook of Developmental Biology*,
https://doi.org/10.1007/978-981-95-7269-4_10

Aim To understand the varied types of placentas based on different parameters.

Learning Objectives
After studying this chapter, students will be able to:

- Define and explain the key physiological roles of the placenta in supporting fetal development.
- Classify placentas based on morphological features and distribution of chorionic villi and histological layers separating maternal and fetal blood.
- Differentiate between the types of placental structures (diffuse, cotyledonary, zonary, discoid) and identify examples of each.
- Compare the histological types of placenta (epitheliochorial, endotheliochorial, hemochorial) based on the number of retained maternal layers.
- Evaluate the evolutionary and functional significance of placental diversity across mammalian species.

10.1 Introduction

The placenta is a temporary organ in viviparous animals, where maternal and embryonic tissues are apposed to each other, facilitating the exchange of nutrients and gases, as well as the removal of body waste. The word "placenta" is derived from the Greek and Latin languages and means "flat" and "cake," respectively (Wang and Zhao 2010). The terminology is derived from the human placenta, which is a round mass with a flat surface that looks like a pancake. The placenta is a transient but multifunctional organ that is exclusively present in eutherian (placental) mammals. It plays an indispensable role in the full-term growth and development of the fetus. The placenta forms soon after the implantation of the fertilized egg and creates an intricate vascular and biochemical interface between the mother and fetus. The process of formation of the placenta and its fusion with the maternal uterus wall is termed "placentation." The maternal-fetal connection allows entry of nutrients, exchange of gases, and the exit of waste products. It also helps in the secretion of hormones and building immunity. Hence, the placenta acts as a short-term lung, kidney, intestine, and endocrine organ for the embryo.

Although all placentas sustain fetal life, the difference in their structure and morphology is astonishing. These variations are the basis for preliminary placental classification:

- The overall shape of the placenta and the distribution of villi and tissues contact between the maternal endometrium and fetal membranes.
- The number of tissue layers separating the fetal and maternal blood.
- The type of tissue layers participating in the formation of tissue layers.

The placenta formation is contributed to by both maternal and fetal tissues. From the fetal side, the trophoblast layer of the blastocyst and the extraembryonic mesoderm take part in its formation, while from the maternal side, the uterine endometrium becomes modified to form the decidua part of the placenta. The early development of the placenta involves the emergence of chorionic villi from the trophoblast layer, which invades the endometrium layer of the maternal side, forming the foundation of tissues for maternal-fetal exchange. The degree of intimacy between maternal and fetal tissues determines the type and structure of the placenta formed.

A mature placenta consists of chorionic villi, which contain fetal blood vessels responsible for exchange. In the hemochorial type of placenta, these villi are surrounded by maternal blood within the intervillous space. The placenta is anchored to the uterus wall of the mother by a basal plate. In addition to these, the placenta also contains the amnion, chorion, allantois, and yolk sac, which support and protect the fetus.

10.2 Requirements

- Photomicrographs/pictures and labelled diagrams of different types of placenta.
- Observation record sheet/auxiliary notebook and pencil.

10.3 Methodology/Observation

Observe the photographs of different types of placenta. Draw diagrams of various types of placenta and compare them in terms of evolutionary significance and species-specific development.

10.4 Description of Different Types of Placenta

As observed in the photographs, the placenta exhibits various forms across mammals. To understand its evolutionary significance based on species-specific development, a thorough understanding of placental structure is necessary. There are several classifications of the placenta based on various aspects, as discussed in the following sections.

10.4.1 Classification Based on the Degree of Association Between Fetal and Maternal Tissue (Eutherian Mammals)

Placental types in eutherian mammals can be categorized based on the degree of association and intimacy between fetal and maternal tissues during implantation and parturition. Three major types are recognized in this category (Arora and Grover 2018):

(i) *Non-deciduate (non-deciduous) or non-contact placenta:* In this type, fetal and maternal tissues do not come into close contact with each other. The chorionic epithelium of the fetus does not deeply penetrate the uterine epithelium, making the embryo's implantation superficial. Therefore, at parturition, the chorionic villi of the fetus are withdrawn from the maternal tissues without causing damage or bleeding. In this case, placental separation is clean, and there is minimal bleeding during parturition. This type of placenta is mainly observed in ruminants such as pigs, cattle, and horses.

(ii) *Deciduate (deciduous) placenta:* In this type of placenta, the connection between maternal and fetal tissues is more intimate, and the villi are embedded deeper into maternal tissues such as connective tissues or maternal blood. Therefore, during delivery, when the fetal chorionic villi are withdrawn, excessive bleeding can occur. When the placenta is separated and expelled at birth, the lining of the uterus also tears away with the blood, and this tissue is called the decidua. This type of placenta is found in higher eutherians such as humans, rabbits, dogs, and cats.

(iii) *Contra-deciduate placenta:* This type of placenta is a deciduous placenta with an intimate connection between maternal and fetal tissues. However, maternal leucocytes digest and eliminate the associated tissues at the implantation site during delivery. Therefore, no bleeding occurs at parturition. Examples of this type of placenta include *Parameles* and the mole (*Talpa*).

10.4.2 Types of Placenta Based on the Source of Vascular Supply

Placentae in mammals can also be classified based on the embryonic structures that contribute to the fetal vascular supply. These are the two main types, depending on whether the yolk sac (vitelline circulation) or the allantois (allantoic circulation) is contributing to the formation of the placenta (Parihar et al. 2019) (Fig. 10.1).

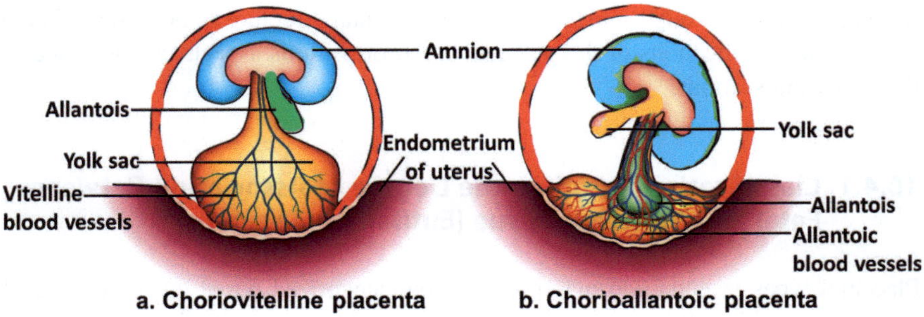

Fig. 10.1 Types of placenta based on vascular supply

(i) *Choriovitelline (Yolk Sac) placenta:* In this type of placenta, the yolk sac with its vitelline blood vessels makes contact with the chorion and forms the surface of the placenta. No involvement of the chorion is seen. This type of placenta can sustain an embryo during the early stages of embryonic development. This arrangement is primarily found in metatherian mammals, such as marsupials, e.g., *Didelphis* (opossum) and *Macropus* (kangaroo).

(ii) *Chorioallantoic placenta:* This is a more complex type of placenta. Here, a highly vascularized embryonic structure, the allantois, fuses with the chorion to create the placenta. The chorion forms a rough surface by forming villi, which are embedded deep into the mother's endothelium. Since such a placenta is more invaded, it allows efficient nutrient and gas exchange. This is found in all eutherian (placental) mammals. Also, some marsupials show this type of placentation during later stages of pregnancy.

10.4.3 Classification Based on Mode of Implantation

The placenta can be classified based on how the embryo implants into the uterine wall. This classification is primarily histological and describes how deeply the embryo embeds itself into the maternal endometrium. Based on the mode of implantation, the placenta is classified into three main types (Arora and Grover 2018) (Fig. 10.2).

(i) *Superficial implantation:* It occurs when the embryo attaches to the surface of the maternal endometrium without deeply penetrating it. The blastocyst remains within the uterine cavity, and the placenta develops externally on the endometrial lining. This is the most common type of implantation and is seen in domestic animals such as pigs, horses, cows, sheep, and dogs. In this type of placenta, the connection between the embryo and maternal endometrium is relatively loose, and the embryo does not become fully surrounded by maternal tissue.

(ii) *Eccentric implantation:* In this type of implantation, the embryo becomes partially embedded in the uterine wall and lies within a crypt or depression in

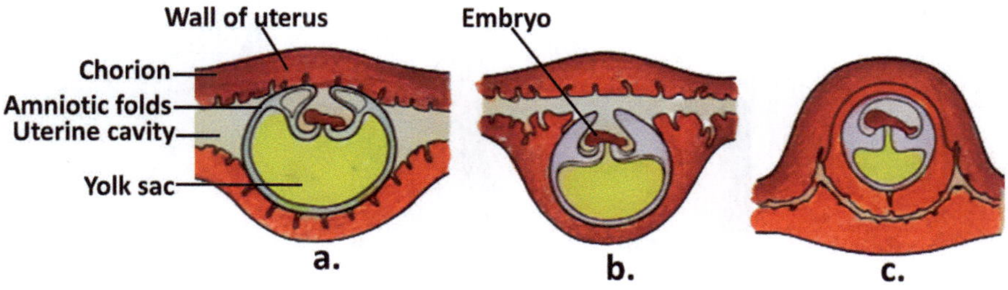

Fig. 10.2 Different types of placenta based on implantation. (**a**) Superficial implantation. (**b**) Eccentric implantation. (**c**) Interstitial implantation

the endometrium. Although the embryo is surrounded on some sides by maternal tissue, it is not fully enclosed. This type of implantation is typical of certain rodents like rats, squirrels, mice, and hamsters, where the embryo burrows into a fold of the endometrium but still extends into the uterine cavity.

(iii) *Interstitial implantation:* It involves deep penetration of the embryo into the endometrial tissue of the mother. The blastocyst becomes entirely embedded within the uterine wall and is completely surrounded by maternal tissue. This type of implantation is more invasive and is observed in humans, chimpanzees, and guinea pigs. The close connection allows for efficient exchange between the maternal and fetal blood systems due to the proximity of the tissues.

10.4.4 Classification Based on Placental Shape and Distribution of Villi

Placenta can also be categorized on the basis of morphology and the distribution pattern of chorionic villi, where the fetal and maternal tissues interface (Furukawa et al. 2014). This classification includes the following main types (Figs. 10.3 and 10.4):

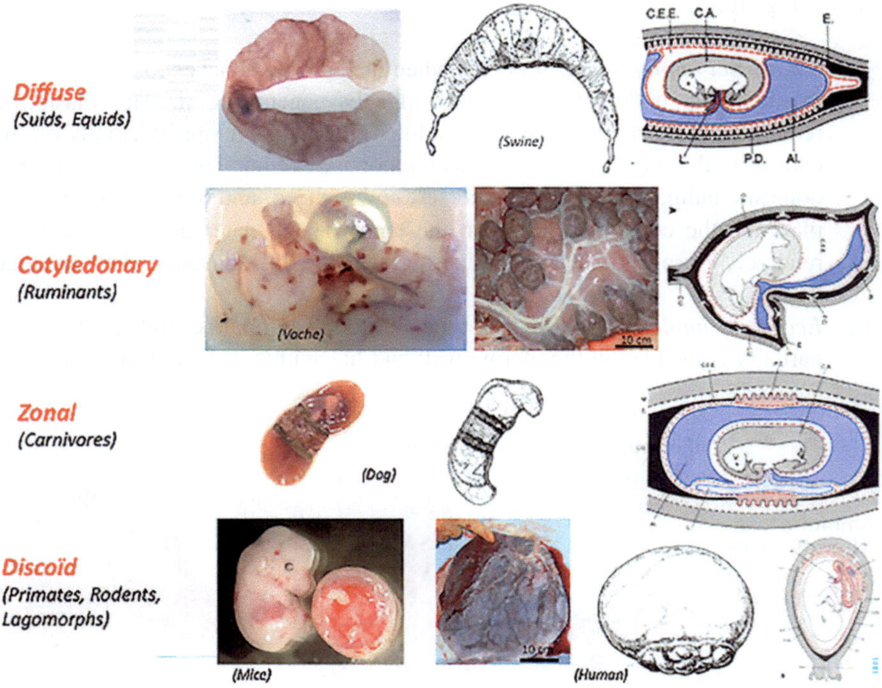

Fig. 10.3 Classification based on placenta shape and villi distribution. (Adapted from Chavatte-Palmer and Tarrade (2016))

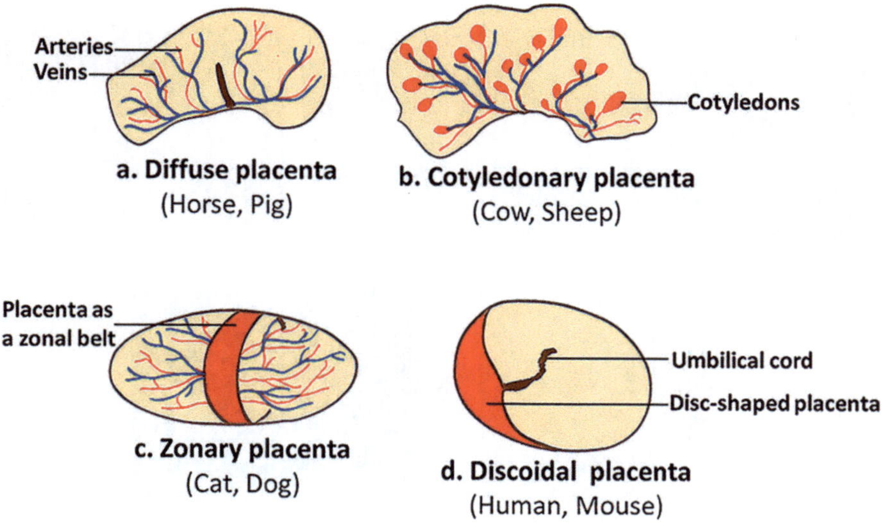

Fig. 10.4 Schematic view of placenta classification based on shape and villi distribution (Modified from Khorami-Sarvestani et al. (2024))

(i) *Diffuse placenta:* In this type, villi are evenly spread on the entire surface of the chorion. This distribution ensures widespread contact with the maternal endometrium for efficient exchange of materials. This type is found in horses, pigs, and mares.

(ii) *Cotyledonary placenta:* Here, villi are distributed into multiple isolated circular clusters, giving an appearance of cotyledons. The fetal cotyledons interface with maternal structures called caruncles, and together they form specialized exchange units called placentomes. Examples of this type of placenta include rudimentary animals such as cows, sheep, and goats.

(iii) *Zonary placenta:* In this form, the villi form a continuous band or girdle around the embryo. The band provides a focused zone of exchange. Such a placenta is found in carnivores such as dogs and cats, as well as in bears and elephants.

(iv) *Discoidal placenta:* In this type of placenta, villi are concentrated in a small disc-shaped area that facilitates highly efficient exchange. It can exist as a single (monodiscoidal) or double (bidiscoidal). It is typical of primates (including humans and apes), bats, and rodents. Monodiscoidal and bidiscoidal structures are found in humans and monkeys, respectively.

10.4.5 Classification Based on Histological Layers Between Maternal and Fetal Blood

The placenta forms through a close connection between maternal and fetal tissues. Another common way to classify placentas is by the number and type of histological layers that separate maternal and fetal blood. Initially, six distinct tissue layers exist

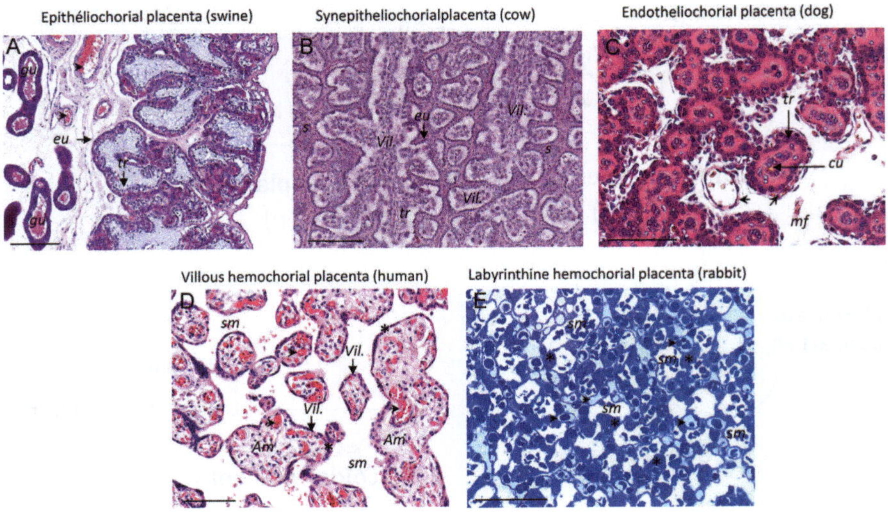

Fig. 10.5 Histological classification of placentas. The trophoblast is denoted by "tr," uterine epithelium as "eu," maternal blood as "sm," blood vessels by arrows, uterine glands as "gu" and stroma as "s," chorionic villi as "Vil," uterine capillaries as "cu," fetal mesenchyme as "mf," and mesenchymal axis as "Am." (Adapted from Chavatte-Palmer and Tarrade (2016))

between the two circulatory systems, but different degrees of erosion or retention of maternal tissues during placental development lead to five main types of placenta. These six layers are composed of both fetal and maternal components. The fetal tissues (extraembryonic membranes) are generally retained across all types and include:

- Fetal endothelium, which lines the capillaries.
- Fetal connective tissue, derived from the chorioallantoic mesoderm.
- Chorionic epithelium, formed by the trophoblast.

In contrast, the maternal components (uterine endometrium), comprising the endometrial epithelium, endometrial connective tissue, and maternal endothelium (blood capillaries), can be lost partially or completely, depending on the degree of placental invasiveness (Montiel et al. 2013; Furukawa et al. 2014). The degree to which these maternal layers are retained forms the basis for classifying the placenta into five types (Figs. 10.5 and 10.6):

(i) *Epitheliochorial placenta:* In this type, all six tissue layers, i.e., three maternal (uterine epithelium, connective tissue, and endothelium) and three fetal (chorionic epithelium, connective tissue, and endothelium) are retained. It is the most primitive and least invasive type of placenta. This type is found in marsupials and ungulates (pigs and horses). In such a placenta, nutrient exchange

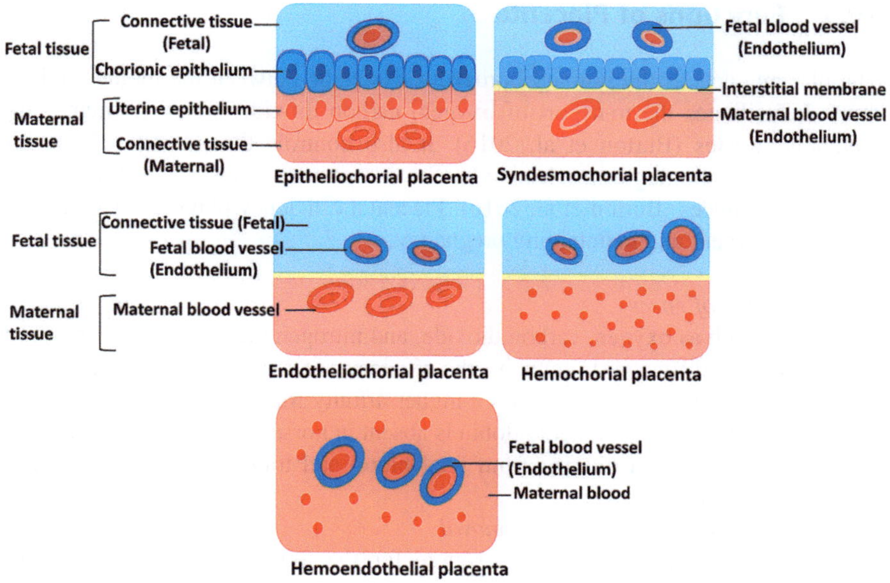

Fig. 10.6 Schematic view of types of placenta based on fetal and maternal tissue layers involved

efficiency is relatively low; however, the long gestation period allows sufficient time for fetal development.

(ii) *Syndesmochorial placenta:* Here, the chorionic villi on the fetal side damage the uterine epithelium, resulting in direct contact with uterine connective tissue. Therefore, only five tissue layers are present between maternal and fetal blood in this case, and it is found in ruminant ungulates (cattle and sheep).

(iii) *Endotheliochorial placenta:* In this type, all fetal layers are present, but only the endothelium of the maternal blood vessels remains on the maternal side. The maternal epithelium and connective tissue are lost due to the deeper penetration of the chorionic epithelium into the uterine walls. This type of placental formation is seen in carnivores such as dogs and cats. The efficiency of maternal-fetal exchange in the endotheliochorial placenta is higher compared to the epitheliochorial type.

(iv) *Hemochorial placenta:* In this type of placenta, the epithelium of maternal capillaries is absent, allowing fetal villi epithelium to be in direct contact with maternal blood. As a result, only three tissue layers are present, and all are fetal layers. Due to its high invasiveness, this placenta enables highly efficient and rapid exchange of gases, nutrients, and waste products, supporting faster fetal growth and development. This type occurs in humans, primates, and rodents.

(v) *Hemoendothelial placenta:* In this type, chorionic villi of fetal layers lose their epithelium and connective tissue, becoming the most invasive form of placenta, where fetal villi are directly bathed in maternal blood. This type is primarily found in rodents such as rabbits, rats, and mice.

10.5 Functions of Placenta

The placenta acts as the primary border for exchange between the mother and the fetus. It facilitates the transport of oxygen, carbon dioxide, water, nutrients, and metabolic wastes (Burton et al. 2016). It also controls the passage of specific substances, functioning as a selective barrier that permits or restricts the passage of specific substances (Burton et al. 2016). Placental efficiency plays a crucial role in fetal development and maintaining pregnancy.

(i) *Exchange of gases*
Gases such as oxygen, carbon dioxide, and nitrogen pass through the placenta by passive diffusion. In humans, the efficiency of oxygen transfer is enhanced by fetal hemoglobin, which has a higher affinity for oxygen than adult hemoglobin. However, fetal hemoglobin is absent in horses, dogs, rabbits, and mice. It removes carbon dioxide from fetal blood and transfers it to the mother's circulation for removal.

(ii) *Nutrient transfer and waste removal*
Glucose, the principal energy source for the fetus, is transported through facilitated diffusion involving specific solute carrier transporters. The types and distribution of these transporters vary between species, with the highest glucose transfer efficiency seen in hemochorial placentas. Maternal proteins do not cross the placental barrier, but amino acids are transported actively by sodium-dependent and sodium-independent transporters. These transporters are mostly located in the syncytiotrophoblast. Fatty acids are transported after hydrolysis of maternal triglycerides at the apical surface of the syncytiotrophoblast. They are then either stored, oxidized for energy, or converted into bioactive molecules like eicosanoids. Cholesterol is transferred via lipoprotein receptors and exported to the fetus through ABC transporters. The placenta removes urea, uric acid, and creatinine from the fetus to the mother's excretory system (Chavatte-Palmer and Tarrade 2016)

(iii) *Barrier function*
It acts as a selective barrier, protecting the fetus from many harmful substances (e.g., alcohol, nicotine, some viruses).

(iv) *Endocrine function*
The placenta has strong endocrine functions and produces various hormones essential for maintaining pregnancy, adapting the mother's body, supporting fetal development, and initiating labor. These hormones are mainly secreted by the trophoblast, especially the syncytiotrophoblast. The primary steroid hormones produced are progesterone and estrogens. Initially, progesterone is secreted by the corpus luteum, but later, the placenta takes over this role, maintaining uterine relaxation and supporting early pregnancy in all mammals. In humans, the placenta begins producing progesterone after six weeks. In pigs, goats, and rabbits, the corpus luteum remains vital throughout pregnancy, whereas in cattle, sheep, and horses, the placenta also makes a significant contribution. Estrogens promote the growth of the endometrium and

myometrium. Their levels rise throughout pregnancy and peak before labor. In cows and sheep, estrone sulfate levels serve as indicators of placental health and fetal well-being. In addition to these, the placenta also secretes proteinaceous hormones such as human chorionic gonadotropin (hCG), placental lactogen (hPL), and placental growth hormone. The hCG is specific to primates and horses, while placental lactogens help develop the mammary glands and support fetal growth by improving maternal nutrient use. Prolactin-related proteins (PRPs), secreted by trophoblast cells in ruminants and rodents, assist with implantation and placental remodeling. In humans and sheep, placental growth hormone replaces pituitary growth hormone during mid-pregnancy and supports placental development (Chavatte-Palmer and Tarrade 2016).

(v) *Immunological role*

The placenta also allows the acceptance of the semi-allogenic fetus by modulating the maternal immune response. Immunoglobulins (especially IgG) are transferred by transcytosis in species with hemochorial and endotheliochorial placentas, such as humans and dogs (Ander et al. 2019). In contrast, in species with epitheliochorial or synepitheliochorial placentas, such as ruminants, immunoglobulins do not cross the placenta. In such cases, colostrum (first milk after birth) is vital for passive immunity transfer after birth.

(vi) *Thermoregulation*

The placenta helps in maintaining a stable temperature for the developing fetus.

10.6 Significance and Research Application

Due to the key role of the placenta in pregnancy and fetal development, it has clinical importance in both human and veterinary medicine. A successful pregnancy is mainly dependent on a normally developed placenta. Various pregnancy-related conditions are diagnosed during fetal development, and an abnormal placenta can result in complications such as placenta previa, placental abruption, preeclampsia, and intrauterine growth restriction (IUGR). All these complications can adversely affect fetal health. Placental insufficiency, where the placenta fails to provide adequate support to the fetus, is a major cause of low birth weight, preterm labor, and even stillbirth (Redline et al. 2023). The placenta also helps in preventing the rejection of the fetus by the maternal immune system. The imbalance in immunological interaction can result in pregnancy loss or autoimmune complications, such as antiphospholipid syndrome.

Placental classification helps predict the efficiency of nutrient, gas, and waste exchange. For example, hemochorial placentas (as seen in humans and rodents) are highly efficient due to the intimate contact between maternal blood and fetal tissues, but they are also more prone to complications such as preeclampsia, placental abruption, and immune challenges. In contrast, epitheliochorial placentas (in pigs, horses, and ruminants), which retain all maternal tissue layers, are less invasive but

Table 10.1 Different types of placenta with their characteristics and examples

S. No.	Placental type	Shape/villous distribution	Histological type	Maternal layers retained	Species	Exchange efficiency	Functional/clinical notes
1	Diffuse	Villi uniformly scattered across chorion	Epitheliochorial	All 3 maternal layers (epithelium, connective tissue, endothelium)	Pig, horse	Low–moderate	Long gestation supports fetal growth
2	Cotyledonary	Villi grouped into multiple patches (cotyledons)	Epitheliochorial	All 3 maternal layers	Cow, sheep, goat	Low–moderate	Specialized exchange units (placentomes); clean separation at birth
3	Zonary	Villi arranged in a band or girdle around the embryo	Endotheliochorial	Only maternal endothelium retained	Dog, cat	Moderate	Intimate connection allows moderate nutrient flow
4	Discoid	Single disc-shaped region of villi	Hemochorial	No maternal layers retained	Human, ape, rodents	High	Most efficient exchange; invasive implantation; greater immune regulation
5	Choriovitelline	Poorly developed; yolk sac contributes to exchange	Varies (simple structure)	Varies; often fewer layers	Marsupials (e.g., opossum)	Low	Temporary; functional in early embryonic stages
6	Chorioallantoic	Fully developed with vascularized allantois	Epitheliochorial to hemochorial	Varies (3–6 layers)	All eutherians (most mammals)	Low to high (depends on species)	Main placenta type in placental mammals; supports full gestation

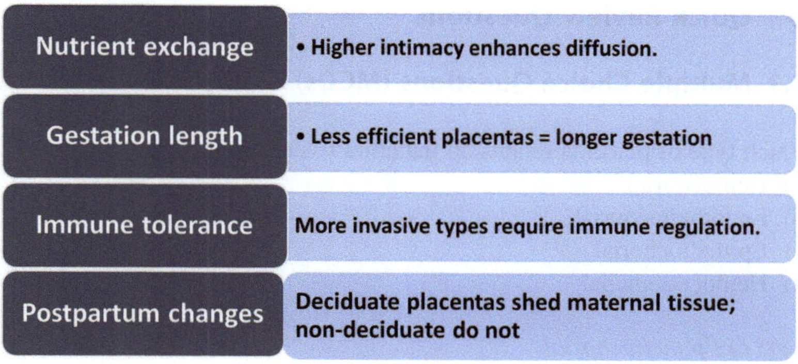

Fig. 10.7 The influence of the placenta on physiological and clinical outcomes

have lower exchange efficiency, often requiring longer gestation periods to ensure proper fetal development (Table 10.1).

Besides its structural and immune roles, the placenta functions as an endocrine organ by producing hormones such as human chorionic gonadotropin (hCG), progesterone, estrogen, and placental lactogen. Any abnormal levels of these hormones can indicate conditions like ectopic pregnancy, trophoblastic diseases, or fetal distress. In assisted reproductive technologies like in vitro fertilization (IVF), successful implantation and proper placental development are crucial for achieving and maintaining a full-term pregnancy. During childbirth, the complete and timely separation of the placenta is essential, as retained tissue can lead to serious complications such as hemorrhage or infection, which may be life-threatening. In veterinary science, understanding the different types of placenta across species is important to manage parturition and prevent problems like retained placenta or dystocia. Additionally, examining the placenta after delivery can offer valuable diagnostic information in cases of fetal infection, congenital disorders, or unexplained stillbirth. Therefore, understanding placental structure and function is vital for improving maternal and fetal health outcomes (Redline et al. 2023).

The placenta, though temporary, is central to mammalian reproduction. Its structural and functional diversity reflects adaptations to different reproductive strategies. Classifying placentas based on their shape, histological structure, and degree of invasiveness provides critical insight into maternal-fetal interactions and guides medical decisions in obstetrics and veterinary medicine (Fig. 10.7). Thus, classifying placentas is not just of academic interest but a practical tool in clinical practices to ensure safer pregnancies and healthier outcomes.

10.7 Precautions

- Record all observations carefully
- Handle the photographs with care

10.8 Quick Review Questions

10.8.1 Multiple Choice Questions (MCQs)

1. Which type of placenta retains all maternal tissue layers?
   (a) Hemochorial
   (b) Endotheliochorial
   (c) Epitheliochorial
   (d) Hemoendothelial

Answer: c) Epitheliochorial

2. Which animal shows a cotyledonary placenta?
   (a) Dog
   (b) Cow
   (c) Human
   (d) Horse

Answer: b) Cow

3. The term "placenta" is derived from the Greek word meaning:
   (a) Layer
   (b) Nutrient
   (c) Flat cake
   (d) Cushion

Answer: c) Flat cake

4. Which of the following is the most invasive type of placenta?
   (a) Epitheliochorial
   (b) Syndesmochorial
   (c) Hemochorial
   (d) Hemoendothelial

Answer: d) Hemoendothelial

5. In which group is the choriovitelline placenta primarily found?
   (a) Primates
   (b) Carnivores
   (c) Marsupials
   (d) Ungulates

Answer: c) Marsupials

6. What type of placenta do dogs have?
   (a) Cotyledonary
   (b) Zonary
   (c) Discoidal
   (d) Diffuse

Answer: b) Zonary

7. Which layer is always retained in all placental types?
   (a) Maternal connective tissue
   (b) Chorionic epithelium
   (c) Uterine epithelium
   (d) Maternal endothelium

Answer: b) Chorionic epithelium

8. Which of the following statements about the placenta is incorrect?
   (a) It functions as a temporary lung, kidney, intestine, and endocrine organ for the fetus.
   (b) It forms only from fetal tissues, without maternal contribution.
   (c) It facilitates the exchange of gases, nutrients, and waste products.
   (d) It helps in immunological protection of the semi-allogenic fetus.

Answer: b) It forms only from fetal tissues, without maternal contribution

9. In which type of placenta are fetal chorionic villi directly bathed in maternal blood, making exchange most efficient?
   (a) Epitheliochorial
   (b) Syndesmochorial
   (c) Hemochorial
   (d) Endotheliochorial

Answer: c) Hemochorial

10. The placental lactogen (hPL) hormone is primarily responsible for:
    (a) Initiating parturition
    (b) Maintaining uterine contraction
    (c) Promoting mammary gland development and nutrient partitioning
    (d) Triggering ovulation

Answer: c) Promoting mammary gland development and nutrient partitioning

10.8.2 Short Answer Questions

(a) Define placentation.
(b) What is the function of the placenta in fetal development?
(c) Name the types of placenta based on histological layers.
(d) List the components of a mature placenta.
(e) Differentiate between choriovitelline and chorioallantoic placenta.
(f) Why is the placenta considered an endocrine organ?
(g) What are placentomes?
(h) What is the role of the placenta in fetal immunity?
(i) Differentiate between deciduate and non-deciduate placenta.
(j) Name different types of placenta based on placental shape and distribution of villi.

Glossary

ABC transporters A membrane protein that uses ATP energy to transport various molecules across cell membranes.

Allantois An embryonic membrane that fuses with the chorion to form the chorio-allantoic placenta.

Amnion A membrane forming the amniotic sac that surrounds and protects the fetus.

Autoimmune A condition in which the body's immune system mistakenly attacks its own cells, tissues, or organs.

Bidiscoidal Refers to a placenta type where two discs are formed instead of a single disc (seen rarely in humans and some primates).

Blastocyst An early stage of mammalian embryo (about 5–6 days after fertilization in humans) consisting of an outer trophoblast layer, an inner cell mass, and a fluid-filled cavity.

Carnivores Animals that primarily feed on other animals (meat-eaters); examples include cats, dogs, lions, and tigers.

Caruncle The maternal part of the placentome in ruminants, located on the uterine wall.

Cattle Large domesticated ruminant animals of the genus Bos, commonly kept for milk, meat, hides, and as draft animals (e.g., cows, bulls, oxen).

Cholesterol A waxy, fat-like sterol found in animal cell membranes and blood. It is essential for cell structure, hormone synthesis, bile salts, and vitamin D production, but excess can lead to health problems.

Chorioallantoic mesoderm The mesodermal tissue layer of the chorioallantoic membrane (formed by the fusion of the chorion and allantois in embryos). It contains blood vessels that facilitate nutrient and gas exchange between the embryo and the mother.

Chorioallantoic placenta A placenta formed by the fusion of the chorion and allantois; predominant in eutherian mammals.

Chorion The outermost fetal membrane that contributes to the formation of the placenta.

Chorionic villi Microscopic finger-like projections of the chorion that invade maternal tissue to facilitate exchange.

Choriovitelline placenta A placenta formed by the chorion and yolk sac, characteristic of marsupials.

Connective tissue A body tissue that supports, binds, and protects organs and other tissues; includes bone, cartilage, blood, and adipose tissue.

Contra-deciduate placenta A rare condition where fetal tissue invades and absorbs maternal tissue.

Corpus luteum A temporary endocrine structure formed in the ovary after ovulation from the ruptured follicle; it secretes progesterone to support pregnancy.

Cotyledon The fetal part of the placentome that interlocks with the maternal caruncle.

Cotyledonary placenta A placenta with multiple button-like structures (cotyledons) that connect with maternal caruncles to form placentomes (e.g., cows, sheep).

Creatinine A waste product formed from muscle metabolism (breakdown of creatine phosphate), excreted in urine; used as a marker of kidney function.

Decidua The modified endometrial lining of the uterus during pregnancy that forms the maternal part of the placenta.

Deciduate placenta A type of placenta where maternal tissue is lost at birth (e.g., humans).

Diffuse placenta A placenta where villi are distributed over the entire surface of the chorion (e.g., pigs, horses).

Discoidal placenta A placenta where villi are confined to a disc-shaped area (e.g., humans, rodents).

Dystocia Difficult or abnormal labor and delivery, often due to fetal or maternal causes.

Ectopic pregnancy A pregnancy in which the embryo implants outside the uterine cavity, most commonly in the fallopian tube.

Eicosanoids Biologically active lipid compounds derived from arachidonic acid (e.g., prostaglandins, thromboxanes, leukotrienes) involved in inflammation, immunity, and reproductive processes.

Embryonic membranes The membranes (amnion, chorion, yolk sac, allantois) that surround the embryo and form the placenta.

Endocrine organ A gland or tissue that secretes hormones directly into the bloodstream to regulate physiological functions (e.g., pituitary, thyroid, adrenal glands).

Endometrium The inner lining of the uterus where the placenta attaches during pregnancy.

Endotheliochorial placenta A placenta where maternal epithelium and connective tissue are lost, leaving the maternal endothelium in contact with the fetal chorion (e.g., dogs, cats).

Endothelium A thin layer of specialized simple squamous epithelial cells that lines the interior surface of blood vessels, lymphatic vessels, and the heart, playing a key role in vascular function, permeability, and blood flow regulation.

Epitheliochorial placenta A type of placenta where all six histological layers between maternal and fetal blood are retained (e.g., pigs, horses).

Epithelium A tissue made up of tightly packed cells that line body surfaces, cavities, and organs, functioning in protection, secretion, absorption, and sensation.

Estrogens A group of steroid hormones (mainly estradiol, estrone, and estriol) produced primarily by the ovaries; they regulate female reproductive development, menstrual cycle, and secondary sexual characteristics.

Eutherian A subclass of mammals, also called placental mammals, characterized by the presence of a well-developed placenta that supports prolonged gestation (e.g., humans, dogs, cows).

Eutherian mammals Mammals that possess a complex placenta and give birth to live young.

Extraembryonic mesoderm The mesodermal tissue that develops outside the embryo proper and contributes to supporting structures like the chorion, yolk sac, and allantois.

Fetal Pertaining to the fetus; the developmental stage in mammals from the end of the embryonic period until birth.

Gestation period The duration of pregnancy, measured from fertilization (or conception) to birth; in humans, about 38–40 weeks

Hemochorial placenta A highly invasive placenta where the fetal chorion is in direct contact with maternal blood (e.g., humans, rodents).

Hemoendothelial placenta A placental type in which the fetal endothelium directly bathes in maternal blood (e.g., rabbits).

Hemorrhage Excessive or uncontrolled bleeding due to rupture or injury of blood vessels.

Hormones Chemical messengers secreted by endocrine glands into the bloodstream that regulate physiological processes and maintain homeostasis.

Human chorionic gonadotropin (hCG) A glycoprotein hormone secreted by the placenta during early pregnancy; it supports the corpus luteum to maintain progesterone production.

Immunoglobulins Also called antibodies; glycoproteins produced by B-lymphocytes and plasma cells that recognize and neutralize antigens as part of the immune response.

Implantation The process by which the embryo attaches to the uterine wall and begins placental formation.

Intrauterine growth restriction (IUGR) A condition in which the fetus does not achieve its expected growth potential in the uterus, often due to placental insufficiency or maternal/fetal factors.

In vitro fertilization (IVF) An assisted reproductive technology (ART) where eggs are fertilized with sperm outside the body in a laboratory, and resulting embryos are transferred to the uterus.

Lactogens Hormones (e.g., placental lactogen, prolactin) that stimulate mammary gland development and milk production.

Marsupials A group of mammals (e.g., kangaroos, opossums) that give birth to relatively undeveloped young, which continue development in a pouch.

Monodiscoidal Refers to a type of placenta that has a single disc-like structure, as seen in humans and rodents.

Myometrium The thick smooth muscle layer of the uterus that contracts during menstruation and childbirth.

Nicotine An alkaloid compound found in tobacco; it is addictive and harmful, with adverse effects on cardiovascular, respiratory, and reproductive systems.

Non-deciduate placenta A placenta where fetal and maternal tissues separate cleanly at birth without tissue loss (e.g., pigs).

Obstetrics The branch of medicine concerned with pregnancy, childbirth, and the postpartum period.

Parameles A genus of marsupials commonly known as bandicoots, found in Australia and New Guinea, characterized by short forelimbs and elongated hindlimbs.

Parturition The process of giving birth; the act of labor and delivery of the fetus.

Pituitary growth hormone Also called somatotropin; a peptide hormone secreted by the anterior pituitary gland that stimulates growth, cell reproduction, and metabolism.

Placenta A temporary organ that forms during pregnancy to support the developing fetus by facilitating nutrient, gas, and waste exchange with the mother.

Placenta previa A pregnancy complication where the placenta implants in the lower uterine segment, partially or completely covering the cervix, leading to bleeding risks during pregnancy and delivery.

Placental abruption A serious condition in which the placenta separates prematurely from the uterine wall, causing bleeding and reduced oxygen/nutrient supply to the fetus.

Placental growth hormone A variant of growth hormone produced by the placenta during pregnancy that regulates maternal metabolism and supports fetal growth.

Placental lactogen (hPL) A peptide hormone secreted by the placenta that promotes mammary gland development and modifies maternal metabolism to ensure nutrient availability for the fetus.

Placentation The process of formation and attachment of the placenta to the uterine wall.

Placentome A structure formed by the union of a fetal cotyledon and a maternal caruncle in ruminants.

Preeclampsia A pregnancy complication often linked with placental dysfunction, characterized by high blood pressure.

Primates An order of mammals including humans, apes, monkeys, and prosimians, characterized by forward-facing eyes, grasping hands, and complex brain development.

Progesterone A steroid hormone mainly produced by the corpus luteum and placenta; it maintains the uterine lining for pregnancy and inhibits uterine contractions.

Rodents A diverse order of mammals (e.g., rats, mice, guinea pigs) with continuously growing incisors, often used as laboratory models.

Ruminants Hooved herbivorous mammals (e.g., cows, sheep, goats) that digest plant material through foregut fermentation in a multi-chambered stomach.

Semi-allogenic fetus The fetus that carries both maternal and paternal antigens; genetically distinct from the mother but tolerated by her immune system.

Sodium-dependent transporters Membrane transport proteins that move molecules (e.g., glucose, amino acids) into cells using the energy from sodium ion gradients.

Sodium-independent transporters Membrane transport proteins that transport molecules without relying on sodium ion gradients, often using facilitated diffusion.

Stillbirth The delivery of a baby that shows no signs of life after 20 weeks of gestation (in humans).

Syncytiotrophoblast The outer multinucleated cell layer of the trophoblast in the placenta; it invades maternal tissue and facilitates nutrient, gas, and hormone exchange.

Transcytosis A cellular transport process in which molecules are taken up on one side of a cell, transported across the cytoplasm in vesicles, and released on the opposite side.

Trophoblast The outer layer of the blastocyst that forms part of the placenta and contributes to implantation.

Ungulates Hoofed mammals, including both even-toed (Artiodactyla like cattle, sheep, pigs) and odd-toed (Perissodactyla like horses, rhinos). Many are herbivores and important domestic animals.

Urea A nitrogenous waste product formed in the liver from protein metabolism, excreted mainly through urine; an important marker of kidney function.

Uric acid An end product of purine metabolism, excreted in urine. In birds and reptiles, it is the main nitrogenous waste, excreted as a paste to conserve water.

Uterine epithelium The cellular lining of the uterus that plays a key role in implantation, protection, and interaction with the embryo during pregnancy.

Veterinary Relating to the medical, surgical, and scientific care of animals.

Viruses Infectious agents consisting of nucleic acid (DNA or RNA) enclosed in a protein coat; they replicate only inside living host cells.

Yolk sac A membrane involved in early embryonic blood supply and nutrient absorption; forms the choriovitelline placenta in some species.

Zonary placenta A placenta with a band-like zone of chorionic villi (e.g., dogs, cats).

References

Ander SE, Diamond MS, Coyne CB (2019) Immune responses at the maternal-fetal interface. Sci Immunol 4(31):eaat6114. https://doi.org/10.1126/sciimmunol.aat6114

Arora R, Grover A (2018) Developmental Biology Principles and Concepts. R.Chand & Co.

Burton GJ, Fowden AL, Thornburg KL (2016) Placental origins of chronic disease. Physiol Rev. https://doi.org/10.1152/physrev.00029.2015

Chavatte-Palmer P, Tarrade A (2016) Placentation in different mammalian species. Ann Endocrinol (Paris). https://doi.org/10.1016/j.ando.2016.04.006

Furukawa S, Kuroda Y, Sugiyama A (2014) A comparison of the histological structure of the placenta in experimental animals. J Toxicol Pathol 27(1):11–18. https://doi.org/10.1293/tox.2013-0060

Khorami-Sarvestani S, Vanaki N, Shojaeian S, Zarnani K, Stensballe A, Jeddi-Tehrani M, Zarnani H (2024) Placenta: an old organ with new functions. Front Immunol 15:1385762. https://doi.org/10.3389/fimmu.2024.1385762

Montiel JF, Kaune H, Maliqueo M (2013) Maternal-fetal unit interactions and eutherian neocortical development and evolution. Front Neuroanat 7:22. https://doi.org/10.3389/fnana.2013.00022

Parihar R, Upadhyay S, Dhiman U (2019) Developmental Biology & Evolution. Vikas Publishing House Pvt. Ltd

Redline RW, Roberts DJ, Parast MM, et al (2023) Placental pathology is necessary to understand common pregnancy complications and achieve an improved taxonomy of obstetrical disease. Am J Obstet Gynecol 228(2):187–202. https://doi.org/10.1016/j.ajog.2022.08.010

Wang Y, Zhao S (2010) Chapter 1, Introduction. In: Vascular biology of the placenta. Morgan & Claypool Life Sciences, San Rafael (CA)

Study of Human/Mammalian Embryonic Development Through Diagrams/Photomicrographs

11

Tanya Tripathi, Anna Senrung, and Alok Chandra Bharti

Abstract

Mammalian embryonic development is a complex, highly regulated process extending from fertilization to the formation of a fully developed fetus. This process involves critical stages such as cleavage, blastocyst formation, gastrulation, neurulation, somite development, and organogenesis. In mammals, including humans, development occurs in vivo, limiting direct observation of intermediate stages due to technical and ethical challenges. While animal models offer foundational insights, direct study of human embryogenesis remains crucial for understanding congenital disorders and advancing reproductive and regenerative medicine. This chapter provides a comprehensive overview of human embryonic development through diagrams and photomicrographs to illustrate key stages, based on the standardized Carnegie staging system, spanning 23 stages across the first eight weeks post-fertilization. It provides a standardized morphological framework to describe human embryonic development. Early stages cover cleavage and blastocyst formation, implantation, and bilaminar disc formation. Mid-stages mark the onset of gastrulation, neural induction, and somite segmentation, while later stages involve organogenesis, including brain development, cardiac formation, limb buds, and facial features. By Stage 23, major organ systems are formed, signaling the transition to the fetal period. Visual tools such as photomicrographs and diagrams serve as indispensable educational resources, enabling learners to identify key structures and developmental timelines while circumventing ethical constraints. Studying these stages enhances understanding of normal development and congenital anomalies influenced by

T. Tripathi · A. C. Bharti (✉)
Molecular Oncology Lab, Department of Zoology, University of Delhi, New Delhi, India

A. Senrung
Department of Zoology, Daulat Ram College, University of Delhi, Delhi, India

A. C. Bharti et al. (eds.), *Textbook of Developmental Biology*,
https://doi.org/10.1007/978-981-95-7269-4_11

genetic and environmental factors. This knowledge supports early diagnosis, improves assisted reproductive technologies, and informs regenerative therapies, ultimately contributing to better reproductive health and management of developmental disorders.

Keywords

Human embryonic development · Mammalian embryogenesis · Carnegie staging system · Photomicrographs · Developmental stages

Aim This chapter aims to enable students to identify, differentiate, and interpret the key stages of human embryonic development from fertilization to early organ formation using diagrams and photomicrographs. It also facilitates understanding of how structural changes during development relate to underlying processes and how their disruption can lead to abnormal embryonic development.

Learning Objectives

Upon completing the experiment, students will be able to:

- Understand the sequence of developmental events from fertilization to early organ formation in the human embryo.
- Identify and describe the key stages of human embryonic development using diagrams and photomicrographs.
- Differentiate between various human embryonic stages such as zygote, cleavage, morula, blastula/blastocyst, gastrula, and organogenesis.
- Interpret important morphological features observed in photomicrographs of embryonic tissues and structures of the human embryo.
- Relate structural changes observed in the embryo to functional developmental processes.

11.1 Introduction

Mammals are a diverse group of vertebrates characterized by several distinctive features, most notably the presence of mammary glands, which enable mothers to nourish their young with milk. In addition to this defining trait, mammals are warm-blooded, typically covered with hair or fur, and exhibit complex brain development and social behaviors. Among the numerous biological processes unique to mammals, embryonic development holds special importance in the field of developmental biology.

Mammalian embryonic development refers to the highly regulated and intricate sequence of events that occur from fertilization to the formation of a fully developed fetus. These stages include fertilization, cleavage, blastocyst formation, gastrulation, and organogenesis. Understanding these stages is essential not only for grasping the fundamental principles of life but also for exploring broader biological themes such as genetics, evolution, and medicine. When viewed through the lens of human development, this knowledge becomes even more impactful, laying the groundwork for research in reproductive health, congenital disorders, and regenerative therapies.

There are several established approaches to studying mammalian embryonic development. One such method involves the use of in vitro culture techniques, which allow observation of embryos under controlled laboratory conditions. However, these approaches are often limited due to technical constraints, ethical considerations, and the species-specific requirements of embryonic development. In most mammals, except egg-laying species like monotremes, embryonic development occurs in vivo, within the mother's uterus, making direct observation at various stages challenging. Additionally, ethical restrictions further limit the use of live mammalian embryos or pregnant animals for educational or research purposes.

To overcome these challenges, the use of diagrams and photomicrographs has emerged as a practical and ethically sound alternative. Diagrams and photomicrographs provide clearly labeled conceptual representations and high-resolution microscopic images that depict actual embryonic structures, respectively. These visual resources are invaluable in helping students and researchers visualize, analyze, and understand developmental processes without the need for invasive procedures. As such, they have become essential tools in developmental biology, especially in academic and training settings.

In this practical exercise, the focus is on the study of human embryonic developmental stages. Observing these stages provides critical insight into the complex biological processes that transform a single fertilized egg into a fully developed, functional organism. This transformation is orchestrated by a coordinated genetic program functioning within a supportive intrauterine environment, resulting in the formation of genetically unique individuals even among identical twins. By using detailed diagrams and photomicrographs, learners are able to visualize and comprehend the morphological changes, tissue differentiation, and organ formation that occur throughout embryonic development. These visual tools help in identifying key embryological structures, understanding developmental timelines, and correlating anatomical changes with physiological function. Such an integrated understanding not only enriches our knowledge of normal human development but also aids in recognizing developmental anomalies and understanding their biomedical implications.

11.2 Requirements

- Photomicrographs/photographs of embryonic stages.
- Observation notebook/copy.
- Pencil and pen for sketching and note-taking.

11.3 Methodology/Observation

- *Detailed observation:* Carefully observe the photographs/images provided, focusing on the structural features visible in each stage of embryonic development.
- *Identification of key features:* Note the important morphological characteristics specific to each Carnegie stage.
- *Sequential study:* Follow the sequence of development beginning from the single-cell stage, progressing through multicellular stages as per the Carnegie staging system.
- *Stage correlation:* Match observed features with the standard criteria outlined in the Carnegie stages to accurately identify each developmental phase.

11.4 Description of the Human Embryonic Development Stages

Human development begins with fertilization, the fusion of two haploid gametes (sperm and ovum) to form a diploid (2n) zygote. This marks the onset of the embryonic stage and initiates the formation of a new individual. A key feature of this process is the transformation of a single cell into a multicellular embryo through rapid mitotic divisions, known as cleavage. Fertilization typically occurs in the ampulla, the outer third, of the fallopian tube (Fig. 11.1). Once a sperm successfully penetrates the ovum (Figure 11.2a), the oocyte becomes impenetrable to prevent polyspermy.

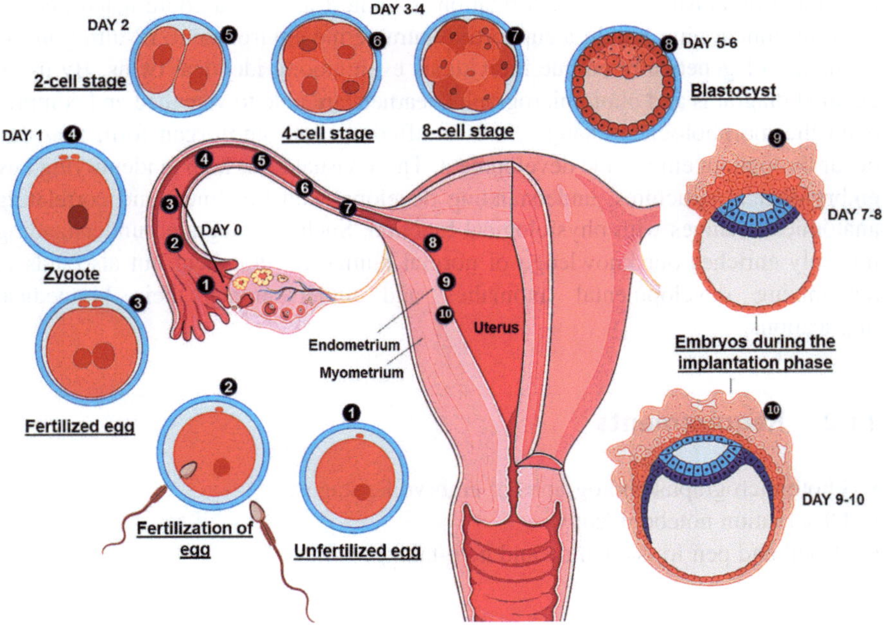

Fig. 11.1 An overview of early human embryonic development, including ovulation, fertilization, cleavage, and implantation. (Created in https://BioRender.com)

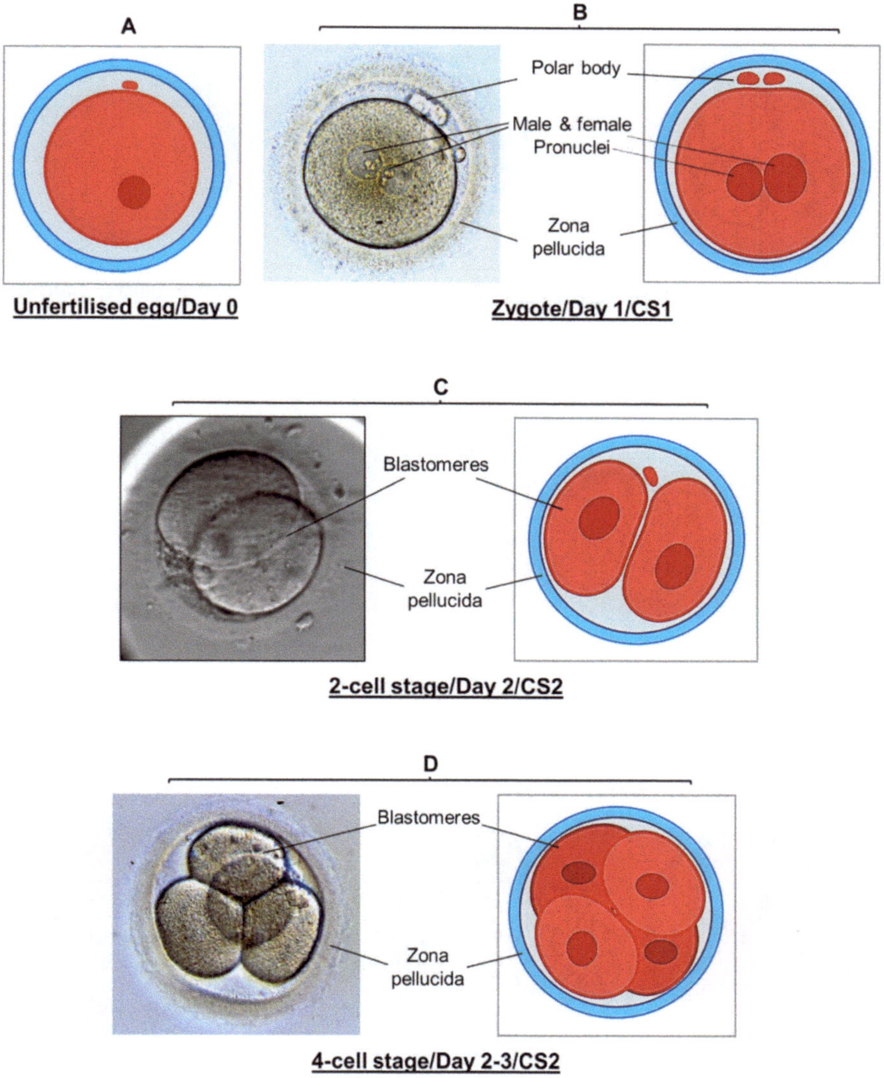

Fig. 11.2 Images and schematic representation of the early human embryonic cleavage (Carnegie stages 1–3). Unfertilized egg at Day 0 (a). The Images were sourced from the following:Fertilized egg from Эмбрион человека первые сутки развития (**b**); The 2-cell stage image was adapted from (Sharma et al. 2023) (**c**)**;** 4-cell stage from Эмбриончеловека 2-е сутки развития (**d**); 8-cell and 16-cell stage from Эмбрион человека 3-и сутки развития (**e** and **f**); Compact morula from Эмбрион человека на 4-е сутки развития (**g**); Early blastocyst from Бластоциста человека 5-е сутки развития (**h**). Blastocyst hatching from zona pellucida (**i**). Free Blastocyst (**j**). (All images are licensed under CC BY-NC-SA 4.0. Figures created in https://BioRender.com)

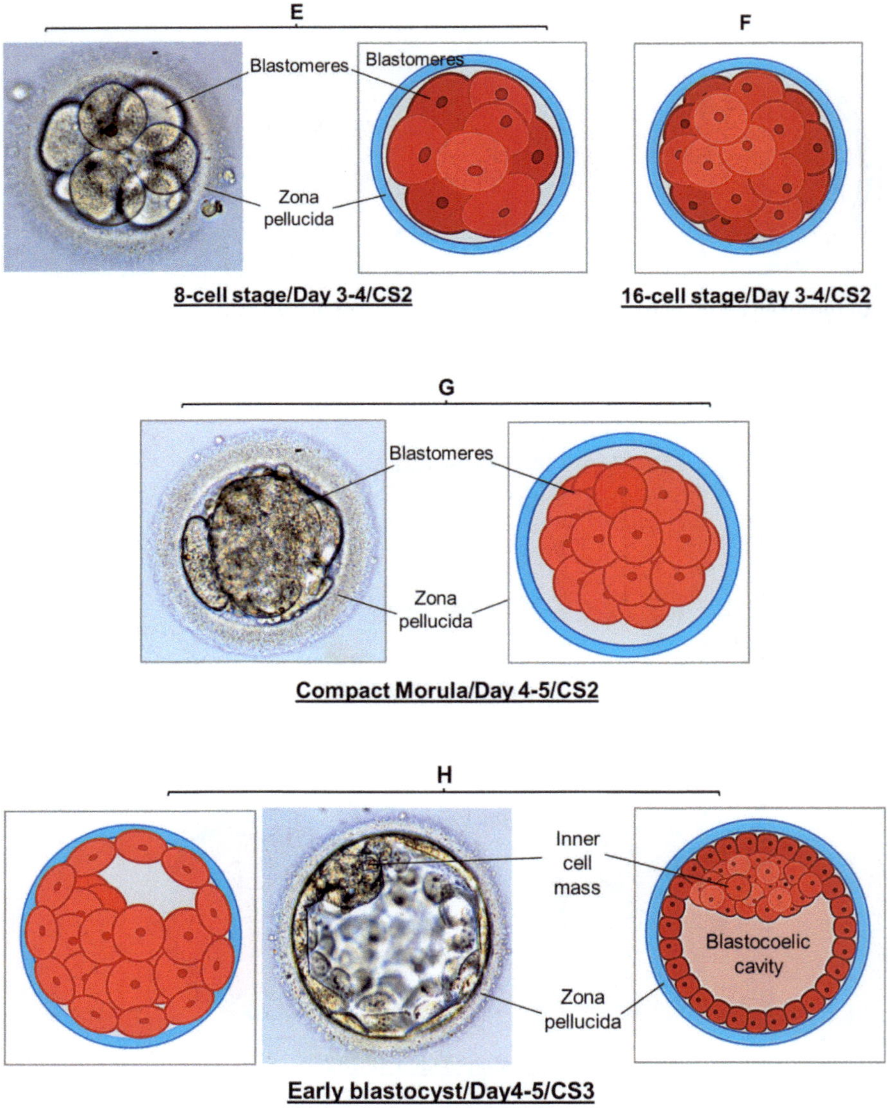

Fig. 11.2 (continued)

Around 16 days after fertilization, gastrulation occurs, during which the implanted blastocyst gives rise to the three primary germ layers: ectoderm, mesoderm, and endoderm. The sequence of events from fertilization to implantation is commonly referred to as conception.

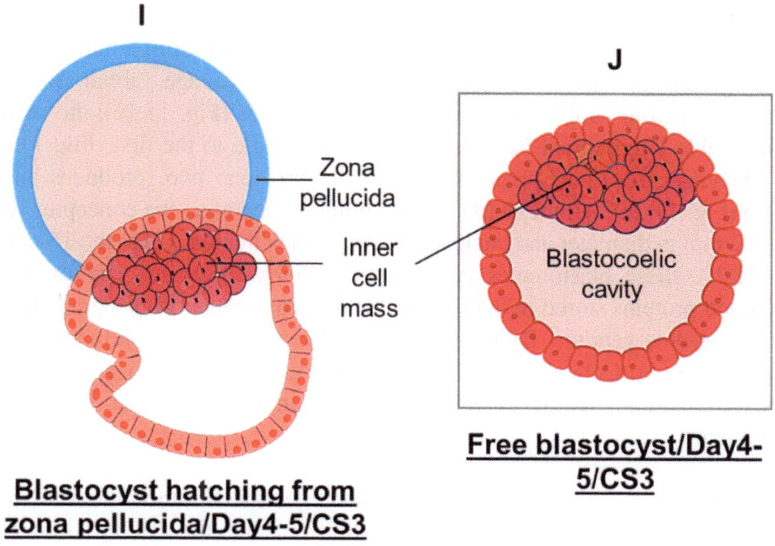

Fig. 11.2 (continued)

11.4.1 Carnegie Stages of Human Embryos

The Carnegie stages represent a standardized system in embryology used to describe vertebrate embryonic development based on distinct morphological characteristics, rather than chronological age or size (Harkness and Baird 1997). This system is specifically applied to human embryos during the first eight weeks of development (Flierman et al. 2023). Numbered from 1 to 23, the Carnegie stages span approximately the first 60 days post-fertilization and are defined by the appearance of key structures such as somites, neural folds, and limb buds. This framework facilitates consistent and comparative analysis of human embryogenesis. Following Stage 23, the embryo transitions into the fetal stage, characterized by continued growth and the maturation of organs beyond the early developmental period.

Stage 1 This stage marks the formation of the primordial cell, the zygote, and is considered Day 1 of embryonic development (Fig. 11.2b). The zygote measures approximately 0.1–0.15 mm in diameter and is unicellular. This stage occurs in the ampullary-isthmic junction of the fallopian tube (Coy et al. 2012). The zygote is surrounded by a protective glycoprotein layer zona pellucida. There has been mixing of maternal and paternal chromosomes during the process of fertilization; hence, the zygote is diploid. The zygote here contains all the genetic information necessary for the formation of an individual.

Stage 2 The second Carnegie stage of embryonic development extends from the 2-cell stage to the formation of the blastocystic cavity, occurring roughly 2–3 days after fertilization. During this interval, the zygote undergoes rapid, holoblastic cleavage which is non-synchronous mitotic divisions that partition the entire zygote into progressively smaller blastomeres. The first cleavage, about 30 h post-fertilization, is longitudinal and yields two blastomeres (Fig. 11.2c); the second, at ~40 h, is also longitudinal but oriented at a right angle to the first (Fig. 11.2d). A third cleavage at ~72 h, perpendicular to the previous two, produces an 8-cell embryo (Fig. 11.2e). Although blastomere number increases, the conceptus remains ~0.1–0.2 mm in diameter and enclosed by the zona pellucida. Up to the 16-cell stage (Fig. 11.2f), these blastomeres are totipotent. Early cleavages rely on maternal mRNA and proteins stored in the oocyte, but developmental control gradually shifts to the embryonic genome. By the stage's end, the embryo is a solid cluster of cells known as morula (Coticchio et al. 2019) (Fig. 11.2g). Formation of the blastocoel cavity signals completion of this phase and transition to the next stage of development.

Stage 3 This stage occurs between days 4 and 5 post-fertilization and is marked by the formation of the blastocystic cavity, transforming the embryo into a "free blastocyst" (Rossant and Tam 2022), measuring approximately 0.1–0.2 mm (Fig. 11.2h). At this point, the embryo's cells begin to differentiate into two distinct populations: the trophoblast cells, which will contribute to the placenta, and the inner cell mass (ICM), which gives rise to the fetus. The trophoblast further divides into polar trophoblasts (adjacent to the ICM) and mural trophoblasts (not in contact with the ICM). The ICM differentiates into the hypoblast, and the remaining formative cells will later form the epiblast, aligning into the embryonic germ disc. Several key processes define this stage: cavitation, expansion/collapse, and hatching. Cavitation begins after the fifth cleavage (32-cell stage), when blastomeres secrete fluid, forming small cavities that merge to create the blastocoel. This transition marks the shift from morula to blastocyst (Watson 1992). The blastocyst then undergoes cyclical collapse and re-expansion, also called "breathing." Collapse is rapid, reducing volume and surface area in under five minutes, while re-expansion is slower, taking hours as fluid reaccumulates. These mechanical cycles exert pressure on the zona pellucida, gradually weakening it. By days 5–6, the blastocyst emerges from the zona pellucida through a process known as hatching (Fig. 11.2i), which is essential for uterine attachment, as the intact zona pellucida inhibits implantation. By the end of this stage, the embryo (Fig. 11.2j) is ready to implant into the uterine epithelium, transitioning into the next phase of development.

Stage 4 This stage is marked by the attachment of the free blastocyst to the uterine luminal epithelium via the process known as implantation (Kim and Kim 2017) (Fig. 11.3) typically around days 5–6 post-fertilization. At this stage, the blastocyst, now called the conceptus and measuring about 0.1–0.2 mm in diameter (Fig. 11.4a), undergoes adplantation, which is the initial phase of implantation. The trophoblast differentiates into two layers: the outer *syncytiotrophoblast*, which invades the endometrial tissue by secreting matrix-degrading enzymes (Ruane et al. 2022), and the inner *cytotrophoblast* (Fig. 11.4b). A coagulation plug seals the site after

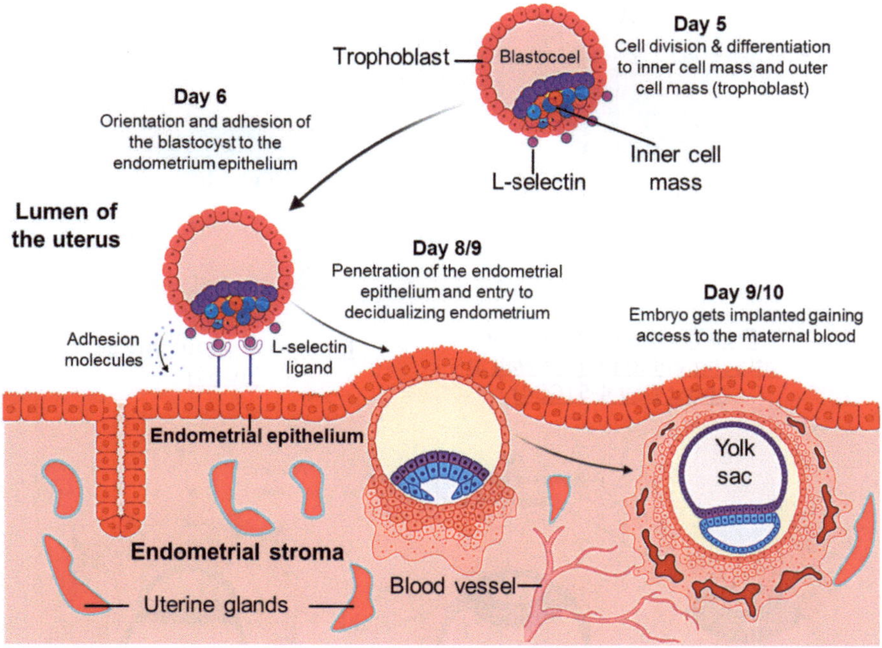

Fig. 11.3 Schematic representation of early human embryo implantation into the uterine wall. After the blastocyst differentiates into the inner cell mass and trophectoderm, it orients and attaches to the uterine luminal epithelium. The trophectoderm (trophoblast) then penetrates the epithelium and invades the decidualizing endometrium. As implantation advances, the embryo becomes fully embedded within the endometrial tissue, establishing early contact with the maternal blood supply essential for continued development and nourishment. (Created in https://BioRender.com)

successful implantation, embedding the embryo within the uterine wall. Concurrently, the inner cell mass (ICM) forms a bilaminar disc composed of epiblast and hypoblast, setting the stage for germ layer development. The conceptus now includes both embryonic and extra-embryonic components, such as the fetal membranes and placenta. Implantation involves adhesion, penetration, proliferation, and placental development. It is regulated by numerous molecular mediators, including cytokines (e.g., LIF, IL-6, IL-1) and adhesion molecules (e.g., integrins, cadherins, Mucin-1). Impaired adhesion, especially involving L-selectin, may contribute to infertility and early pregnancy loss (Ochoa-Bernal and Fazleabas 2020).

Stage 5 This stage represents early post-implantation human embryos, approximately 7–12 days old and measuring about 0.1–0.2 mm in diameter. Although implantation is complete, chorionic villi have not yet formed (Fig. 11.4c). A key developmental event during this period is the formation of the amniotic cavity between the embryonic disc and the surrounding chorion (Dard et al. 2008; Mole et al. 2020). Between days 8 and 9 post-fertilization, implantation progresses further as the embryo becomes more deeply embedded in the decidualizing endometrium

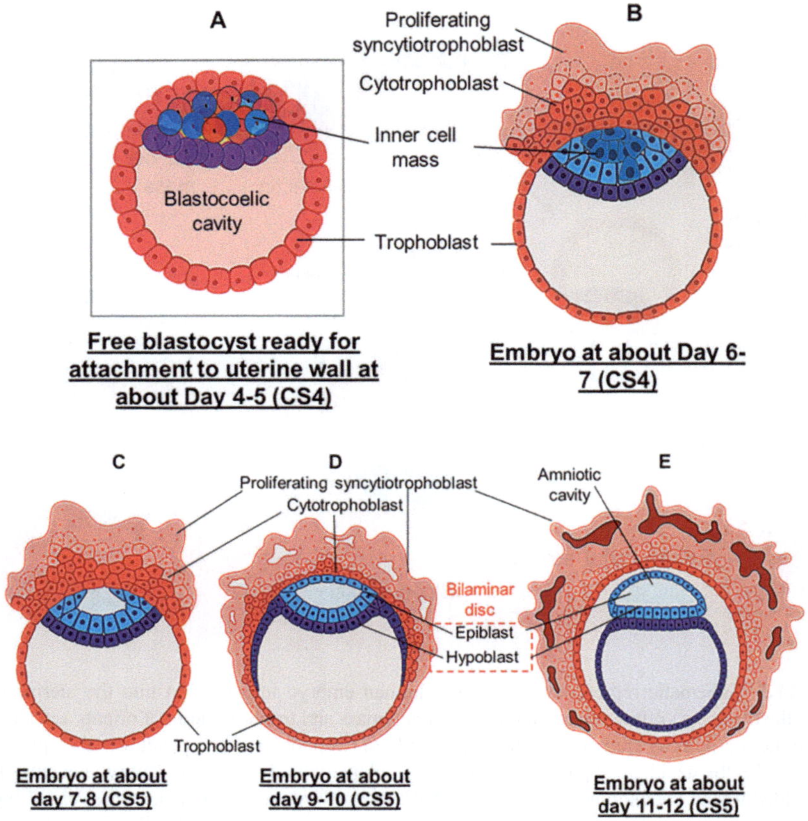

Fig. 11.4 Images and schematic representation of the early Human embryonic cleavage (Carnegie stages 4–5). Carnegie stages 4 (**a**, **b**) and 5 (**c**–**e**). (Created in https://BioRender.com)

(Fig. 11.4d). Endoblast formation begins along the inner surface of the trophoblast. The embryonic disc, now around 0.1 mm in diameter, differentiates into two layers: a thicker epiblast and a thinner hypoblast. The amniotic cavity begins to form as the epiblast arches upward. By days 9–10, the syncytiotrophoblast develops interconnected lacunae, marking the onset of uteroplacental circulation. At the same time, cytotrophoblastic projections emerge, initiating the formation of future chorionic villi. Extraembryonic mesoblasts begin to line the inner surface of the trophoblastic shell. The amniotic cavity continues to expand, and the bilaminar embryonic disc adopts a more defined, oval shape. The primary yolk sac forms, enclosed by the exocoelomic (Heuser's) membrane. By days 11–12, implantation is fully complete, with the embryo entirely embedded in maternal tissue (Fig. 11.4e). The syncytiotrophoblast forms a complex network of large, irregular, and interconnecting lacunae that fill with maternal blood, marking the early establishment of placental circulation (Gauster et al. 2022; Hamilton and Boyd 1960). These blood-filled spaces may appear as a discontinuous red ring on the surface of the endometrium. Concurrently, the cytotrophoblast continues to proliferate, supporting the growth and expansion of the trophoblast. The primary yolk sac becomes well-developed, and the oval-shaped bilaminar disc, seen dorsally, lies between the yolk sac and the enlarging amniotic cavity. The primitive streak and Hensen's node (primitive node) begin to appear, serving as crucial organizing centers for subsequent embryonic development.

Stage 6 This stage occurs approximately on day 13 post-ovulation, with the embryo measuring about 0.2 mm in length. A key feature of this stage is the appearance of chorionic villi, which play a crucial role in establishing maternal-fetal exchange (Kojima et al. 2022). The primitive streak forms at the caudal end of the embryonic disc, marking the onset of bilateral symmetry and the establishment of the body axis (Sheng et al. 2021) (Fig. 11.5a). During this stage, a well-defined amniotic cavity and yolk sac are present. Additionally, the secondary umbilical vesicle develops as a result of the expansion of the space enclosed by the chorion. This period is identified as the prosomite stage, with the embryonic disc size ranging from 0.15 to 0.5 mm. Gastrulation begins, characterized by the migration of epiblast cells through the primitive streak to give rise to the three germ layers.

Stage 7 This stage occurs at approximately 16 days post-ovulation, with the embryo measuring around 0.4 mm in length (Fig. 11.5b). A defining feature of this stage is the appearance of the notochordal process, which will later develop into the notochord, a key structural element in the early embryo (Fig. 11.5c). The primitive streak becomes more prominent and elongated, extending along the midline of the embryonic disc. Chorionic villi begin to exhibit branching, indicating further placental development. The process of gastrulation continues, giving rise to the three primary germ layers-ectoderm, mesoderm, and endoderm (Fig. 11.5c). As a result of

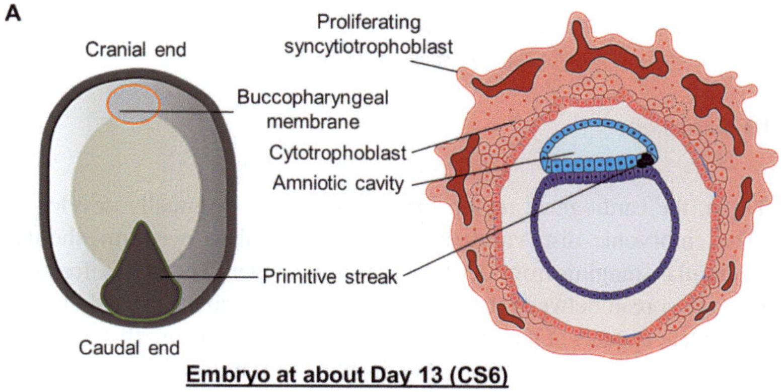

Fig. 11.5 Schematic representation highlighting the key changes of the early Human embryonic development (Carnegie stages 6–7). Carnegie stages 6 (**a**) and 7 (**b**, **c**). (Created in https://BioRender.com). Diagram of the transverse section of the embryo through primitive streak (c) has been redrawn from (Sadler, T.W. 2023)

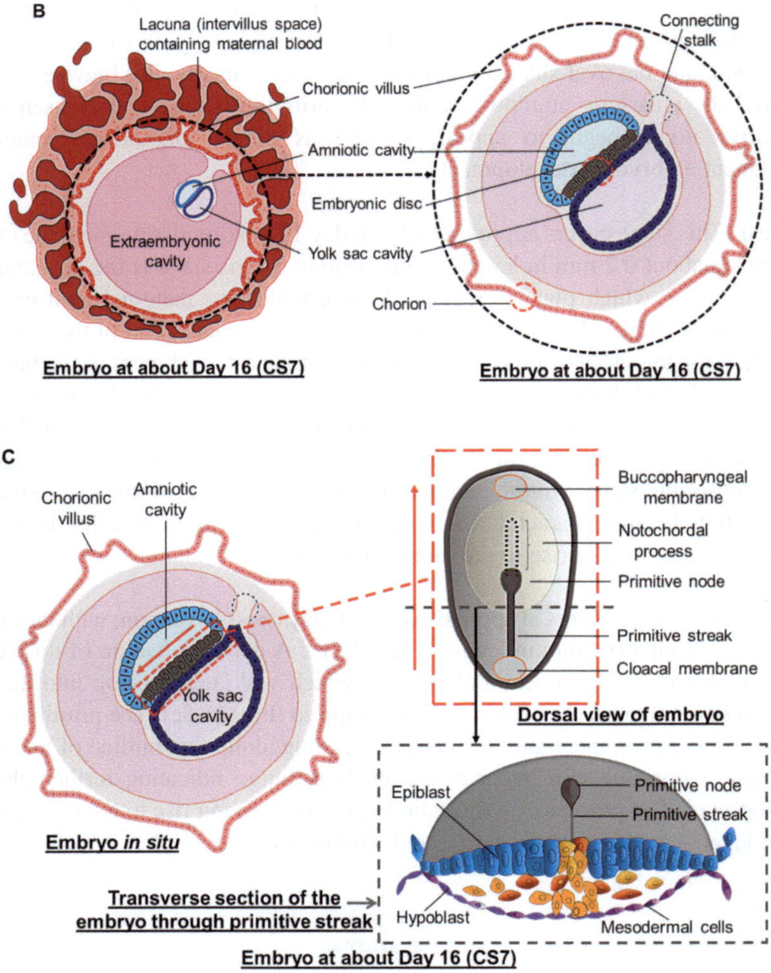

Fig. 11.5 (continued)

gastrulation, the cardiogenic region forms, which will eventually develop into the heart. The embryonic disc varies in shape from oval to pyriform, though often remains circular, ranging from 0.1 to 1 mm in diameter. The newly formed mesoderm starts to spread between the ectoderm and endoderm layers. Additionally, the prechordal plate appears at the cranial end of the disc, a structure that will contribute to the development of the forebrain. Meanwhile, both the amniotic cavity and yolk sac continue to expand. The earliest blood cells begin to form within the yolk sac, marking the onset of primitive hematopoiesis (Palis and Yoder 2001). This stage establishes the foundational body plan and sets the stage for neurulation and organogenesis in subsequent phases of development.

Stage 8 This developmental stage corresponds to approximately 19 days post-ovulation, with the embryo measuring between 1 and 1.5 mm in length. Key morphological features of this stage include the formation of the primitive pit, notochordal canal, and neurenteric canal, which collectively signal the initiation of the neurulation process (Smith and Schoenwolf 1997). The lateral edges of the neural plate begin to elevate, forming neural folds that will later fuse to give rise to the neural tube. The notochordal process plays a critical inductive role in the early development of the central nervous system (Fig. 11.6). The complete process of neurulation is given in Fig. 11.7. At this point, the embryo is in the presomitic phase, and its overall shape may appear ovoid, circular, or pyriform, typically ranging in size from 0.5 to 2 mm (Fig. 11.8a). This stage occurs during the third week of embryogenesis, a period marked by the establishment of utero-placental circulation and full development of the decidua. Interestingly, a high incidence of sex chromatin has been observed in both the notochordal process and the umbilical vesicle, suggesting early epigenetic activity or sex-specific gene expression patterns (Park 1957).

Stage 9 This stage marks a pivotal transition in human embryonic development, signifying the completion of gastrulation and the onset of organogenesis (Yuan et al. 2025). It occurs around day 20 post-ovulation, with the embryo measuring approximately 1.5 to 2.5 mm in length. A defining characteristic of this stage is the

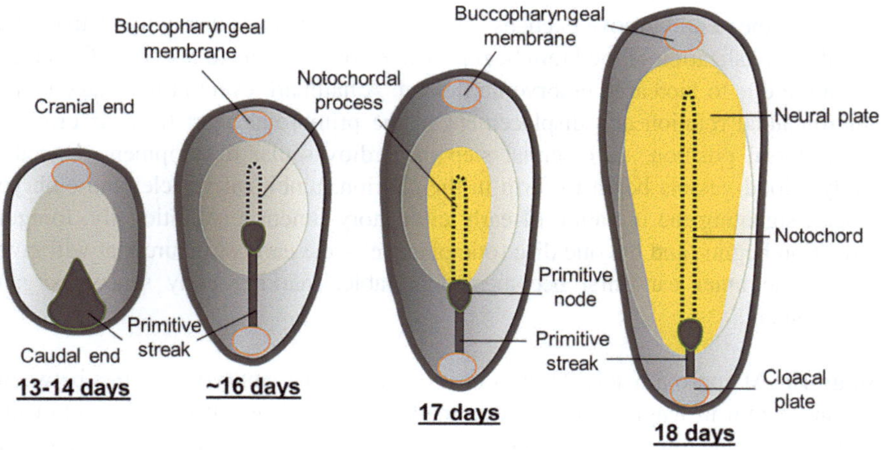

Fig. 11.6 Schematic representation of neurulation induction. Neurulation, or the formation of the neural tube, is initiated from the ectoderm overlying the mesodermal structure known as the notochord. This process begins around day 16 of embryonic development and involves crucial interactions between the ectoderm and the underlying mesoderm. Inductive signals are released from the Hensen's node, triggering ectodermal cell proliferation. This leads to the extension, bending, and folding of the neural plate, the embryonic precursor to both the brain and spinal cord, resulting in significant morphological changes. Upon the notochord formation and elongation, the neural plate expands and extends beyond its lateral boundaries, laying the groundwork for central nervous system development (Acharya et al. 2017). (Created in https://BioRender.com)

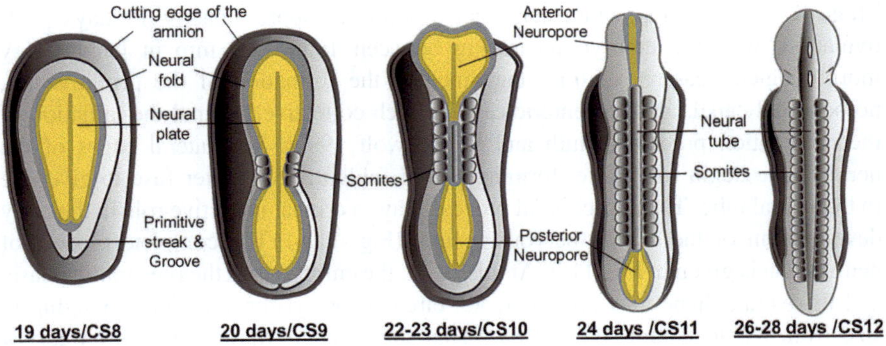

Fig. 11.7 Schematic representation of neurulation. The neural plate begins to invaginate along the central axis, forming a longitudinal median groove flanked by neural folds. In the region destined to become the spinal cord, these neural folds rise and gradually move toward each other, eventually fusing around day 22 of gestation to form the neural tube. This fusion occurs at multiple sites along the rostral-caudal axis and progresses bidirectionally. Closure of the anterior neuropore is typically completed by day 24, while the posterior neuropore closes by day 28, marking the end of primary neurulation. Secondary neurulation then gives rise to the caudal spinal cord structures, including the conus medullaris, lower lumbar, and sacrococcygeal segments (Acharya et al. 2017). *CS* Carnegie stage. (Created in https://BioRender.com)

appearance of 1 to 3 pairs of somites, segmented blocks of paraxial mesoderm that will later develop into vertebrae, ribs, and skeletal muscles (Buckingham et al. 2003) (Fig. 11.8b). Neural development advances as the neural folds become more elevated and the neural groove deepens. The head fold becomes distinctly visible, and initial regionalization of the brain begins. The embryo begins to take on a C-shaped curvature due to increased embryonic flexion. A hallmark event at this stage is the ventro-caudal rotation and displacement of the primordial heart from its original rostro-dorsal position, an essential step in cardiovascular development. Concurrently, blood vessels begin to form in the chorion, umbilical vesicle, and embryo proper, signaling the initiation of early circulatory function. Additionally, foregut formation begins, and the otic disc (otic placode)—the early structure that will give rise to the inner ear—first becomes identifiable, marking early sensory organ development.

Stage 10 At approximately 23 days post-ovulation, the embryo measures between 1.5 and 3 mm in length. This stage is marked by rapid neural tube development, continued somite formation, and the onset of organogenesis. Typically, 4–12 pairs of somites are present (Fig. 11.8c). The neural folds elevate and fuse in a zipper-like manner, with the five major brain subdivisions, including the prosencephalon, mesencephalon, and rhombencephalon, becoming distinguishable (Noel 2024). The forming neural tube still features open rostral and caudal neuropores. The heart tube begins faint contractions, and the presumptive atrial and ventricular regions can be identified, alongside the initiation of cardiac looping (Hikspoors et al. 2022). Additional key features include the formation of the optic sulcus with

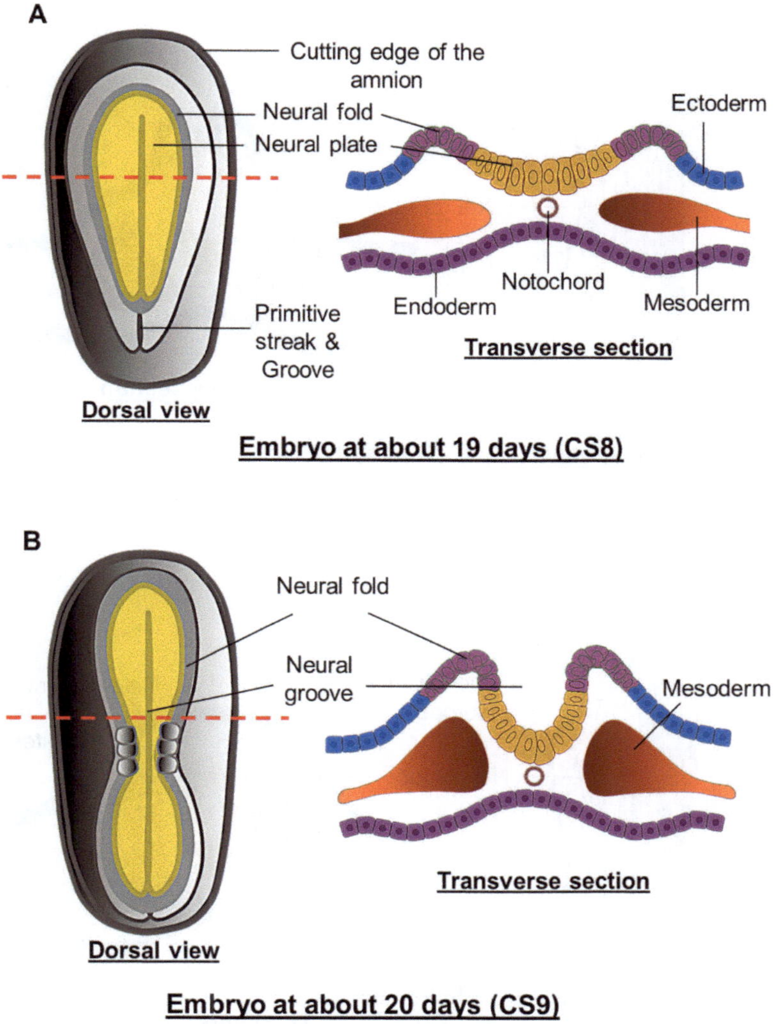

Fig. 11.8 Schematic representation highlighting the key changes of the early human embryonic development (Carnegie stages 8–12). Carnegie stages 8 (**a**), 9 (**b**), 10 (**c**), 11 (**d**), 12 (**e**, **f**). (Created in https://BioRender.com)

a midline chiasmatic plate, the emergence of the first pharyngeal arch, and the development of the laryngotracheal sulcus. The stomodeum, future mouth opening, is visible, as are the precursors to the thyroid gland and hepatic plate. The intermediate mesoderm becomes apparent, and the intraembryonic coelomic cavity begins to form within the lateral plate mesoderm, marking the early development of the body cavities.

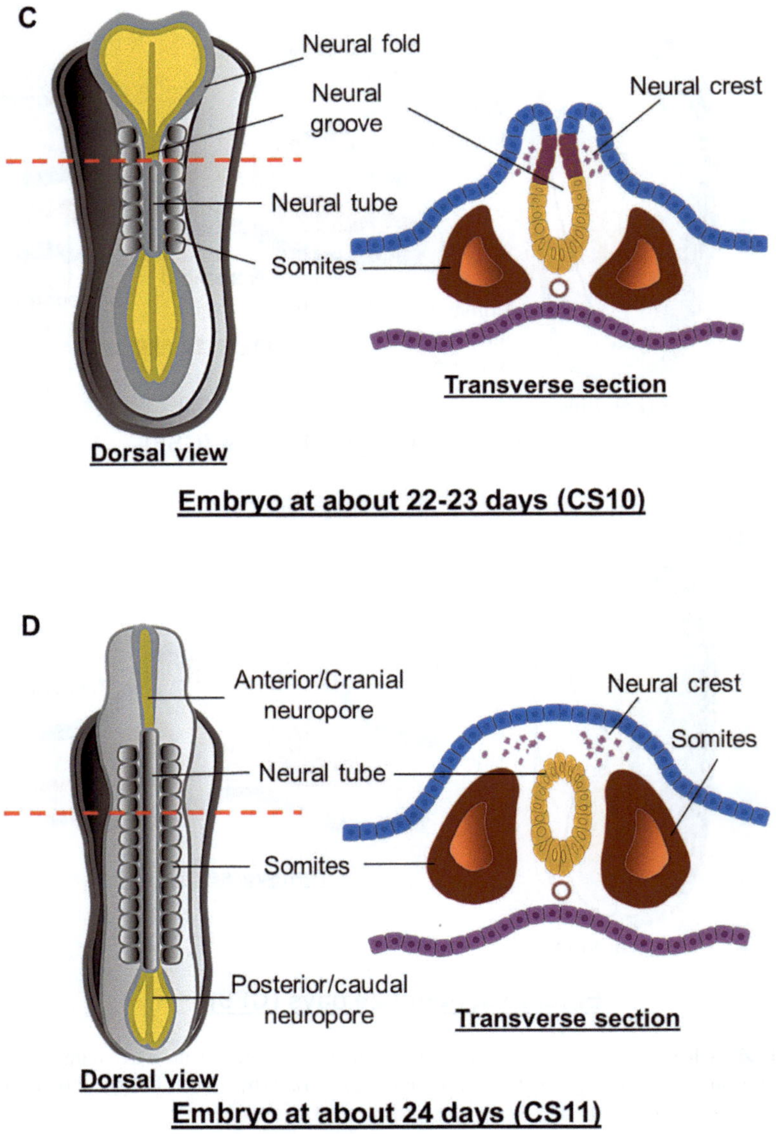

Fig. 11.8 (continued)

Stage 11 This stage occurs at approximately 24 post-ovulatory days, with the embryo measuring around 2.5–4.5 mm in length. It is characterized by the presence of 13–20 pairs of somites (Fig. 11.8d). Additional distinguishing features include the closure of the rostral neuropore at the anterior end of the neural tube. The hindbrain begins to show segmentation into rhombomeres. Optic sulci, or eye grooves, are visible on either side of the forebrain. The first and second pharyngeal arches: the

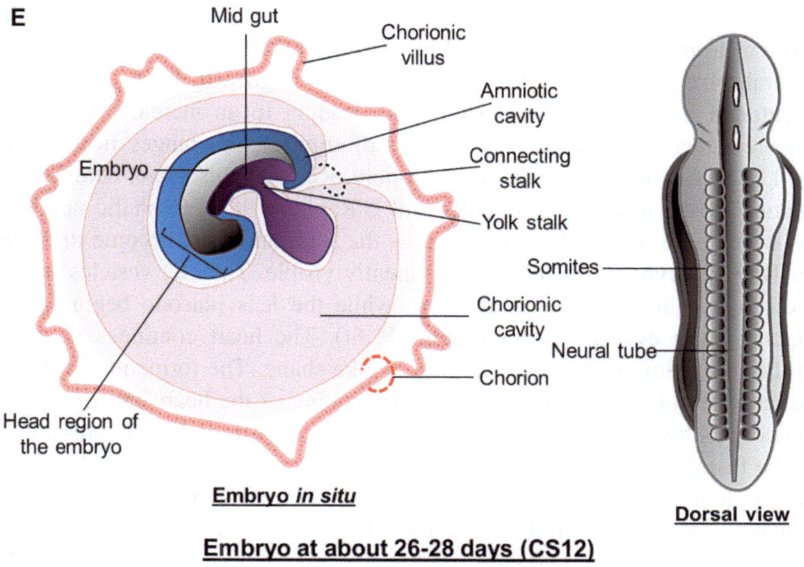

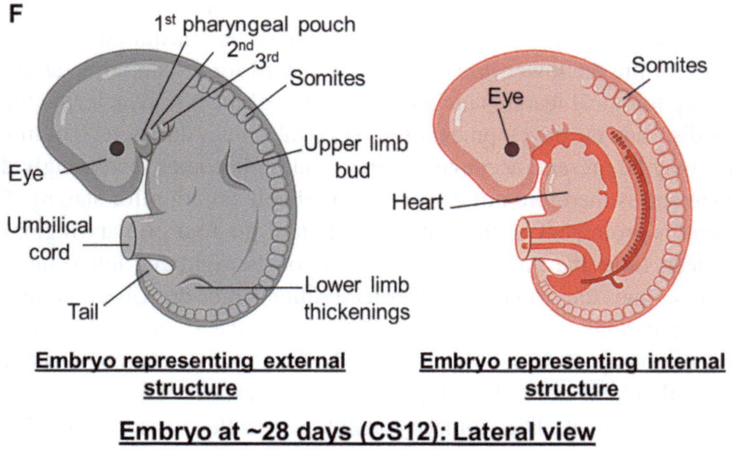

Fig. 11.8 (continued)

mandibular and hyoid arches become visible, which will contribute to the development of the face, neck, and pharynx (Fabik et al. 2021). Coelom formation, which began at stage 10, becomes more prominent at this point. The heart starts exhibiting peristaltic blood flow, and the development of the sinus venosus is underway (Hikspoors et al. 2022). The oropharyngeal membrane, separating the primitive digestive tract from the oral cavity, begins to rupture. Meanwhile, the mesonephric ducts and tubules, involved in early kidney development, start to form.

Stage 12 This stage occurs between 26 and 28 post-ovulatory days, with the embryo measuring approximately 3–4 mm in length. There is no significant increase in overall size compared to the previous stage, as growth during this period is concentrated in the thoraco-lumbar region, leading to an increase in embryonic curvature rather than length. The number of somite pairs ranges from 21 to 29 (Fig. 11.8e). Closure of the caudal neuropore occurs at this stage, completing the formation of the neural tube (Muller and O'Rahilly 1987). Cephalic neural crest cells, which give rise to structures such as the skull and face, become distinguishable. Three pharyngeal arches are prominently visible. The otic vesicles are in the process of closing but remain attached, while the lens placode begins to form, indicating early eye development (Fig. 11.8f). The heart continues to develop, with the four primary chambers starting to take shape. The formation of the interventricular septum begins, marking the partitioning of the heart into right and left circulatory pathways. The atrio-ventricular canal and aortic arches are present, along with the fusion of the dorsal aortae. Other notable developments include the emergence of lung buds (Sutliff and Hutchins 1994), the appearance of the dorsal pancreatic primordium, and the initial formation of limb buds.

Stage 13 At this stage of embryonic development, the embryo measures approximately 46 mm in length and is around 28 days post-ovulation, corresponding to the fifth week of gestation (Fig. 11.9a). All four limb buds are present. The upper limb buds, which initially appeared during stage 12, have now acquired a more defined morphology, while the lower limb buds are observed for the first time at this stage. Notable morphological features include the closure of the otic vesicle, the formation of the lens disc, and the development of the cervical sinus. Although 30 or more pairs of somites may be present, accurate somite counting becomes increasingly difficult at this stage and is therefore not considered a reliable criterion for staging. Cardiac development is prominent, with clear delineation of the four primary regions of the primitive heart tube: the trabeculated segments of the right and left ventricles, the conus cordis, and the truncus arteriosus (Gittenberger-de Groot et al. 2005). Although the cardiac valves have not yet formed, the initial appearance of the common pulmonary vein can be identified. The cardiac chambers are notably distended with blood cells. Development of the gastrointestinal and respiratory systems is also evident. The stomach exhibits a spindle-shaped morphology, and both the pancreas and liver are undergoing differentiation. The right and left lung buds are distinctly visible. Furthermore, glomerular structures are beginning to form within the mesonephros, and the neural tube is closed.

Stage 14 At this stage of development, the embryo measures approximately 5–7 mm in length and is about 32 days post-ovulation. A defining feature is the invagination of the lens disc, resulting in an open lens pit, within which the lens will subsequently develop (Fig. 11.9b). A characteristic dorsal indentation, located at the level of sclerotomes 5 and 6 and known as the Nackengrube of His, is now consistently observed. The upper limb buds are more prominent, extending outward and curving ventrally, with the distal tapering end that will later form the hand plate.

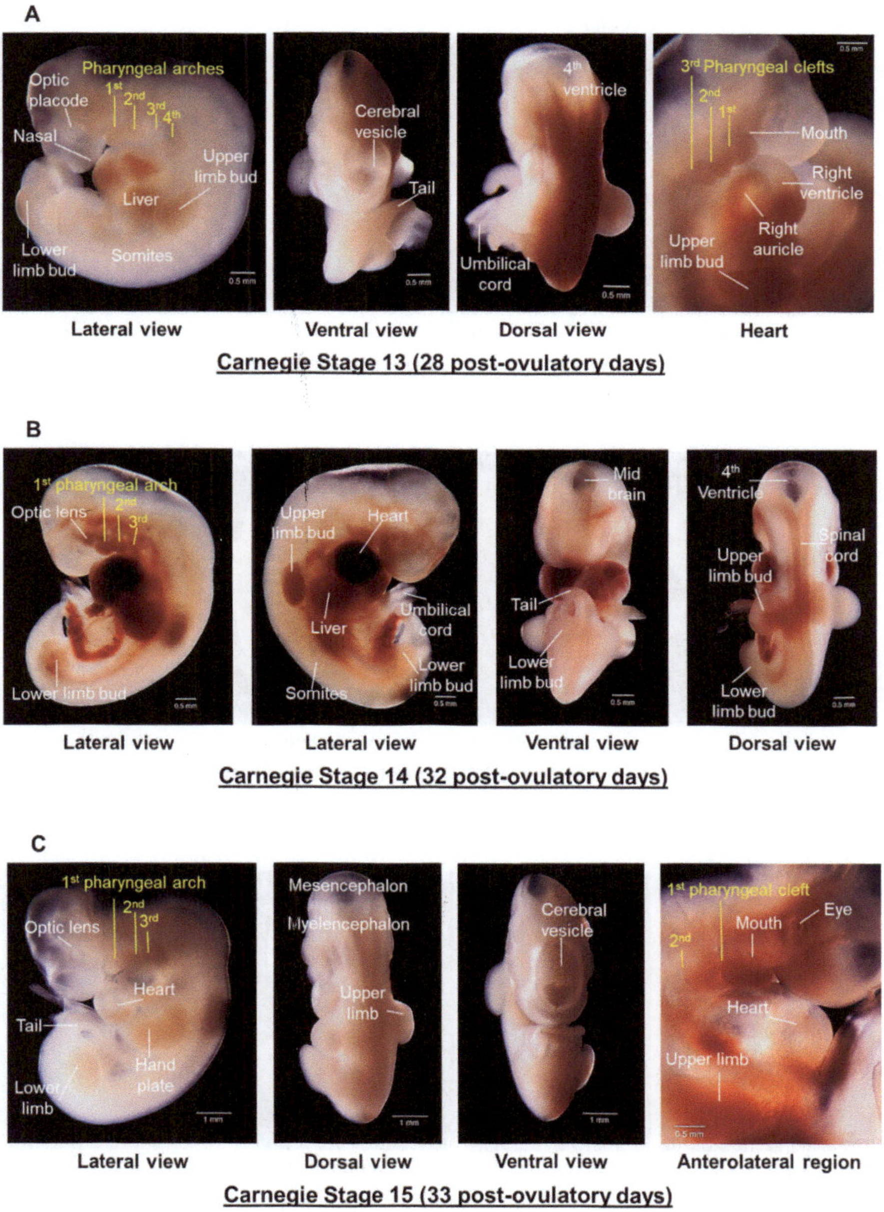

Fig. 11.9 Photographs of human embryos (Carnegie stages 13–23). Carnegie stages 13 (**a**), 14 (**b**), 15 (**c**), 16 (**d**), 17 (**e**), 18 (**f**), 19 (**g**), 20 (**h**), 22 (**i**), 23 (**j**). (The images were sourced from Multi-dimensional Human Embryo, Carnegie Stages, and are licensed under CC BY-NC-SA 4.0)

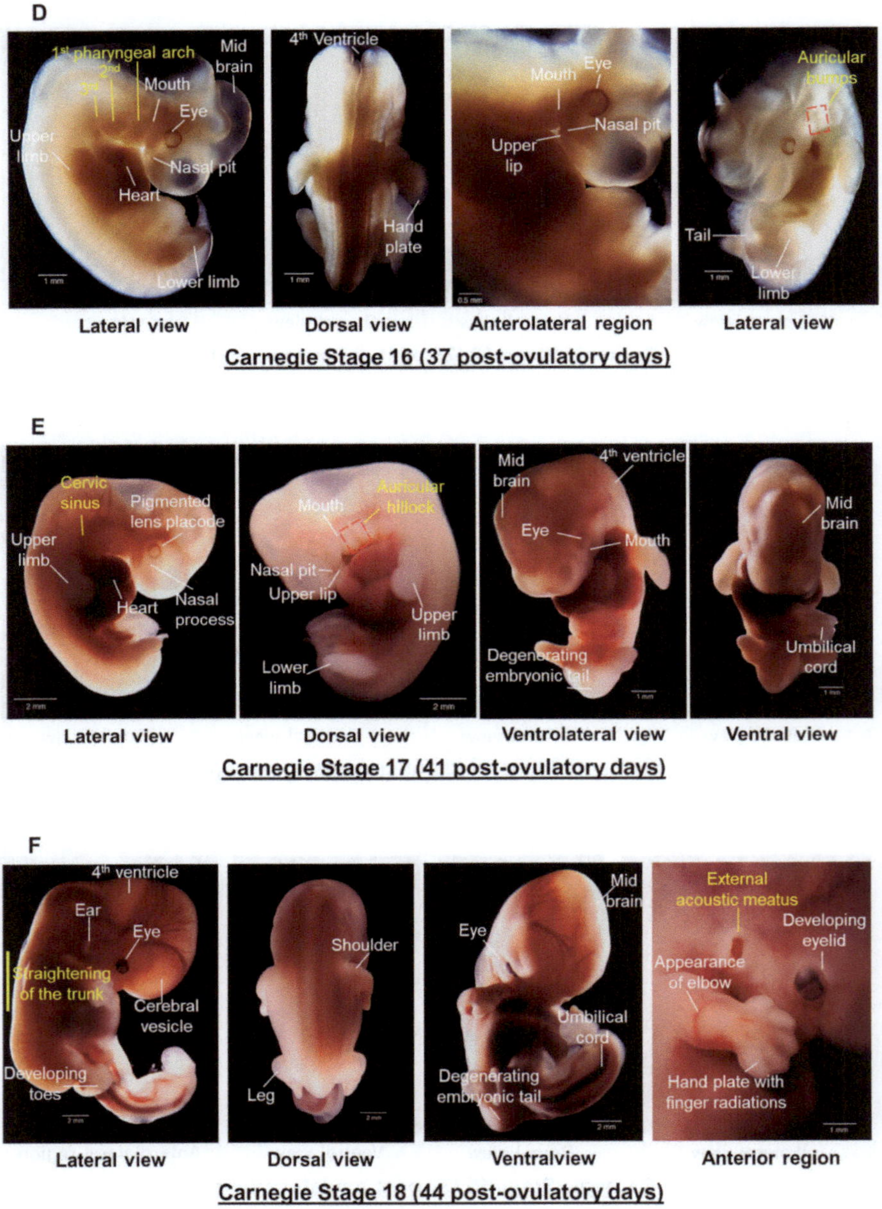

Fig. 11.9 (continued)

In contrast, the lower limb buds remain comparatively less developed. The endolymphatic ducts also begin to form during this stage. In the cardiovascular system, the four main subdivisions of the heart initially observed in the previous stage are now well-defined, and the sinoatrial node becomes identifiable (Sano and Yamagishi

G

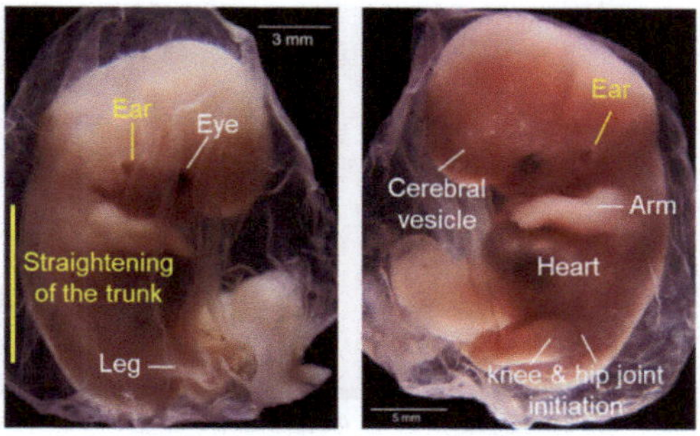

Lateral view Lateral view

Carnegie Stage 19 (47 post-ovulatory days)

H

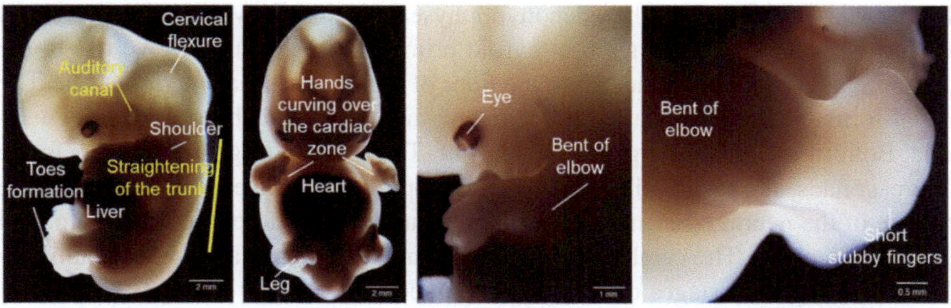

Lateral view Ventral view Dorsal view Forelimbs with short, stubby fingers

Carnegie Stage 20 (50 post-ovulatory days)

I

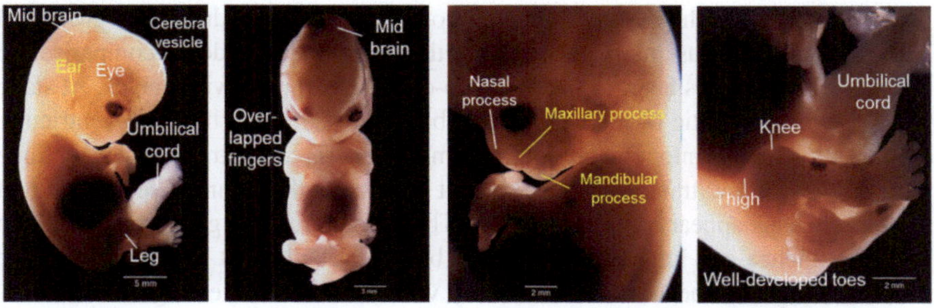

Lateral view Front view Anterolateral region Posterior region of the embryo

Carnegie Stage 22 (54 post-ovulatory days)

Fig. 11.9 (continued)

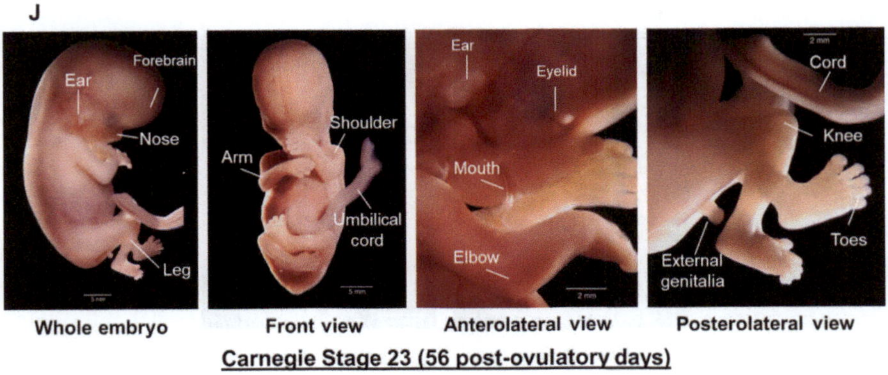

Fig. 11.9 (continued)

1965). The emergence of the adenohypophyseal (craniopharyngeal) structure is a key developmental milestone. The thyroid primordium, which will give rise to the thyroid gland, remains connected to the floor of the pharynx via a narrow stalk. The initial formation of intestinal coils is evident, and the trachea becomes distinguishable. Both right and left lung sacs are present. The gonadal ridges, precursors to the future gonads, are also visible. Neural development continues with the complete closure of the neural tube and the clear demarcation of the forebrain, midbrain, and hindbrain regions. The cerebral hemispheres and cerebellar plates are beginning to form, marking early stages of brain differentiation.

Stage 15 Embryos at this stage correspond to ~33 postovulatory days and are 7–9 mm in length. The characteristic external features observed at this stage are the closing of lens vesicles, the appearance of nasal pits, and the formation of hand plates (Fig. 11.9c). The width of the embryo's trunk increases due to the growth of spinal ganglia and muscular plates, making the embryo look wider in comparison to stage 14. The bulbous heart changes to two-chambered pump with valves. At stage 15, a key characteristic of the heart is that blood flow through the atrio-ventricular canal is already split into left and right streams. These separate streams then continue through the outflow tract and aortic sac. At this stage, the foramen secundum and semilunar cusps appear at the same time. The laryngotracheal groove lengthens to form the trachea and lung buds, while the bronchial buds start to branch out. The intestinal loop and caecum become clearly visible; lobar buds emerge in the bronchial tree; the ureteric pelvis begins to develop, and the primary urogenital sinus takes shape. The outlines of the future cerebral hemispheres become more distinct, the early paleostriatum and neostriatum can be identified, the primordium of the pineal gland becomes apparent, and retinal pigmentation starts to appear.

Stage 16 This stage occurs at approximately 37 post-ovulatory days with embryo measuring around 11–14 mm. This stage is characterized by various external and internal features. The forebrain is expanding, and the retinal pigmentation becomes visible (Fig. 11.9d). The nasal pits start to face ventrally. The second pharyngeal arch enlarges and becomes conspicuous. Auricular hillocks begin to appear. There is appearance of thickenings for semi-circular ducts. The foot plates are developing in the upper limb, whereas the thigh, leg, and foot are becoming distinguishable in the lower limb. The heart is prominent with the presence of the foramen secundum and visible chamber formation (Dhanantwari et al. 2009). The intestinal rotation has started. A definitive mesentery has also formed. The primordial muscle and skeletal system begin to organize at this stage.

Stage 17 At approximately 41 post-ovulatory days, the embryo corresponds to Carnegie stage 17 and measures about 11–14 mm in length. The embryonic axis is becoming more linear, although a slight curvature in the lumbar region may still be observed. The nasal pits deepen, and the upper lip begins to form (Fig. 11.9e). All six auricular hillocks are present on the first and second pharyngeal arches, indicating the early development of the external ear. Eyelid folds begin to appear as localized thickenings of the surface ectoderm. The hand plates now display distinct digital rays, while the foot plates take on a rounded shape with early digit formation. In the heart, the semilunar cusps are forming. The palate appears, and the vermiform appendix becomes distinguishable. The dorsal and ventral pancreatic buds begin to fuse. Additional features include the appearance of the auditory ossicles and the initial formation of the olfactory bulbs. Early ossification centers are evident in the humerus and radius, marking the onset of skeletal development in the upper limb.

Stage 18 At Carnegie stage 18, the embryo measures approximately 13–17 mm in length and is around 44–48 days post-ovulation. This stage is marked by the emergence and refinement of several key anatomical structures. In the limbs, digital rays become more pronounced in the hand plate, with intermetacarpal and intermetatarsal notches visible at the peripheries of the hand and foot plates (Fig. 11.9f). The metacarpals and metatarsals appear as precartilaginous condensations or cartilage, and many previously undifferentiated mesenchymal blastema now begin to form recognizable muscles and cartilages. The upper and lower limb buds, wrists, and feet are clearly defined. Craniofacial development advances significantly: the tip of the nose becomes distinguishable, along with the vomeronasal organ, eyelids, and nasolacrimal ducts. The auricular hillocks begin to merge, forming the external ear, and three pharyngeal arches (1st–3rd) are evident. Internally, the heart shows further maturation with the appearance of the second interatrial septum and the foramen ovale. The left postcardinal vein regresses near the heart, with its distal segment draining into the newly formed inferior vena cava. The diaphragm continues forming via peripheral growth from the septum transversum (Merrell and Kardon 2013), and the tertiary segmental bronchi are identifiable in the lungs. Additional features include the development of the phallus and genital tubercle (early glans penis or clitoris), the initial appearance of mammary glands on the thoracic ectoderm, and the formation of the paramesonephric ducts lateral to the cranial mesonephros.

Stage 19 At this stage, the embryo is estimated to be approximately 47–48 post-ovulatory days old, with a crown-rump length of about 17–20 mm. The trunk begins to elongate and straighten, resulting in the head no longer forming a right angle with the dorsal surface. Limb development progresses, with prominent toe rays visible on the foot plate, although interdigital notching has not yet occurred (Fig. 11.9g). The nasal pits deepen, and the initial formation of the upper lip becomes evident (Som and Naidich 2013). The auricular hillocks begin to merge, contributing to the development of the definitive external ear. Eyelid fold formation also commences during this period. Approximately 44–46 pairs of fully segmented somites are present. In the central nervous system, the cerebral hemispheres continue to expand, reflecting ongoing brain development. Prominent cardiac structures are visible, and the liver and umbilical cord show further differentiation.

Stage 20 Embryos at approximately 50–51 post-ovulatory days correspond to Carnegie stage 20 and typically measure between 21 and 23 mm in crown-rump length. From this stage onward, surface examination of internal structures becomes increasingly challenging due to mesodermal thickening. Consequently, only well-preserved fresh or formalin-fixed specimens are suitable for detailed anatomical analysis. The forelimbs show noticeable growth and exhibit a slight bend at the elbow. Fingers are now fully separated, with a significant reduction in inter-digital webbing (Fig. 11.9h). In contrast, the hindlimbs still display prominent webbing, and the toes are not yet completely separated. The eyelid folds have matured into definitive structures, and a few optic nerve fibers are now visible. The auricular hillocks have fused to form a distinct external ear (pinna), while in the inner ear, the cochlear duct begins to exhibit coiling (Hans et al. 2007). Cranial flexure persists, with ongoing expansion of the forebrain. The trunk continues to straighten as axial development progresses. At this stage, 47–49 pairs of somites can typically be identified. Gonadal differentiation also becomes more apparent. The testes appear elliptical with a smooth surface and contain straight seminiferous tubules. The ovaries, identified for the first time during this stage, appear cylindrical and are composed of undifferentiated oogonia.

Stage 21 At approximately 52 post-ovulatory days, the embryo reaches a crown-rump length of about 22–24 mm. At this stage, the fingers are fully separated, and the thumb becomes clearly distinguishable. The hands exhibit a slight flexion at the wrists. The feet begin to move toward the midline, and in some cases, the toes of opposite feet may come into contact. Auricular development progresses as the pinna becomes more elevated, gains defined contours, and begins to migrate upward from its initial low position. The upper lip is now completely formed, and the nasal openings have established external communication. The trunk continues to straighten, and the head assumes a more rounded appearance. The brain shows ongoing expansion, and for the first time, cortical plates appear within the walls of the cerebral hemispheres (Hadders-Algra 2018). Optic nerve fibers, which began forming in earlier stages, are now observed reaching the brain.

Stage 22 This stage occurs approximately between 54 and 56 post-ovulatory days, with the crown-rump length measuring around 25–27 mm. At this point in development, the head remains disproportionately large relative to the body, although the trunk begins to show signs of straightening (Fig. 11.9i). The eyelids, now thickened, cover about half of the visible portion of the eyes. The external ears are becoming more defined, with clearer differentiation of the tragus and antitragus (Honkura et al. 2020). Limb development is notable, with marked elongation and separation of the digits, though residual webbing between the fingers may still be present. The hands show increasing mobility, and in some cases, the fingers of one hand may overlap those of the other. Salivary gland development progresses, with the appearance of secondary branching. Skeletal development enters the cartilaginous phase, and early signs of ossification are observed. The brain continues to enlarge, and cortical plates begin to form in the cerebral hemispheres. The superior and inferior colliculi of the midbrain exhibit initial lamination, and the cranial nerves become distinguishable. In the cardiovascular system, the formation of the chordae tendineae within the developing heart begins (Buijtendijk et al. 2020). Overall, this stage is characterized by significant morphogenesis and increasing internal reorganization, laying the groundwork for the subsequent stage.

Stage 23 This stage, occurring around 56–57 post-ovulatory days, marks the transition from the embryonic to the fetal period. It corresponds to the end of the eighth week of gestation. At this point, the crown-rump length measures approximately 27–31 mm. Most organ systems are already formed, although they remain in an immature state. The head appears distinctly rounded, and the eyelids have begun to separate (Fig. 11.9j). External ears are clearly visible and prominent. The limbs are elongated, fully segmented, and display well-defined fingers and toes, with the early development of nails. Ossification centers have begun to appear in the skeletal system, indicating the onset of bone formation. Significant brain development is also evident. The rhombencephalon shows advanced organization, and cerebral commissures have emerged. The brainstem is beginning to display nuclei, fiber tracts, and commissures (Stiles and Jernigan 2010). Around 33–34 cartilaginous vertebrae are present, although the spinous processes have not yet formed. Joint cavitation has also begun. The heart is now fully partitioned and functioning, with clearly identifiable arteries. The superior vena cava is obliterated at this stage. While the external genitalia are well developed, they are not yet sexually differentiated. Testicular tubules can also be identified. By the end of this stage, the basic structures for all major organ systems are in place, laying the groundwork for continued growth and maturation during the fetal period. The preserved human fetal specimens displayed in museum collections are shown in Fig. 11.10.

11.4.2 Abnormal Embryos

Abnormal embryos refer to embryos that exhibit deviations from normal development due to genetic, environmental, or unknown factors. These abnormalities can manifest at various stages of embryogenesis and may affect cell division,

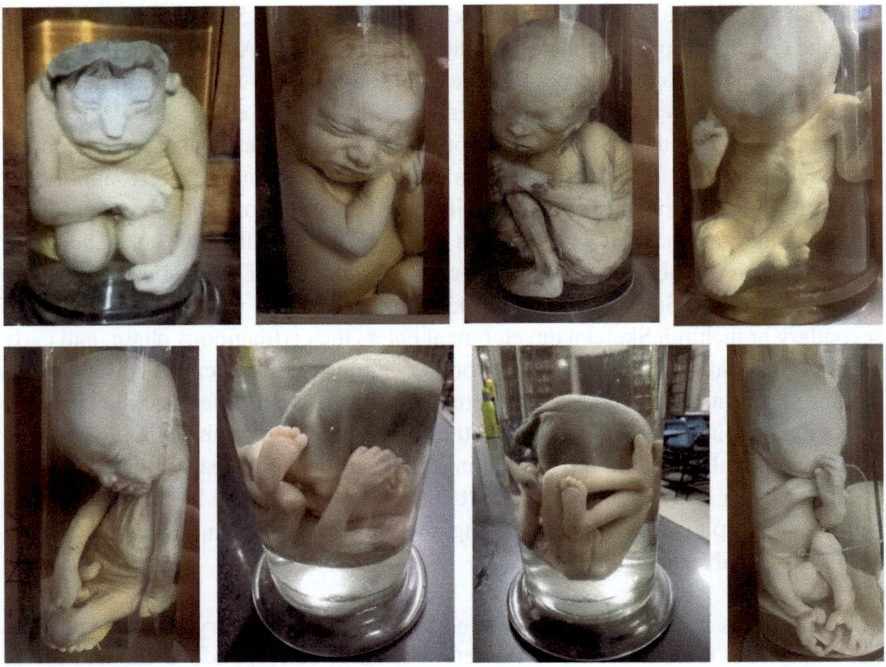

Fig. 11.10 Photographs of human fetus. (Source: Museum, Department of Zoology, University of Delhi, and Department of Zoology, Daulat Ram College, University of Delhi, India)

differentiation, morphogenesis, or organ formation. Many embryos with abnormalities fail to implant or are lost early in pregnancy, while embryos with serious defects often result in spontaneous abortion (Fig. 11.11). Structural anomalies in non-vital organs like limbs or facial features may not affect fetal survival but can pose risks after birth. Approximately 5% of newborns have detectable mild to severe malformations (Verma 2021). The abnormalities could range from numerical chromosomal abnormalities (e.g., Down syndrome, Turner syndrome) and structural chromosomal abnormalities to structural birth defects (limb malformations, congenital diaphragmatic hernia, neural tube defects, abdominal wall defects (Bence and Wagner 2021), heart defects, cleft lip or palate).

The cause of abnormalities in embryos could be from varied sources, including lack of folic acid whose deficiency could lead to neural tube defects (e.g., anencephaly, pina bifida), infections caught by the mother during pregnancy (e.g., Infection with Zika virus results in microcephaly, rubella leads to heart defects and deafness in developing embryo), medication or drugs taken by the mother during pregnancy (e.g., isotretinoin present in acne drugs increases the risk of physical deformities in more than 35% growing embryo), exposure to environmental pollutants (e.g., phthalates, heavy metals -arsenic, cadmium), weight of the mother (obesity is linked with hormonal imbalance that could lead to congenital heart defects and cleft palate in the embryo), and the maternal age (advanced maternal age increases the risk of chromosomal non-disjunction during meiosis, genetic disorders such as Down

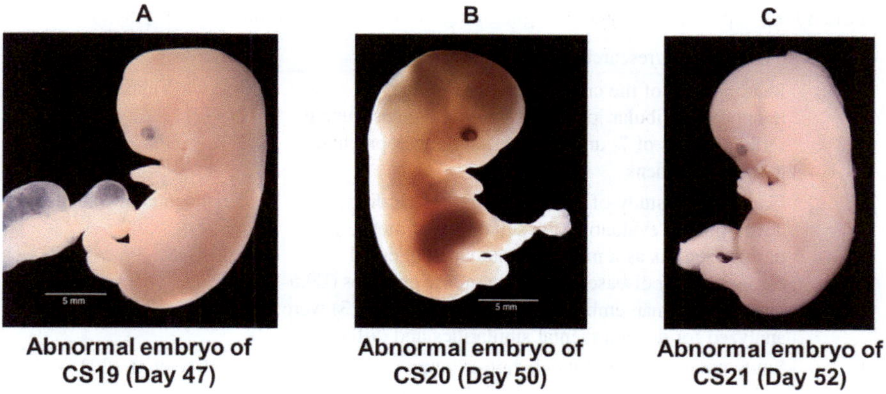

Fig. 11.11 Photographs of abnormal human embryos. Carnegie stages 19 (**a**), 20 (**b**), and 21 (**c**). (The images were sourced from Multi-dimensional Human Embryo, Carnegie Stages, and are licensed under CC BY-NC-SA 4.0)

syndrome and Turner syndrome). Consumption of alcohol and smoking during pregnancy could also lead to fetal alcohol syndrome and increased risk of problems such as defective lung development, low birth weight, and increased risk of death in infancy.

11.5 Significance/Research Applications

Understanding human embryonic development is vital for uncovering the origins of congenital anomalies. Studying normal developmental progression enables researchers and healthcare professionals to identify the timing and causes of disruptions that lead to structural or functional abnormalities. This knowledge underpins early detection, preventive strategies, and treatment of conditions such as neural tube defects, cardiac malformations, and limb deformities.

Beyond clinical relevance, insights into embryonic development significantly advance stem cell research by clarifying the mechanisms of stem cell fate determination and differentiation, and by identifying factors that guide embryogenesis. Such knowledge allows researchers to mimic developmental processes in vitro, optimize conditions for embryonic growth, and improve assisted reproductive technologies like IVF. Table 11.1 lists a few representative studies of the relevance of human embryology knowledge in biomedical research.

11.6 Precautions

- Handle photomicrographs with care to avoid scratches or damage to visual data.
- Ensure proper labelling and identification of developmental stages to prevent confusion.
- Follow ethical guidelines when discussing human embryonic material.

Table 11.1 Representative studies highlighting the significance of human embryology

S. no.	Type of study/research	Reference
1.	Identification of the critical period of human temporomandibular joint (TMJ) morphogenesis using serial sections of 7- and 17-week-old developmental human specimens	Merida-Velasco et al. (1999)
2.	Comparative study of mouse and human cardiac development, evaluating the suitability of mouse morphogenesis as a model for human congenital cardiovascular disease. Wild-type mouse embryos (E9.5–birth) and human embryos (Carnegie stage 13–23) were analyzed for developmental similarities and differences	Krishnan et al. (2014)
3.	Study the formation of the definitive human notochord to understand the pathogenesis of birth defects associated with notochord using sectioned 17–41 day-old human embryos (Carnegie stage 8–15)	de Bree et al. (2018)
4.	Generation of synthetic embryology	Cornwall-Scoones and Zernicka-Goetz (2021)
5.	Study to understand the cause of mosaicism in human embryos	Capalbo et al. (2021)
6.	To understand the process of neural tube formation as well as tail development in humans and its comparison to mice, a study that could facilitate the interpretation of research findings, including the recapitulation of neurulation events in in vitro models such as stem cell-derived organoids using human embryos (Carnegie stages 10–18)	Santos et al. (2024)

11.7 Quick Review Questions

11.7.1 Multiple Choice Questions

1. The Carnegie staging system is based on:
   (a) Embryo size alone
   (b) Chronological age in days
   (c) Morphological features and developmental landmarks
   (d) Genetic markers
   Answer: c) Morphological features and developmental landmarks
2. The zygote formed after fertilization is:
   (a) Haploid
   (b) Diploid
   (c) Polyploid
   (d) Aneuploid
   Answer: b) Diploid

3. In mammals, the outer layer of blastocyst cells that contributes to placenta formation is:
   (a) Epiblast
   (b) Hypoblast
   (c) Trophoblast
   (d) Inner cell mass
   Answer: c) Trophoblast
4. The inner cell mass (ICM) of the blastocyst develops into:
   (a) Placenta
   (b) Extraembryonic membranes only
   (c) Embryo proper
   (d) Yolk sac only
   Answer: c) Embryo proper
5. Gastrulation in mammals begins with the formation of:
   (a) Neural tube
   (b) Primitive streak
   (c) Somites
   (d) Blastocyst
   Answer: b) Primitive streak
6. Which of the following is a derivative of the endoderm?
   (a) Muscles
   (b) Nervous system
   (c) Lining of the gut
   (d) Skeleton
   Answer: c) Lining of the gut
7. The neural tube in mammalian embryos ultimately gives rise to the:
   (a) Digestive tract
   (b) Brain and spinal cord
   (c) Circulatory system
   (d) Reproductive organs
   Answer: b) Brain and spinal cord
8. In human embryonic development, implantation of the blastocyst into the uterine wall occurs around:
   (a) Day 1–2 after fertilization
   (b) Day 6–7 after fertilization
   (c) Day 14 after fertilization
   (d) Day 28 after fertilization
   Answer: b) Day 6–7 after fertilization
9. Cavitation marks the transition of:
   (a) Zygote to morula
   (b) Morula to blastocyst
   (c) Blastocyst to gastrula
   (d) Oocyte to zygote
   Answer: b) Morula to blastocyst

10. Somites are derived from:
    (a) Ectoderm
    (b) Endoderm
    (c) Paraxial mesoderm
    (d) Intermediate mesoderm
    Answer: c) Paraxial mesoderm

11.7.2 Short Answer Questions

1. Define the Carnegie staging system and state its significance in human embryology.
2. Differentiate between a morula and a blastocyst.
3. Write the importance of the inner cell mass (ICM) in mammalian development.
4. State the role of the trophoblast in implantation.
5. Define neurulation. Which germ layer gives rise to the neural tube?
6. Name the three primary germ layers formed during gastrulation and their main derivatives.
7. Define implantation and mention its usual site in humans.
8. Explain the developmental role of the placenta.
9. At what Carnegie stage does the primitive streak first appear?
10. What is the function of the notochord in embryonic development?

Glossary

Amnion A thin, elastic membrane, forms the amniotic sac filled with fluid, protecting the developing embryo and regulating temperature.

Amniotic cavity Fluid-filled space within the amniotic sac that protects the growing embryo.

Amniotic fluid A watery substance present in the amniotic cavity that nourishes and protects the embryo during pregnancy.

Anencephaly A severe neural tube defect characterized by the absence of major portions of the brain, skull, and scalp due to incomplete closure of the neural tube during early embryonic development.

Aortic Sac An embryonic vascular structure at the distal end of the truncus arteriosus that gives rise to the aortic arches, contributing to the formation of major arteries.

Atrio-ventricular canal An embryonic passage connecting the primitive atrium and ventricle, which later partitions to form the atrioventricular valves and septa of the heart.

Auditory ossicles A chain of three small bones, namely malleus, incus, and stapes, in the middle ear that transmit and amplify sound vibrations from the eardrum to the inner ear.

Auricular hillocks Small mesenchymal swellings that appear around the first and second pharyngeal arches in the embryo and later fuse to form the external ear.

Blastomeres The early embryonic cells formed by the division of a fertilized egg during embryonic development.

Blastocoel Fluid-filled central cavity that forms within the blastula during embryonic development.

Blastocoel cavity The fluid-filled central cavity of the blastula that provides space for cell migration during early embryonic development.

Blastocyst An early-stage embryo consisting of an outer trophoblast layer, an inner cell mass, and a fluid-filled cavity, formed around five to six days after fertilization.

Blastocystic cavity The fluid-filled space within the blastocyst that separates the inner cell mass from the surrounding trophoblast.

Blastomeres The individual cells produced by the early cleavage divisions of a fertilized egg.

Brainstem The lower part of the brain that connects to the spinal cord and controls essential functions such as breathing, heart rate, and reflexes.

Bronchial buds Embryonic outgrowths from the foregut that develop into the bronchi, lungs, and the branching respiratory tree.

Bronchial tree The branching system of airways within the lungs, starting from the trachea and dividing into bronchi and bronchioles, which conduct air to the alveoli.

Cardiac chambers The four compartments of the heart (two atria and two ventricles) that receive blood and pump it to the lungs and the rest of the body.

Cardiac valves Structures within the heart that ensure unidirectional blood flow by opening and closing in response to pressure changes between chambers.

Cardiovascular system The organ system composed of the heart and blood vessels that circulates blood, delivering oxygen and nutrients while removing wastes from the body.

Cavitation The developmental process in which small fluid-filled spaces arise within a solid cell mass, later coalescing to form a central cavity such as the blastocoel.

Cerebellar plates Embryonic thickenings of the alar region in the hindbrain that develop into the cerebellum.

Cerebral commissures Bundles of nerve fibers that connect corresponding regions of the two cerebral hemispheres, enabling communication between them.

Cerebral hemispheres The two symmetrical halves of the cerebrum responsible for higher brain functions such as perception, thought, memory, and voluntary movement.

Cervical sinus A transient embryonic depression on the lateral neck formed by the overgrowth of the second pharyngeal arch, which normally disappears during development.

Chordae tendineae Thin, fibrous cords in the heart that anchor the atrioventricular valve leaflets to the papillary muscles, preventing valve prolapse during ventricular contraction.

Chorionic villi Finger-like projections of the chorion that extend into the uterine lining and facilitate nutrient, gas, and waste exchange between maternal and fetal blood.

Cleavage Repeated mitotic divisions of the zygote that lead to the formation of blastomeres.

Cleft lip A congenital defect caused by incomplete fusion of the maxillary and medial nasal processes, resulting in a gap or opening in the upper lip.

Conceptus The collective term for all structures derived from the zygote, including the embryo or fetus, along with its associated membranes and tissues.

Conception The process by which a sperm cell fertilizes an egg cell, resulting in the formation of a zygote.

Congenital anomalies Structural or functional abnormalities that occur during intrauterine life and present at the time of birth.

Congenital diaphragmatic hernia A birth defect in which an abnormal opening in the diaphragm allows abdominal organs to move into the chest cavity, impairing lung development.

Cortical plates Layers of developing neurons in the embryonic brain that give rise to the cerebral cortex.

Cranial mesonephros The anterior portion of the embryonic mesonephros that contributes to early kidney function before regressing as permanent kidneys develop.

Craniofacial Relating to the bones and soft tissues of the skull and face, particularly in the context of development, anatomy, or congenital anomalies.

Crown-rump The measurement from the top of the embryo or fetus's head (crown) to the base of the buttocks (rump), commonly used to estimate gestational age.

Cytotrophoblastic The inner layer of trophoblast cells in the early embryo that serve as progenitors for the syncytiotrophoblast and contribute to placental development.

Diagrams Labeled visual representations that conceptually illustrate the structure, process, or subject under study.

Digital rays Embryonic mesenchymal condensations within the hand and foot plates that outline the future fingers and toes.

Down syndrome A genetic disorder caused by the presence of an extra copy of chromosome 21 (trisomy 21), characterized by intellectual disability, distinct facial features, and various health complications.

Ectoderm The outermost germ layer of the embryo that gives rise to structures such as the skin, nervous system, and sensory organs.

Elbow The joint connecting the upper arm (humerus) with the forearm (radius and ulna), allowing flexion, extension, and rotational movements.

Embryo The early stages of development in multicellular organisms post fertilization and preceding the fetal stage.

Endoblast The innermost layer of cells in the embryo that gives rise to the endoderm.

Endoderm The innermost germ layer of the embryo that develops into the lining of the digestive tract, respiratory system, and associated organs.

Endolymphatic ducts Embryonic tubular structures of the inner ear that regulate endolymph fluid balance and connect the membranous labyrinth to the endolymphatic sac.

Endometrium The inner mucosal lining of the uterus that undergoes cyclic changes and provides the site for implantation and support of the developing embryo.

Epiblast The upper cell layer of the bilaminar embryonic disc that gives rise to all three germ layers during gastrulation.

Exocoelomic (Heuser's) membrane A thin membrane formed by hypoblast cells lining the blastocyst cavity, contributing to the formation of the primary yolk sac.

Extraembryonic mesoblasts Mesodermal cells that develop outside the embryo proper, forming structures like the chorion, amnion, and connecting stalk to support embryonic development.

Fetus The later stages of human embryonic development from the ninth week of pregnancy till birth.

First pharyngeal arch An embryonic structure that gives rise to the maxilla, mandible, muscles of mastication, and parts of the middle ear.

Foramen ovale An opening in the fetal heart's interatrial septum that allows blood to bypass the lungs by flowing directly from the right atrium to the left atrium.

Foramen secundum An opening in the fetal interatrial septum that permits blood flow between the atria, contributing to normal fetal circulation.

Germ layers The three primary layers of the embryo, ectoderm, mesoderm, and endoderm, that give rise to all tissues and organs of the body.

Gestation The period of development from conception to birth during which the embryo or fetus grows and matures in the uterus.

Hand plate The flattened, paddle-like distal end of the embryonic upper limb bud that gives rise to the fingers.

Holoblastic cleavage Complete cleavage of the egg.

Humerus The long bone of the upper arm extending from the shoulder to the elbow, forming joints with the scapula and the forearm bones.

Hypoblast The lower cell layer of the bilaminar embryonic disc that contributes to the formation of extraembryonic structures like the yolk sac.

Inferior colliculi Paired structures in the midbrain that act as principal auditory centers, integrating and relaying sound information to higher brain regions.

Inferior vena cava The large vein that carries deoxygenated blood from the lower half of the body back to the right atrium of the heart.

Inner cell Mass A cluster of cells inside the blastocyst that gives rise to the embryo proper and some extraembryonic tissues.

Interatrial septum The muscular and membranous wall that separates the right and left atria of the heart.

Inter-digital webbing The presence of skin connecting adjacent fingers or toes, which normally regresses during embryonic development.

Intermetacarpal notches Indentations between the bases of adjacent metacarpal bones that allow articulation and flexibility of the hand.

Intermetatarsal notches Small depressions between the bases of adjacent metatarsal bones that facilitate articulation and stability of the foot.

Lens disc A thickened, circular region of ectoderm in the developing eye that invaginates to form the lens vesicle.

Lens pit The invaginated depression of the lens placode that eventually pinches off to form the lens vesicle during eye development.

Lens vesicles A hollow, spherical structure formed from the invaginated lens pit that develops into the lens of the eye.

Lobar buds Secondary outgrowths from the primary bronchial buds that give rise to the individual lobes of the lungs during embryonic development.

Lumbar region The lower part of the back between the thorax and pelvis, comprising the five lumbar vertebrae.

Mesenchymal blastema A mass of undifferentiated mesenchymal cells in the embryo that differentiates to form connective tissues, bones, or other structures.

Mesentery A double layer of peritoneum that suspends the intestines from the posterior abdominal wall and contains blood vessels, nerves, and lymphatics.

Mesoderm The middle germ layer of the embryo that gives rise to muscles, bones, the cardiovascular system, and connective tissues.

Mesonephros A transient embryonic kidney that functions temporarily before the development of the permanent metanephros.

Microcephaly A congenital condition characterized by an abnormally small head and often associated with impaired brain development.

Morula An early-stage embryo consisting of a solid ball of blastomeres resulting from successive cleavage of the fertilized egg.

Muscular plates Embryonic regions of mesoderm that differentiate to form skeletal muscles in the developing body.

Nasal pits Depressions in the embryonic fronto-nasal prominence that develop into the nostrils and nasal cavities.

Nasolacrimal ducts Tubes that develop from the embryonic nasolacrimal groove, connecting the lacrimal sac to the nasal cavity to drain tears.

Neostriatum A major part of the basal ganglia in the brain, comprising the caudate nucleus and putamen, involved in motor control and learning.

Neurenteric canal A transient embryonic passage connecting the amniotic cavity and yolk sac through the notochordal region, playing a role in early gut and notochord development.

Olfactory bulbs Paired neural structures located at the forebrain's base that receive sensory input from the nasal epithelium and process the sense of smell.

Oocyte A female germ cell or egg cell that undergoes maturation to become capable of fertilization.

Ossification The biological process by which bone tissue is formed from mesenchymal or cartilage templates.

Ossification centers Specific sites in developing bone where bone tissue first begins to form during skeletal development.

Otic vesicle A fluid-filled, sac-like structure derived from the otic placode that gives rise to the inner ear, including the cochlea and vestibular apparatus.

Ovaries Paired female reproductive organs that produce oocytes (eggs) and secrete hormones such as estrogen and progesterone.

Palate The roof of the mouth that separates the oral and nasal cavities, consisting of the hard palate anteriorly and the soft palate posteriorly.

Paleostriatum A region of the basal ganglia in the brain, primarily comprising the globus pallidus, involved in the regulation of voluntary movement.

Paramesonephric ducts Embryonic ducts that develop into the female reproductive tract, including the fallopian tubes, uterus, and part of the vagina.

Pharyngeal arches The temporary structures developed in the early stages of human embryonic development that later form the face and neck.

Pharynx A muscular tube connecting the nasal and oral cavities to the larynx and esophagus, serving both respiratory and digestive functions.

Photomicrographs High-resolution microscopic images showing the actual structures of the specimen under study.

Phthalates A group of chemical compounds used as plasticizers to increase flexibility in plastics, often associated with endocrine-disrupting effects.

Pina bifida A congenital neural tube defect in which the vertebral arches fail to close completely, potentially exposing the spinal cord.

Placenta A temporary organ in pregnancy that facilitates nutrient, gas, and waste exchange between the mother and developing fetus while producing hormones to support gestation.

Polyspermy Penetration and fertilization of an ovum by more than one sperm.

Postcardinal vein An embryonic vein that drains blood from the body wall and contributes to the formation of the inferior vena cava and other major veins.

Post-ovulatory days The days following ovulation during which the ovum is available for fertilization and the endometrium undergoes changes to support potential implantation.

Radius The lateral bone of the forearm, extending from the elbow to the wrist, involved in rotation and movement of the forearm.

Retina The light-sensitive inner layer of the eye that contains photoreceptor cells and converts visual information into nerve signals for the brain.

Rhombencephalon The embryonic hindbrain that develops into the metencephalon (pons and cerebellum) and myelencephalon (medulla oblongata).

Rubella A contagious viral infection, also called German measles, that can cause mild illness in children and serious congenital defects if contracted during pregnancy.

Sclerotomes Portions of the somites in the developing embryo that give rise to the vertebrae and rib cartilage.

Second pharyngeal arch An embryonic structure that contributes to the formation of the stapes, styloid process, part of the hyoid bone, and muscles of facial expression.

Semilunar cusps Crescent-shaped flaps of the aortic and pulmonary valves that prevent backflow of blood from the arteries into the ventricles.

Seminiferous tubules Coiled structures within the testes where sperm production (spermatogenesis) takes place.

Septum transversum A thick mass of mesoderm in the embryo that contributes to the formation of the diaphragm and part of the liver.

Sinoatrial node A specialized group of cardiac cells in the right atrium that generates electrical impulses, acting as the heart's natural pacemaker.

Skeletal System relating to the bones of the body that provide structural support, protection of organs, and attachment points for muscles.

Somites Temporary paraxial mesodermal blocks that form vertebral columns and skeletal muscles in the embryo.

Spinal ganglia Clusters of sensory neuron cell bodies located along the spinal cord that transmit sensory information from the body to the central nervous system.

Spinous processes Bony projections extending posteriorly from each vertebra, serving as attachment points for muscles and ligaments of the spine.

Superior colliculi Paired structures in the midbrain that coordinate visual reflexes and control eye movements.

Superior vena cava A large vein that carries deoxygenated blood from the upper half of the body back to the right atrium of the heart.

Syncytiotrophoblast The outer multinucleated layer of the trophoblast that invades the uterine lining and facilitates nutrient exchange between mother and embryo.

Testes Paired male reproductive organs that produce sperm and secrete male sex hormones, primarily testosterone.

Testicular tubules Coiled tubules within the testes where sperm is produced and begins its maturation process.

Third pharyngeal arch An embryonic structure that forms parts of the hyoid bone and contributes to the development of the stylopharyngeus muscle and portions of the carotid arteries.

Totipotent The potency of cells to form a complete organism even after separation.

Teratogen External agents that cause congenital abnormalities in the fetus during embryonic development.

Turner syndrome A chromosomal disorder in females characterized by the partial or complete absence of one X chromosome, leading to short stature, infertility, and other physical features.

Umbilical cord A flexible cord connecting the fetus to the placenta, containing blood vessels that transport nutrients, oxygen, and waste between mother and fetus.

Ureteric pelvis The funnel-shaped upper part of the ureter that collects urine from the kidney's calyces before passing it to the ureter.

Urogenital sinus An embryonic cavity that gives rise to parts of the urinary and reproductive tracts, including the bladder and urethra.

Uteroplacental Functional connection between the uterus and placenta that facilitates nutrient, gas, and waste exchange between mother and fetus.

Vertebrae The individual bones that form the spinal column, providing structural support, protecting the spinal cord, and allowing flexible movement.

Vomeronasal organ A chemoreceptive structure in the nasal cavity involved in detecting pheromones and mediating certain social and reproductive behaviors.

Zika virus A mosquito-borne virus that can cause mild illness in adults but is associated with severe birth defects, including microcephaly, when infection occurs during pregnancy.

Zona pellucida An outer protective glycoprotein layer of the mammalian oocyte that acts as a selective filter before fertilization.

References

Acharya UV, Pendharkar H, Varma DR, Pruthi N, Varadarajan S (2017) Spinal dysraphism illustrated; Embroyology revisited. *Indian J Radiol Imaging* 27(4):417–426. https://doi.org/10.4103/ijri.IJRI_451_16

Bence CM, Wagner AJ (2021) Abdominal wall defects. *Transl Pediatr* 10(5):1461–1469. https://doi.org/10.21037/tp-20-94

Buckingham M, Bajard L, Chang T, Daubas P, Hadchouel J, Meilhac S, Montarras D, Rocancourt D, Relaix F (2003) The formation of skeletal muscle: from somite to limb. J Anat 202(1):59–68. https://doi.org/10.1046/j.1469-7580.2003.00139.x

Buijtendijk MFJ, Barnett P, van den Hoff MJB (2020) Development of the human heart. Am J Med Genet C Semin Med Genet 184(1):7–22. https://doi.org/10.1002/ajmg.c.31778

Capalbo A, Poli M, Rienzi L, Girardi L, Patassini C, Fabiani M, Cimadomo D, Benini F, Farcomeni A, Cuzzi J, Rubio C, Albani E, Sacchi L, Vaiarelli A, Figliuzzi M, Findikli N, Coban O, Boynukalin FK, Vogel I, Hoffmann E, Livi C, Levi-Setti PE, Ubaldi FM, Simon C (2021) Mosaic human preimplantation embryos and their developmental potential in a prospective, non-selection clinical trial. Am J Hum Genet 108(12):2238–2247. https://doi.org/10.1016/j.ajhg.2021.11.002

Cornwall-Scoones J, Zernicka-Goetz M (2021) Unifying synthetic embryology. Dev Biol 474:1–4. https://doi.org/10.1016/j.ydbio.2021.03.007

Coticchio G, Lagalla C, Sturmey R, Pennetta F, Borini A (2019) The enigmatic morula: mechanisms of development, cell fate determination, self-correction and implications for ART. Hum Reprod Update 25(4):422–438. https://doi.org/10.1093/humupd/dmz008

Coy P, Garcia-Vazquez FA, Visconti PE, Aviles M (2012) Roles of the oviduct in mammalian fertilization. Reproduction 144(6):649–660. https://doi.org/10.1530/REP-12-0279

Dard N, Breuer M, Maro B, Louvet-Vallee S (2008) Morphogenesis of the mammalian blastocyst. Mol Cell Endocrinol 282(1–2):70–77. https://doi.org/10.1016/j.mce.2007.11.004

de Bree K, de Bakker BS, Oostra RJ (2018) The development of the human notochord. PLoS One 13(10):e0205752. https://doi.org/10.1371/journal.pone.0205752

Dhanantwari P, Lee E, Krishnan A, Samtani R, Yamada S, Anderson S, Lockett E, Donofrio M, Shiota K, Leatherbury L, Lo CW (2009) Human cardiac development in the first trimester: a high-resolution magnetic resonance imaging and episcopic fluorescence image capture atlas. Circulation 120(4):343–351. https://doi.org/10.1161/CIRCULATIONAHA.108.796698

Fabik J, Psutkova V, Machon O (2021) The mandibular and hyoid arches-from molecular patterning to shaping bone and cartilage. Int J Mol Sci 22(14). https://doi.org/10.3390/ijms22147529

Flierman S, Tijsterman M, Rousian M, de Bakker BS (2023) Discrepancies in embryonic staging: towards a gold standard. Life (Basel) 13(5). https://doi.org/10.3390/life13051084

Gauster M, Moser G, Wernitznig S, Kupper N, Huppertz B (2022) Early human trophoblast development: from morphology to function. Cell Mol Life Sci 79(6):345. https://doi.org/10.1007/s00018-022-04377-0

Gittenberger-de Groot AC, Bartelings MM, Deruiter MC, Poelmann RE (2005) Basics of cardiac development for the understanding of congenital heart malformations. Pediatr Res 57(2): 169–176. https://doi.org/10.1203/01.PDR.0000148710.69159.61

Hadders-Algra M (2018) Early human brain development: starring the subplate. Neurosci Biobehav Rev 92:276–290. https://doi.org/10.1016/j.neubiorev.2018.06.017

Hamilton WJ, Boyd JD (1960) Development of the human placenta in the first three months of gestation. J Anat 94(Pt 3):297–328. https://www.ncbi.nlm.nih.gov/pubmed/14399291

Hans S, Christison J, Liu D, Westerfield M (2007) Fgf-dependent otic induction requires competence provided by Foxi1 and Dlx3b. BMC Dev Biol 7:5. https://doi.org/10.1186/1471-213X-7-5

Harkness LM, Baird DT (1997) Morphological and molecular characteristics of living human fetuses between Carnegie stages 7 and 23: developmental stages in the post-implantation embryo. Hum Reprod Update 3(1):3–23. https://doi.org/10.1093/humupd/3.1.3

Hikspoors J, Kruepunga N, Mommen GMC, Kohler SE, Anderson RH, Lamers WH (2022) A pictorial account of the human embryonic heart between 3.5 and 8 weeks of development. Commun Biol 5(1):226. https://doi.org/10.1038/s42003-022-03153-x

Honkura Y, Hayashi S, Kim JH, Murakami G, Abe H, Rodriguez-Vazquez JF, Katori Y (2020) Development and growth of auricular cartilage and muscles: a study using human fetuses. Int J Pediatr Otorhinolaryngol 133:109973. https://doi.org/10.1016/j.ijporl.2020.109973

Kim SM, Kim JS (2017) A review of mechanisms of implantation. Dev Reprod 21(4):351–359. https://doi.org/10.12717/DR.2017.21.4.351

Kojima J, Ono M, Kuji N, Nishi H (2022) Human chorionic villous differentiation and placental development. Int J Mol Sci 23(14). https://doi.org/10.3390/ijms23148003

Krishnan A, Samtani R, Dhanantwari P, Lee E, Yamada S, Shiota K, Donofrio MT, Leatherbury L, Lo CW (2014) A detailed comparison of mouse and human cardiac development. Pediatr Res 76(6):500–507. https://doi.org/10.1038/pr.2014.128

Merida-Velasco JR, Rodriguez-Vazquez JF, Merida-Velasco JA, Sanchez-Montesinos I, Espin-Ferra J, Jimenez-Collado J (1999) Development of the human temporomandibular joint. Anat Rec 255(1):20–33. https://doi.org/10.1002/(SICI)1097-0185(19990501)255:1<20::AID-AR4>3.0.CO;2-N

Merrell AJ, Kardon G (2013) Development of the diaphragm—a skeletal muscle essential for mammalian respiration. FEBS J 280(17):4026–4035. https://doi.org/10.1111/febs.12274

Mole MA, Weberling A, Zernicka-Goetz M (2020) Comparative analysis of human and mouse development: from zygote to pre-gastrulation. Curr Top Dev Biol 136:113–138. https://doi.org/10.1016/bs.ctdb.2019.10.002

Muller F, O'Rahilly R (1987) The development of the human brain, the closure of the caudal neuropore, and the beginning of secondary neurulation at stage 12. Anat Embryol (Berl) 176(4): 413–430. https://doi.org/10.1007/BF00310083

Noel ES (2024) Cardiac construction-recent advances in morphological and transcriptional modeling of early heart development. Curr Top Dev Biol 156:121–156. https://doi.org/10.1016/bs.ctdb.2024.02.005

Ochoa-Bernal MA, Fazleabas AT (2020) Physiologic events of embryo implantation and Decidualization in human and non-human primates. Int J Mol Sci 21(6). https://doi.org/10.3390/ijms21061973

Palis J, Yoder MC (2001) Yolk-sac hematopoiesis: the first blood cells of mouse and man. Exp Hematol 29(8):927–936. https://doi.org/10.1016/s0301-472x(01)00669-5

Park WW (1957) The occurrence of sex chromatin in early human and macaque embryos. J Anat 91(3):369–373. https://www.ncbi.nlm.nih.gov/pubmed/13448995

Rossant J, Tam PPL (2022) Early human embryonic development: blastocyst formation to gastrulation. Dev Cell 57(2):152–165. https://doi.org/10.1016/j.devcel.2021.12.022

Ruane PT, Garner T, Parsons L, Babbington PA, Wangsaputra I, Kimber SJ, Stevens A, Westwood M, Brison DR, Aplin JD (2022) Trophectoderm differentiation to invasive syncytiotrophoblast is promoted by endometrial epithelial cells during human embryo implantation. Hum Reprod 37(4):777–792. https://doi.org/10.1093/humrep/deac008

Sadler TW (2023) Langman's Medical Embryology (15th ed.). Wolters Kluwer

Sano T, Yamagishi S (1965) Spread of excitation from the sinus node. Circ Res 16:423–430. https://doi.org/10.1161/01.res.16.5.423

Santos C, Marshall AR, Murray A, Metcalfe K, Narayan P, de Castro SCP, Maniou E, Greene NDE, Galea GL, Copp AJ (2024) Spinal neural tube formation and tail development in human embryos. elife 12. https://doi.org/10.7554/eLife.88584

Sharma A, Ansari AZ, Kakulavarapu R, Stensen MH, Riegler MA, Hammer HL (2023) Predicting cell cleavage timings from time-lapse videos of human embryos. Big Data and Cognitive Computing 7(2):91

Sheng G, Martinez Arias A, Sutherland A (2021) The primitive streak and cellular principles of building an amniote body through gastrulation. Science 374(6572):abg1727. https://doi.org/10.1126/science.abg1727

Smith JL, Schoenwolf GC (1997) Neurulation: coming to closure. Trends Neurosci 20(11):510–517. https://doi.org/10.1016/s0166-2236(97)01121-1

Som PM, Naidich TP (2013) Illustrated review of the embryology and development of the facial region, part 1: early face and lateral nasal cavities. AJNR Am J Neuroradiol 34(12):2233–2240. https://doi.org/10.3174/ajnr.A3415

Stiles J, Jernigan TL (2010) The basics of brain development. Neuropsychol Rev 20(4):327–348. https://doi.org/10.1007/s11065-010-9148-4

Sutliff KS, Hutchins GM (1994) Septation of the respiratory and digestive tracts in human embryos: crucial role of the tracheoesophageal sulcus. Anat Rec 238(2):237–247. https://doi.org/10.1002/ar.1092380210

Verma RP (2021) Evaluation and risk assessment of congenital anomalies in neonates. Children (Basel) 8(10). https://doi.org/10.3390/children8100862

Watson AJ (1992) The cell biology of blastocyst development. Mol Reprod Dev 33(4):492–504. https://doi.org/10.1002/mrd.1080330417

Yuan Y, Wang X, Yan X, He N, Lu X, Yang J, Xie X, Yuan H, Chen N, Liu Y, Ren H, Zhang R, Cui L, Ren P, Lin S, Cheng S, Yang X, Guo Y, Li R, Yan T, Guo J, Xiao Z, Wei Y, Yu L (2025) 3D reconstruction of a human Carnegie stage 9 embryo provides a snapshot of early body plan formation. Cell Stem Cell 32(6):1006–1024 e1005. https://doi.org/10.1016/j.stem.2025.04.007

Microinjection in *Caenorhabditis elegans* 12

Gaurav Sharma, Srianshu Kumar Panda, Mukesh Behera,
Rahul Singh, Aman Kumar, and Narender K. Dhania

Abstract

Caenorhabditis elegans is a small, transparent nematode and a well-known model organism for different biological studies. Its short life cycle and well-defined biology make it easy to study its physiological processes and track the fate of injected samples during microinjection. Microinjection is a very powerful and versatile technique involving incorporation of oligonucleotides, chromogenic dye, or a fluorophore into the worm's body, which can be used for genotypic and phenotypic analysis. In microinjection experiment, a very thin glass needle/micropipette is used to inject the samples into the worm's body. It is also used as a physical method of gene transfer and is generally used in several fields like genetic engineering, molecular cell biology, in vitro fertilization, and generation of transgenic *C. elegans*. It can be used to label biomolecules to study its physiological function. Similarly, oligonucleotides can be injected into the

Supplementary Information The online version contains supplementary material available at https://doi.org/10.1007/978-981-95-7269-4_12.

G. Sharma · S. K. Panda · M. Behera · R. Singh · N. K. Dhania (✉)
Department of Zoology, Faculty of Sciences, Wound Healing Research Lab, University of Delhi, Delhi, India
e-mail: narender@zoology.du.ac.in

A. Kumar
Department of Animal Biotechnology, Laboratory of Molecular Diagnosis, Lala Lajpat Rai University of Veterinary and Animal Sciences, Hisar, Haryana, India

A. C. Bharti et al. (eds.), *Textbook of Developmental Biology*,
https://doi.org/10.1007/978-981-95-7269-4_12

worm's gonads to generate transgenic generations using knock-in or knock-out approaches. Microinjection requires skilled personnels, preciseness, and specific volume transfer, making it a robust biological technique.

Keywords

Microinjection · Microscope · Microneedle · *C. elegans*

Aim

To perform Microinjection in *Caenorhabditis elegans.*

Learning Objectives

Upon completing the experiment, students will be able to:

- Prepare microneedles of different diameter using needle puller.
- Prepare agarose bed and pick up single worm for microinjection.
- Preparation of microinjector and microinjection in the worms via micropipette.

12.1 Introduction

C. elegans is a soil nematode and one of the most common invertebrates used as a model organism for research. Sydney Brenner was the first person to use *C. elegans* as a model organism to study nervous system development in the 1960s. Since then, the round worm has been extensively used to study different aspects of biological processes related to humans. In the last decades, many studies have been performed on *C. elegans* to reveal the mechanism of wound healing. This encompassess transcription factors (cnc, nlp, etc.) that are activated during cuticular injury of the worm and generate immune responses. It shows a certain analogy to the human wound response. Establishing the role of mitochondria in nematodes wound healing was one of the major revolutionary studies (Silverman et al. 2009; Pouget et al. 2022). Recent studies have also shown the role of specific extracellular matrix proteins involved in wound healing (Zhu et al. 2023; Sundaram et al. 2024). Adult *C. elegans* is about 1 mm in length, and it is easy to culture in the lab that feeds on different strains of *E. coli* bacteria. *C. elegans* are either male having single X chromosome (XO) or hermaphrodite having two X chromosome (XX). Generally, the presence of hermaphrodites is huge in numbers, and the individual males are rare in number (one in 600 worms) (Meneely et al. 2019). After mating, eggs are laid which hatch after 12 h. The newly hatched larva called L1 is 1/6th length of an adult. It is almost transparent and molts to form L2 by replacing the cuticle. L2 larva further

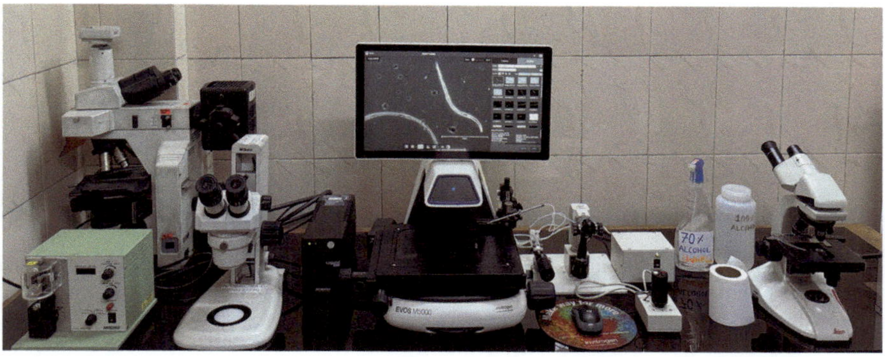

Fig. 12.1 Microinjection setup

molts into L3 and L4 and then to an adult after about 10 h. The adult starts to lay eggs again and the worm completes these stages in 3–4 days in laboratory conditions. Under unfavorable conditions like starving or overcrowding state, the L2 larva molts into a different stage called dauer stage (Fig. 12.1). Once the conditions become favorable, the dauer larva molts directly into L4, and the L3 stage is skipped. Low temperature of incubation prolongs the developmental stages (Byerly et al. 1976). Males can be distinguished from hermaphrodites as they have a fin-like tail. The genome of *C. elegans* consists of six chromosomes having 100,286,401 base pairs, out of which 19,984 are coding genes and 24,812 are non-coding genes. Most of the *C. elegans* are hermaphrodites and consist of 959 somatic cells from which various tissues and germ layers develop.

12.2 Requirements

Chemicals: Agarose, high growth media (HGM) or nematode growth media (NGMA) plates, 70% ethanol, M9 buffer, wax, mineral oil.

Apparatus/Instruments: Microinjector (IM-9B), three-axis joystick, oil hydraulic micromanipulator (MO-202 U), needle puller (PC-100), micropipette (Narishige 1 mm), microscope (fluorescence and stereo zoom microscope) (Fig. 12.1), glass slide, toothpicks, eyelashes.

12.3 Methodology (Video 12.1 Methodology)

12.3.1 Preparation of Agarose Bed for Microinjection

Agarose bed (2%) is prepared for microinjection. Agarose bed provides a substrate for the worm to attach and prevent its movement while performing microinjection. Agarose bed should be made a day prior to the microinjection experiment. Once the agarose bed is made, it can be stored and used multiple times.

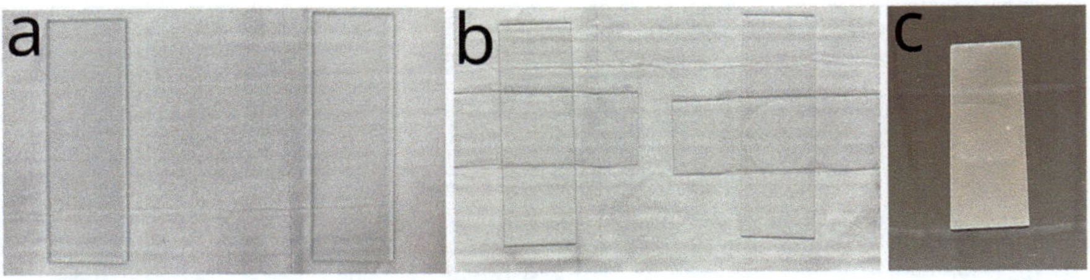

Fig. 12.2 Preparation of agarose bed for microinjection. (**a**) Glass slides containing 50 μl of agarose. (**b**) Perpendicular slide places on 50 μl agarose allowing its even spreading. (**c**) Dried agarose bed for microinjection

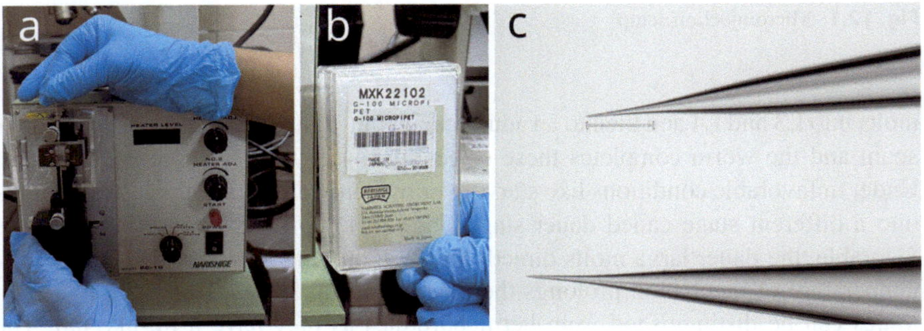

Fig. 12.3 Preparation of microneedle using needle puller. (**a**) Needle puller PC-100. (**b**) Micropipette (Narishige 1 mm). (**c**) Needles of different size and shape formed by the needle puller

(i) Weigh 0.2 g agarose and add 10 ml of distilled water.
(ii) Heat the solution for 30 s in a microwave to dissolve the agarose in water.
(iii) Place 5–6 cleaned glass slides on a clean surface (Fig. 12.2a).
(iv) Transfer 50 μl of agarose solution onto one slide and quickly place another slide perpendicularly on top of the slide (Fig. 12.2b).
(v) Repeat this step for all the slides and wait for 5 min and let the agarose dry.
(vi) After 5 min, remove the top slides carefully and store the agarose bed slides in a slide box and let it dry for one day prior to the experiment (Fig. 12.2c).

12.3.2 Preparation of Microneedle Using Needle Puller

Needle with appropriate pore size is essential for injecting sample into the worm's body. Needles having larger pore size causes damage to the worm, whereas preparation of very fine needle may block its pore; therefore, it is necessary to use the appropriate pore sized needles by adjusting the temperature and weights on needle puller.

(i) For microinjecting sample into the worm, a single-step needle pulling will be performed using two heavy weights and two light weights (Fig. 12.3a).
(ii) *Needle placement:* Place the 1 mm micropipette in the heating chamber of the needle puller such that the heating element is at the center of the capillary. Tighten the screws to prevent the slippage of capillary (top screw first) (Fig. 12.3b).
(iii) Temperature adjustment: Switch on the power button and adjust the temperature to 62.7 °C.
(iv) Needle removal: Once the capillary is pulled slowly, remove the microneedle without touching the tip.
(v) Storage of needle: Microneedles should be stored in clean space as dirt can also clog the needle; it should be stored on a foam pad carefully so that any movement should prevent the damage to the tip (Fig. 12.3c).

12.3.3 Worm Picking and Transfer to Agarose Bed

C. elegans can be cultured on HGM or NGMA plates, and single plates may contain thousands of *C. elegans*. For microinjection in *C. elegans*, a single adult worm is required (Fig. 12.4a). Adult hermaphrodite can be identified by its large size and developed gonads.

Procedure for Worm Picking

(i) Worm picker preparation: For preparing a worm picker, gently pull your eyelash and place it on a white sheet. Attach the eyelash to the toothpick tip using tape or melted wax (dip the toothpick in melted wax and attach it to the base on eyelash) (Fig. 12.4b).
(ii) Worm selection: Place the worm plate in stereo zoom microscope and switch on the bottom light source. Locate an adult hermaphrodite *C. elegans* (Fig. 12.4c).

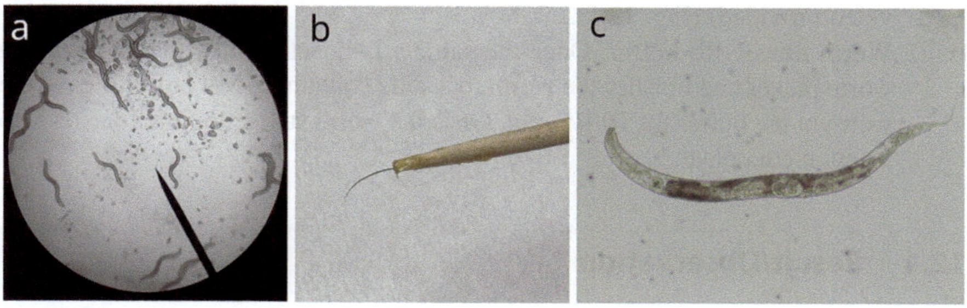

Fig. 12.4 Worm picking and transfer to agarose bed. (**a**) Stereo zoom microscopic view of *C. elegans* on HGM plates. (**b**) Worm picker (attachment of eyelash to toothpick using wax). (**c**) Worm attached to agarose pad

(iii) Worm picking: Sterilize the worm picker using 70% ethanol and place the worm picker under the worm and quickly lift the worm on the eyelash.
(iv) Oil dipping: Add a drop of mineral oil on the corner of a cleaned glass slide having agarose bed (make sure that drop do not touch the agarose bed). Place the worm in the oil for 1–2 min to remove bacteria or the agar attached to surface of the worm and prevent desiccation of the worm.
(v) Placement on the agarose bed: Pick the worm from mineral oil using worm picker and place it straight on agarose bed and gently press the worm body with the eyelash (make sure worm do not bend) and add a small drop of mineral oil to prevent desiccation of the worm.

12.3.4 Microinjection of Sample in *C. elegans*

(i) When *C. elegans* is placed in mineral oil, the following steps should be performed:
(ii) Needle adjustment: Place the needle by loosening the knob of microinjector, and once the needle is adjusted, tighten the knob (make sure the needle is tight).
(iii) Oil filling: Fill the needle with mineral oil along its tip using a micromanipulator.
(iv) Sample filling: Add a drop of sample (fluorophore) on the glass slide and move the microneedle toward the slide. Dip the microneedle into the surface of the sample drop and aspirate the sample using micromanipulator (rotate anticlockwise).
(v) Worm visualization: Visualize the worm in 10× under the fluorescence microscope (make sure worm is attached to agarose bed).
(vi) Needle visualization: Visualize the needle in 10× and place the needle on the same plane as of *C. elegans* using the joystick and X, Y, and Z axis of the micromanipulator.
(vii) Sample injection: For injecting the sample, adjust the microscope to 60X, reduce the joystick speed to its lowest, and inject the sample slowly using a micromanipulator (rotate clockwise) at your desired location (refer to the video provided) (Fig. 12.5).
(viii) Worm transfer to buffer: Once the sample is injected, pick the worm using worm picker and transfer the worm to a slide containing M9 buffer. Leave the worm in the buffer for 10–15 min. Once the worm starts moving, transfer it to a new media plate.

12.4 Result/Observation

Observe your fluorescent-labeled sample in worm's body. If the sample is injected in gonads, look for the offsprings that are fluorescently labeled. Depending upon the site of microinjection, the result duration may vary:

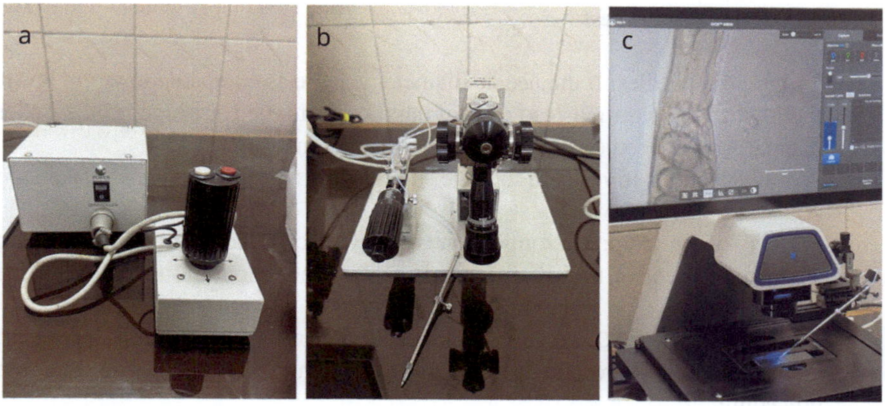

Fig. 12.5 Microinjection of sample in *C. elegans*. (**a**) Three-axis joystick for movement of microinjector. (**b**) Oil hydraulic micromanipulator (MO-202 U) for sample transfer into the worm. (**c**) Microinjection of sample in *C. elegans*

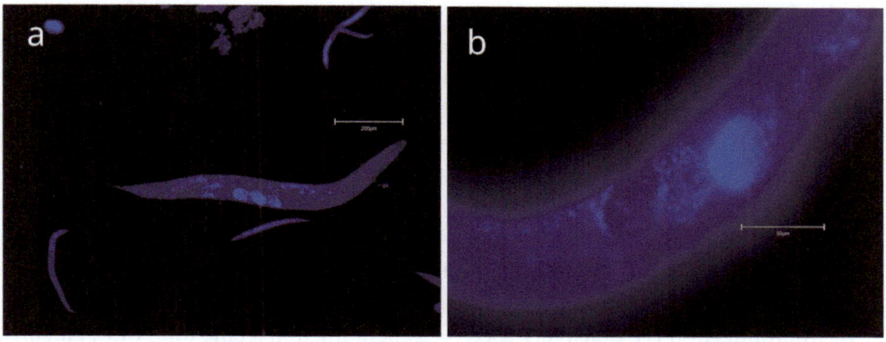

Fig. 12.6 Visualization of fluorescent-labeled *C. elegans* under fluorescence microscope. (**a**) 10× magnification; (**b**) 60× magnification

Microinjection into the vitals: The sample can be visualized immediately after the microinjection has been performed, and the movement of sample can be tracked in real time.

Microinjection in eggs/embryo: If the sample is injected into the egg or embryo, transfer the worm to a new plate with no worm and visualize the plate the next day or until the eggs hatch.

Microinjection in gonads: If the sample is injected in gonads (ventral or distal), track the plate every day, as out of thousands of newly hatched worms, only a few may show fluorescence.

For our experiment, we injected Hoechst fluorescence dye into the worm's gonads. The worm was incubated in M9 buffer for 15 min at 22 °C and then transferred into a new growth plate. The worms were picked using a worm picker and visualized under the fluorescence microscope using DAPI filter (Fig. 12.6).

Errors During the Process

(i) Larger pore needles: If the needle diameter is large as compared to worm body, it will push the worm instead of piercing it (refer to the Narishige needle puller manual for the needle's pore size as per the organism).
(ii) Worm selection: If you aim to target the gonads of the *C. elegans*, make sure you select the adult worm.
(iii) Stability of the setup: Microinjection is a very precise and controlled technique; slight movement of the setup can cause large inflection and breakage of the needle.
(iv) Worm desiccation: *C. elegans* desiccate easily once they are exposed to air; therefore, make sure to place them in mineral oil while performing the procedure.
(v) Needle placement on microinjector: While placing the needle in the microinjector, make sure the needled is fixed tight in the gasket and the knob is tight. The loose needles often eject out during the process.

Media for *C. elegans*
C. elegans are maintained in two types of growth media plates. HGM (NaCl, bactopeptone, bactoagar) or NGMA (peptone, NaCl, agarose, and agar). Once the media is autoclaved, cholesterol, MgSO4, KH2PO4, and CaCl2 are added in the media and poured in the plates. The worms are fed by OP50 strain of *E. coli,* and the culture is maintained at 22 °C.

12.5 Significance/Research Applications

(i) Microinjection is extensively used for generating transgenic lines of *C. elegans*, allowing the incorporation of transgenes of interest or modifying genetic sequences. This can be utilized for studying the function of genes, specifically in mutant strains in which functional genes are knocked out. This technique is also used in precise genome editing tools such as CRISPR/Cas-9 systems.
(ii) Microinjection is used to introduce reporter genes, such as GFP or lacZ, to visualize gene expression patterns in different tissues and developmental stages. By microinjecting constructs that express proteins with specific tags or modifications, researchers can study protein structure, function, and interactions.
(iii) Microinjection is used to deliver double-stranded RNA (dsRNA) molecules that target specific genes, allowing for the study of gene function and pathways through RNAi- mediated gene silencing. Microinjection can be used to deliver drugs or chemicals into the worm's body, allowing researchers to assess their effects on development, physiology, and lifespan.
(iv) Microinjection allows researchers to study the role of specific genes and pathways in development, including embryonic development, tissue formation, and organogenesis.

(v) *C. elegans* is a model organism for studying aging, and microinjection is used to study the effects of genes and pathways on lifespan and age-related changes.
(vi) *C. elegans* can be used to study the molecular and cellular mechanisms of human diseases, and microinjection is used to model these diseases and assess the effects of potential therapies.

12.6 Precautions

(i) Needles should be carefully placed in the needle puller. The acrylic hood of the needle puller should be closed while the heating filament is on. Carefully handle the microneedles as slight shock or touch on its tip may cause breakage.
(ii) Ensure that the worm is always placed in mineral oil if it is out of the media. Exposure of worm to air causes its rapid desiccation.
(iii) For precise injection of sample into the worm's body, use 60X magnification.
(iv) When injecting the sample into the worm in 60X magnification, reduce the speed of joystick movement to its minimum.
(v) Ensure that the small amount of sample is injected in the worm as releasing larger volume may cause bursting of the worm.
(vi) Once injection in completed, quickly transfer the worm to the slide with M9 buffer and place the slide in incubator (22 °C) for 10–15 min; make sure the amount of M9 buffer is sufficient such that it does not dry up.

12.7 Quick Review Questions

12.7.1 Multiple Choice Questions

1. What is the purpose of microinjection in *C. elegans*?
   (a) Observe feeding behavior of worm
   (b) Delivery of DNA/sample into the worm
   (c) Study mating behavior of worm
   (d) None of the above

Answer: b) Delivery of DNA/sample into the worm.

2. What is the ideal temperature for *C. elegans* growth?
   (a) 20 °C ± 2
   (b) >26 °C ± 5
   (c) <35 °C ± 1
   (d) 37 °C ± 2

Answer: a) 200C ± 2

3. Which of the following is not part of *C. elegans* culture media?
   (a) NaCl
   (b) Bactopeptone
   (c) Bactoagar
   (d) Borax

Answer: d) Borax

4. Microinjection in *C. elegans* is typically performed into which structure?
   (a) Intestine
   (b) Gonads
   (c) Syncytial gonad
   (d) Gut

Answer: c) Syncytial gonad

5. What is the advantage of syncytial gonad of the *C. elegans* for microinjection?
   (a) Allows for rapid digestion of DNA
   (b) It prevents DNA expression
   (c) It allows DNA to be distributed to multiple germ nuclei
   (d) It stops the formation of transgenic lines

Answer: c) It permits DNA to be distributed to multiple germ nuclei

6. What is the most commonly used stage of *C. elegans* for microinjection?
   (a) L1 larva
   (b) L2 larva
   (c) Dauer stage
   (d) Adult worm

Answer: d) Adult worm

7. Which equipment is used for microinjection in *C. elegans*?
   (a) Thermal cycler
   (b) Electroporator
   (c) Micromanipulator
   (d) Microplate reader

Answer: c) Micromanipulator

8. What is the volume range of the sample injected into C. *elegans during microinjection?*
   (a) 1–10 μL
   (b) 10–100 nL
   (c) 10–100 μL
   (d) 0.1–1 μL.

Answer: a) 1–10 μL.

9. *Which of the following is NOT a typical component of a microinjection mix in C. elegans?*
   (a) Buffer
   (b) Target DNA sequence
   (c) Antibiotic
   (d) Fluorescent dye

Answer: c) Antibiotic

10. What is a common reason for low survival rate of worms after microinjection?
    (a) Incorrect DNA concentration
    (b) Injection into the body cavity instead of the gonad
    (c) Presence of GFP in the plasmid
    (d) Use of extrachromosomal arrays

Answer: b) Injection into the body cavity instead of the gonad

12.7.2 Short Answer Questions

(i) What percentage of agarose is used for preparation of agarose bed for microinjection in *C. elegans*?
(ii) How many weights are used for preparation of microneedle?
(iii) What is the significance of using mineral oil during the microinjection process?
(iv) How can you successfully visualize the microinjection of sample in *C. elegans*?

Glossary

Buffer A solution that resists change in pH when small amount of acid or base is added. Somatic cell—non-reproductive cells which form the body of an organism.

Cas-9 Endonuclease enzyme that plays a key role in the CRISPR-Cas9 gene editing.

Chromogenic dye Color-producing compound that develops color when it reacts with a specific enzyme, chemical, or biological molecule.

Chromosome A thread-like structure made of DNA and proteins carrying genetic information.

Cnc Transcription factors involved in stress responses, especially oxidative stress.

CRISPR Clustered Regularly Interspaced Short Palindromic Repeats is gene editing tool derived from bacterial immune system.

DAPI DNA-binding fluorescent stain.

Dauer stage Inactive stage of C. elegans to survive unfavorable condition.

Double-stranded RNA (dsRNA) Two complementary strands of DNA bound together.

Extracellular matrix proteins Proteins that are component of extracellular matrix.

Fluorophore Fluorescent chemical compound that can absorb light at one wavelength and emit light at a longer wavelength.

Gene knocked-in Technique in which a specific gene is inserted into a particular locus in the genome.

Gene knocked-out Technique in which a specific gene is deleted from a particular locus in the genome.

Genetic engineering Process of genome editing.

Genome Consist of DNA, gene and non-coding sequences of an organism.

Genotype Genetic makeup of an organism.

Germ layers Three primary cell layers formed during embryonic development which later develop into all tissues and organs of the body.

GFP Green fluorescent protein found in the jellyfish Aequorea victoria.

Gonads Organs that produce gametes.

Hermaphrodite Organism that contains both male and female reproductive organs.

Hoechst Dye used to stain DNA.

Invertebrates Animals lacking backbone or spinal cord.

Lacz Gene from Lac operon of E. coli that produces β-galactosidase and is commonly used as reporter gene in molecular biology.

Micropipette Instrument used to precisely transfer volumes of liquid in the microliter range.

Mutant strains Variant of an organism that has undergone permanent genetic changes (mutations) compared to its wild-type strain.

Nematode Phylum consisting of roundworm or eelworms

Nlp Neuropeptide-like proteins that regulate diverse physiological processes.

Oligonucleotides Short segments of DNA or RNA.

Phenotype Set of observable characteristics or traits of an organism.

RNAi RNA interference is a process where small RNA is used for gene silencing.

Somatic cells All the body's cells except reproductive (germ) cells.

Transcription factors Factors that regulate transcription of gene.

Transgene Insertion of a foreign gene or gene that is not part of an organism using genetic engineering techniques.

Transgenic C. elegans C. elegans in which foreign DNA has been introduced to create genetically modified strain.

Wound healing Process where body repairs dead tissue.

References

Byerly L, Cassada RC, Russell RL (1976) The life cycle of the nematode Caenorhabditis elegans: I. Wild-type growth and reproduction. Dev Biol 51(1):23–33

Meneely, P. M., Dahlberg, C. L., & Rose, J. K. (2019). Working with worms: Caenorhabditis elegans as a model organism. Current Protocols Essential Laboratory Techniques, 19(1), e35

Pouget C, Dunyach-Remy C, Bernardi T, Provot C, Tasse J, Sotto A, Lavigne JP (2022) A relevant wound-like in vitro media to study bacterial cooperation and biofilm in chronic wounds. Front Microbiol 13:705479

Silverman GA, Luke CJ, Bhatia SR, Long OS, Vetica AC, Perlmutter DH, Pak SC (2009) Modeling molecular and cellular aspects of human disease using the nematode Caenorhabditis elegans. Pediatr Res 65(1):10–18

Sundaram MV, Pujol N (2024) The Caenorhabditis elegans cuticle and precuticle: a model for studying dynamic apical extracellular matrices in vivo. Genetics 227(4):iyae072

Zhu Y, Li W, Dong Y, Xia C, Fu R (2023) C. elegans Hemidesmosomes sense collagen damage to trigger innate immune response in the epidermis. Cells 12(18):2223

Chicken Embryo as a Preclinical Model in Biomedical Research

13

Anna Senrung, Pallavi Sethi, and Alok Chandra Bharti

Abstract

The chicken embryo represents one of the most versatile and ethically favorable vertebrate models in preclinical biomedical research. Its external development, cost-effectiveness, and physiological similarities to mammals make it an ideal intermediate system bridging in vitro studies and mammalian models. A major advantage of the chick embryo lies in its extraembryonic chorioallantoic membrane (CAM), a highly vascularized structure that serves as a dynamic platform for studying angiogenesis, tumor biology, and toxicology. The CAM assay allows direct visualization and quantification of vascular responses to pro- or antiangiogenic agents, providing rapid and reproducible results under controlled experimental conditions. Beyond angiogenesis studies, the chick CAM supports the establishment of tumor xenografts derived from cell lines or patient-derived samples. Owing to the natural immunodeficiency of the embryo until day 14 of incubation, tumor grafts can grow without immune rejection, mimicking neovascularization and tumor stroma interactions seen in human cancers. This facilitates the assessment of tumor progression, angiogenic modulation, and therapeutic efficacy of anticancer drugs. Moreover, systemic or localized drug administration on the CAM offers a convenient method for pharmacological screening. The chick embryo model also provides a reliable system for early-stage toxicity and biocompatibility evaluations. Drug-induced alterations in embryonic morphology, organ weight, or histoarchitecture can be rapidly assessed, allowing early identification of compounds with potential adverse effects. Overall, the chicken embryo serves as a robust, accessible, and ethically

A. Senrung · P. Sethi
Department of Zoology, Daulat Ram College, University of Delhi, Delhi, India

A. C. Bharti (✉)
Molecular Oncology Lab., Department of Zoology, University of Delhi, Delhi, India

A. C. Bharti et al. (eds.), *Textbook of Developmental Biology*,
https://doi.org/10.1007/978-981-95-7269-4_13

advantageous preclinical model that combines simplicity with translational relevance. Its applications extend across angiogenesis research, cancer therapy evaluation, and toxicological screening, offering valuable insights into drug efficacy and safety before mammalian testing. This chapter details methodologies for CAM-based assays and emphasizes the significance of the chick embryo model in advancing biomedical and translational research.

Keywords

Chick embryo · Chorioallantoic membrane · Angiogenesis · Tumor xenograft · Cytotoxicity

AIM This exercise aims to familiarize students with the handling of chick embryos and their applications as a preclinical research model in biomedical science.

Learning Objectives

By the end of this experiment, students should be able to:

- Employ the chick embryo as a preclinical model for cancer research.
- Perform angiogenesis assays using the chick chorioallantoic membrane (CAM) to evaluate angiogenesis or antiangiogenic agents.
- Develop tumor xenografts on the chick CAM.
- Assess the biocompatibility of anticancer drugs or therapeutic agents using the chick embryo.

13.1 Introduction

The chicken embryo is a classical model in developmental biology and a valuable preclinical system for biomedical research. Its external development, year-round availability, warm-blooded physiology, and similarities to mammalian developmental processes make it ideal for studying embryogenesis, organogenesis, toxicology, and disease mechanisms.

A key feature that underpins its utility in preclinical studies is the extraembryonic chorioallantoic membrane (CAM) (Anna Senrung et al. 2024b). The CAM is a highly vascularized membrane formed by the fusion of the mesodermal layers of the chorion and allantois, comprising chorionic ectoderm, chorioallantoic mesoderm, and allantoic endoderm. Functioning as a respiratory surface with an extensive capillary network, the CAM is widely used to study angiogenesis in vivo and to evaluate the efficacy of pro- and antiangiogenic compounds. Its vascularization begins around embryonic day 4–5, with maximal responsiveness to angiogenic factors between days 8 and 9, making this period optimal for angiogenesis assays.

The CAM also provides a permissive environment for tumor xenografts, as the chick embryo remains naturally immunodeficient until approximately day 14, preventing immune rejection of grafted cells. Xenografts established on the CAM, whether patient-derived (PDX) or cell line-based, recapitulate tumor vascularization and microenvironmental interactions, supporting studies of tumor-induced angiogenesis and the testing of anticancer therapeutics. The rich CAM vascular plexus ensures adequate nutrient and oxygen supply, allowing reproducible establishment of vascularized tumor models within 2–5 days of inoculation.

The chick embryo model has been extensively applied in preclinical research for: *Angiogenesis studies*, to investigate vascular growth in physiological and pathological conditions such as wound healing, diabetic retinopathy, age-related macular degeneration, and cancer; *Tumor biology*, including evaluation of tumor progression, tumor-induced angiogenesis, efficacy of anticancer agents; *Toxicology*, assessing drug safety and biocompatibility before mammalian testing.

Owing to its accessibility, reproducibility, cost-effectiveness, and ethical advantages, the chicken embryo remains one of the most widely used in vivo systems for translational and biomedical research. This chapter highlights the chick embryo as a preclinical model, with emphasis on its applications in angiogenesis, tumor xenografts, and toxicological studies.

13.2 Requirements

- Fertilized chicken eggs, tumor cells, BOD incubator, polystyrene weigh boat, parafilm strip, plastic tray, spirit lamp, match stick, steel blade, gelatin sponge, alcohol 70%, alcohol 90%, test compounds, muslin cloth, Dulbecco's Modified Eagle Medium/Harry Eagle's Minimal Essential Media (DMEM/MEM) cell culture media, microtips, dispenser, mini electric drill.

13.3 Methodology

13.3.1 CAM Angiogenesis Assay

Angiogenesis in the chick embryo is most effectively studied using the boat method, where the embryo is cultured in a polystyrene weigh boat (Fig. 13.1). This technique provides a larger working surface, allows easy visualization of the CAM, and facilitates superior imaging and manipulation compared to other methods, such as the windowing or hammock techniques (Anna Senrung et al. 2024b). The following protocol describes the steps followed for evaluating angiogenesis using white Leghorn (*Gallus gallus*) embryos, adapted from Senrung et al. (Anna Senrung et al. 2024a; Senrung et al. 2023).

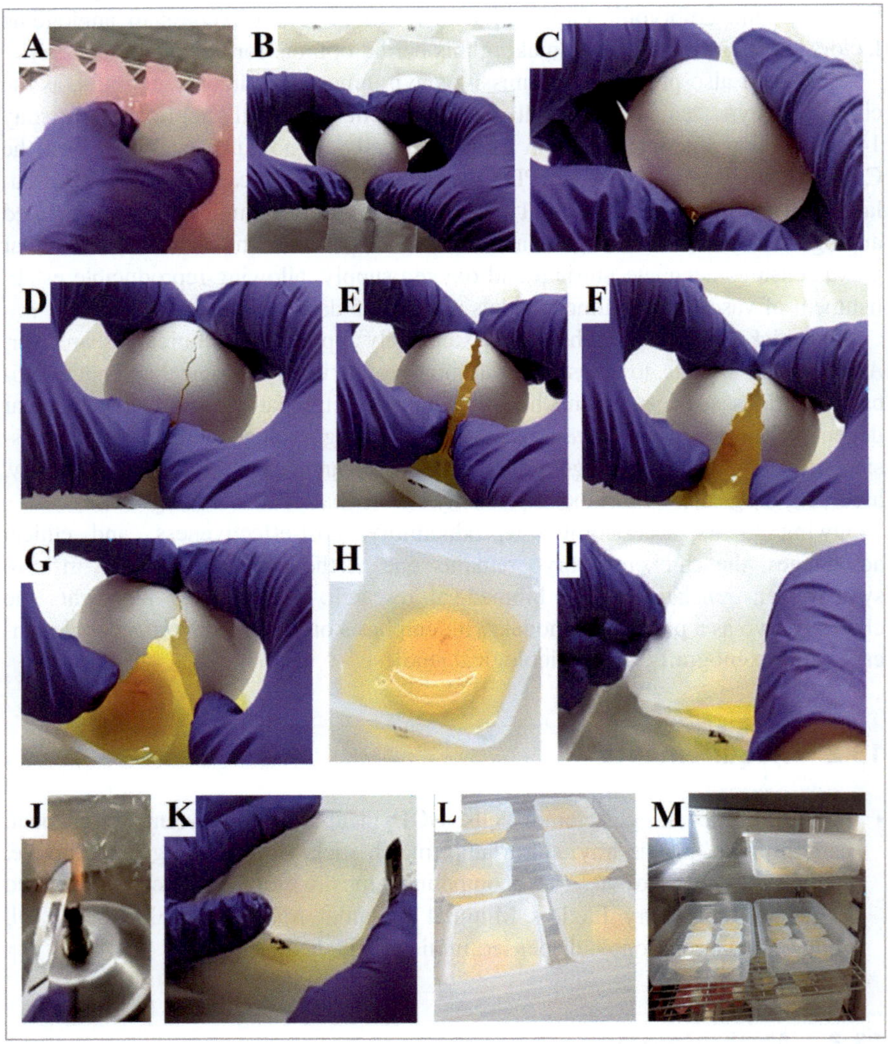

Fig. 13.1 Steps to the setting of the boat method of chick embryo culture

Procedure

(i) Incubate fertilized white Leghorn eggs in an incubator at 37 ± 1 °C and 80 ± 10% or ≥60% relative humidity.

(ii) On embryonic development day 4 (EDD 4), carefully crack open the eggs and transfer the entire contents into a sterile polystyrene weigh boat (Fig. 13.1). Ensure the yolk remains intact.

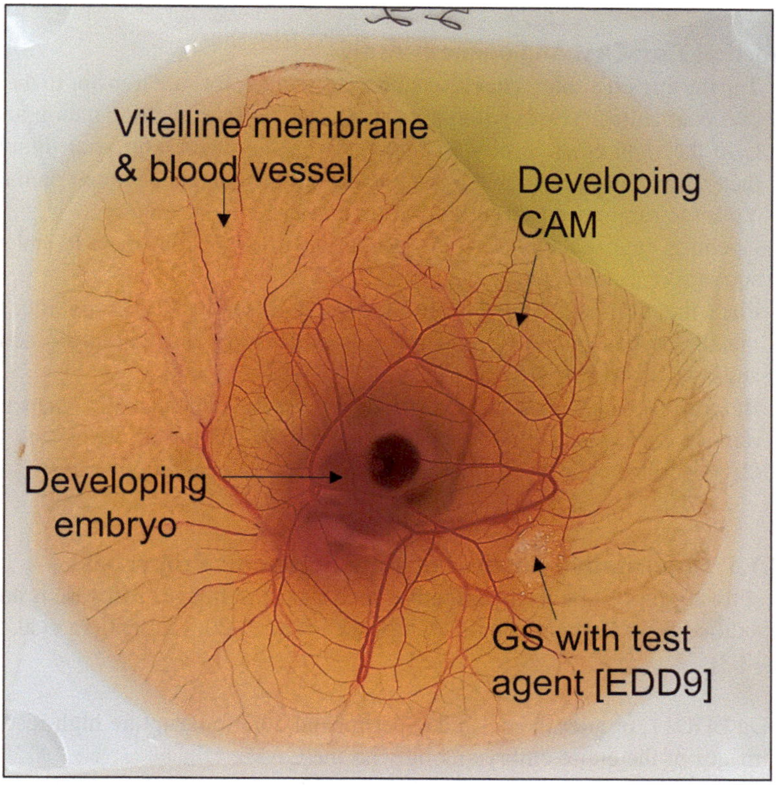

Fig. 13.2 Schematic representation of the chick CAM angiogenesis assay setup

(iii) On EDD 9, prepare small gelatin sponge (GS) cubes pre-soaked in the desired concentrations of the test agent. Using sterile forceps, gently place the GS cube loaded with test agent onto the CAM surface.

(iv) Position the GS-loaded test agent between the branches of the primary allantoic vein, preferably within the zone of active angiogenesis (Fig. 13.2), to ensure optimal vascular response.

(v) Reincubate the embryo and evaluate the CAM vasculature around the test site after desired hours (24, 48, and 72 h) of exposure.

(vi) Observe the vascular density and pattern of blood vessel growth around the site of test agent application under a stereomicroscope or digital imaging setup.

(vii) Capture high-resolution images of the CAM region of test site and analyze the vascular network parameters (e.g., total vessel length, branching, density) using ImageJ software with the Angiogenesis Analyzer plugin (Carpentiera et al. 2012)

Practical Tips: CAM Angiogenesis Assay

- Tap the egg on a sharp edge to create a clean crack for safe transfer to the ex ovo boat culture vessel without rupturing the yolk or vitelline membrane.
- Grip the shell edges firmly on both sides of the crack with your thumbs, then slowly and gently pull apart to pour out the contents without damage.
- Yolk sac rupture causes embryo death and experiment failure.
- Opening eggs on day 1–2 of incubation often results in low embryo survival in ex ovo culture.
- Place the test agent-loaded gelatin sponge (GS) on the CAM only after the membrane has spread beyond the amniotic region and within the active angiogenic zone.
- Before placement, rest the GS briefly on sterile Whatman filter paper to remove excess test solution, preventing it from sinking or depressing the CAM surface over the yolk or along the sides of the amniotic sac.

13.3.1.1 Quantitative Analysis of Antiangiogenic Properties

The following procedure for quantitative analysis using ImageJ software is adapted and modified from Carpentier et al. (Carpentier et al. 2020; Carpentiera et al. 2012) (Fig. 13.3).

(i) On EDD11, or after the desired experimental duration, capture high-resolution images of the entire embryo along with the CAM.
(ii) Select and crop a square region (~3.3 × 3.3 cm) around the site of test agent application from the captured image (Fig. 13.3a and b).
(iii) Convert the cropped image into a binary format for further processing (Fig. 13.3c).
(iv) Process the binary image in ImageJ using the Angiogenesis Analyzer plugin. The analysis generates visual outputs, including a skeletonized version of the image (Fig. 13.3d and e), along with statistical data representing various angiogenic parameters.
(v) The analysis detects parameters such as nodes, segments, branches, extremities, meshes, twigs, and isolated segments (Fig. 13.3d).
(vi) Depending on the study objectives, the most relevant parameters can be selected and graphically represented to assess the antiangiogenic or pro-angiogenic effects of the test compounds.

13.3.2 Generation of CAM Tumor Xenograft (U87 Cell Line)

(i) Load 25 μl of U87 tumor cell suspension (cell density: 1.5×10^6 cells) onto a gelatin sponge (dimensions: 4 mm × 4 mm × 2 mm thickness) and place it on the chick CAM of EDD9 (HH stage 35) (Fig. 13.4). The seeding density may vary depending on the cell type used.

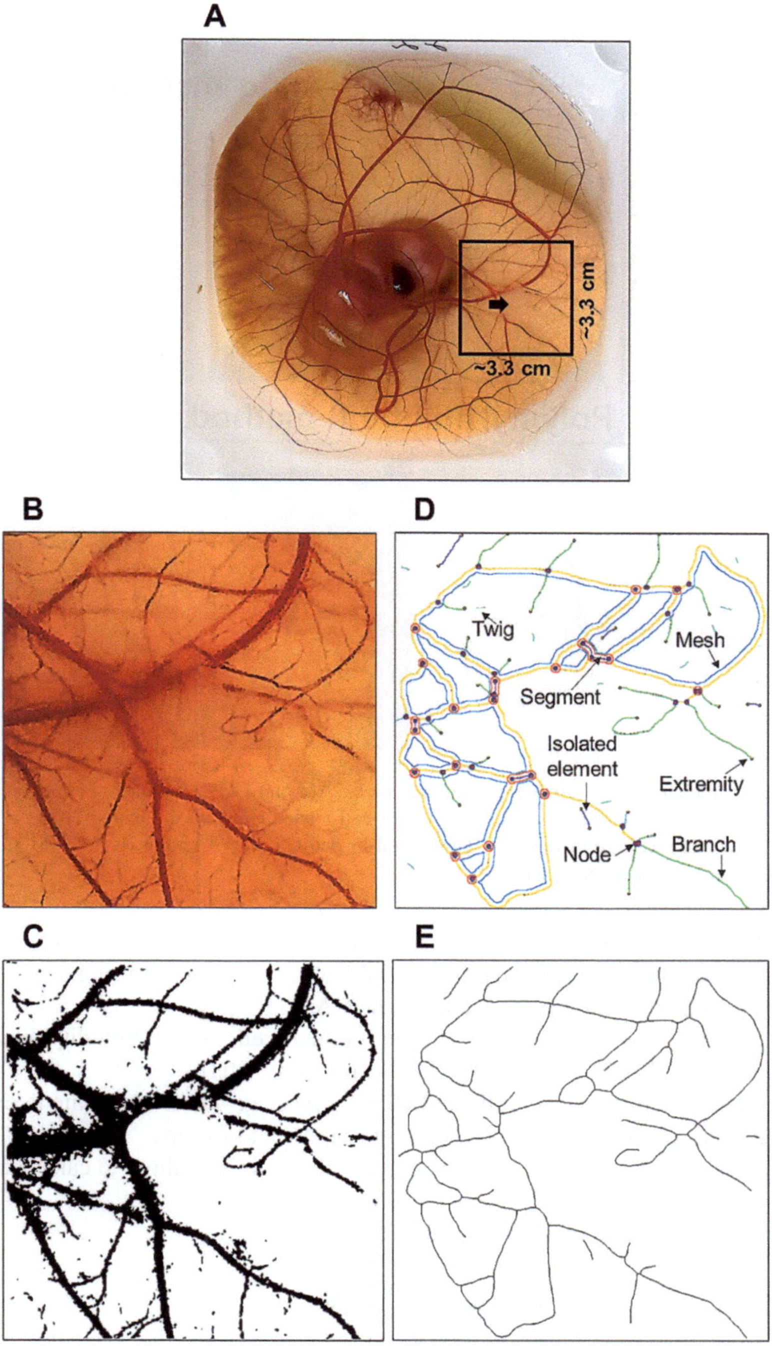

Fig. 13.3 Quantitative analysis of antiangiogenic activity using ImageJ. (Modified from (Carpentier et al. 2020; Carpentiera et al. 2012; A. Senrung et al. 2024a, b). (**a**) Image of Chick embryo on EDD11 with ~3.3 × 3.3 cm square selected around the test agent site. (**b**) Cropped image of the selected area for analysis. (**c**) Binary image of the CAM with the gelatin sponge (GS) cut-out. (**d**) Detection of branches, extremities, isolated segments, meshes, nodes, segments, and twigs. (**e**) Skeletonized image of the binary CAM area for quantitative assessment. (Images are adapted from Senrung et al. (2023))

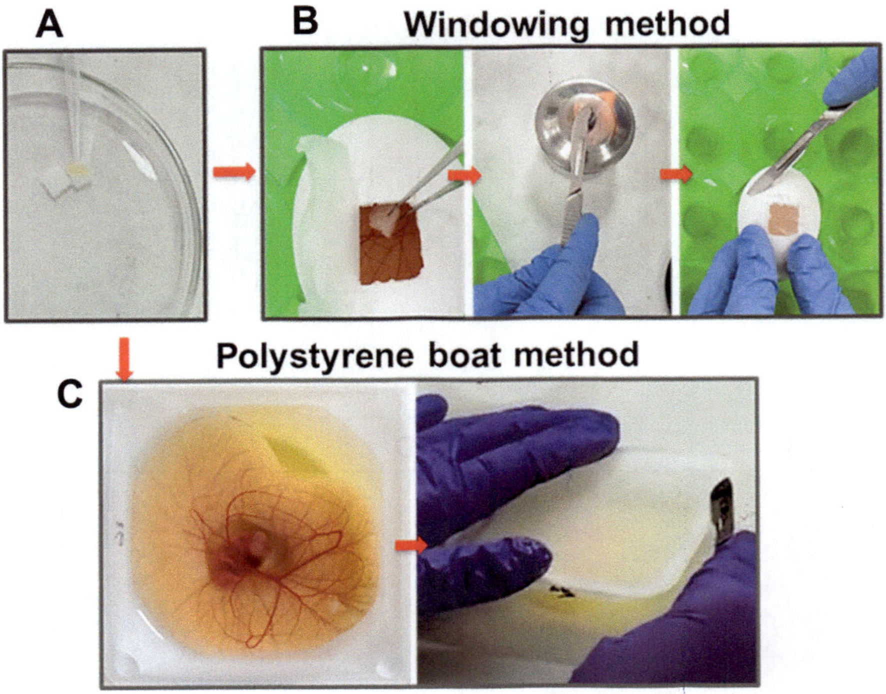

Fig. 13.4 Implantation of tumor cells onto the chick CAM on EDD9. (**a**) Loading tumor cells onto the gelatin sponge (GS). (**b**) Placement of tumor cell-loaded GS onto the chick CAM on EDD9 using the window culture method. (**c**) Tumor cell-loaded GS placed on the chick CAM at EDD9 using the polystyrene culture method

(ii) To maintain uniformity, position the sponges in the region of minor vascular branches between the two primary/major allantoic veins, ensuring that the cell-seeded surface of the sponge faces the CAM.
(iii) Incubate the chick embryos with grafted cells in a humidity chamber maintained at 37 °C and 80% relative humidity (RH).
(iv) Excise the tumor xenografts on EDD14 (HH stage 40) or according to experimental requirements using dissecting scissors and forceps.
(v) Measure the excised tumor xenografts using a digital caliper to calculate their volume.
(vi) Calculate tumor volume using the formula:

$$\text{Volume}\left(\text{mm}^3\right) = \left(W^2 \times L\right)/2,$$

where L = major anteroposterior diameter and W = lateral diameter (Ayers et al. 2010; Chen et al. 2014).
(vii) Capture microscopic images of the excised xenografts for morphological assessment.

Practical Tips: Generation of CAM Tumor Xenograft

- When implanting tumor cell-loaded gelatin sponges onto the CAM, place an additional sponge layer and add extra culture medium to support cell survival until the cells establish contact and integrate with the CAM, which typically takes 1–2 days.

13.3.3 CAM Tumor Xenograft Angiogenesis Assay

Chick CAM-derived xenografts are inherently highly angiogenic, making them particularly suitable for evaluating antiangiogenic agents, key components in cancer therapy, where angiogenesis is a hallmark of tumor progression. The effects of these agents can be assessed by analyzing vascular density, quantifying angiogenesis-related activity, or measuring specific protein expression through immunohistochemistry (e.g., anti-desmin) or biochemical assays.

For measuring vascular density in tumor xenografts, conventional hematoxylin and eosin (H&E)-stained histological sections can be prepared after the tumor xenograft generation and treated with test compounds. In these sections, blood vessels can be distinctly identified to evaluate the promotion or inhibition of tumor-induced angiogenesis by the test agents.

A reduction in vascular density or angiogenic marker expression is indicative of the antiangiogenic efficacy of the tested compound.

13.3.4 Toxicological Study

The biocompatibility or potential toxicity of identified therapeutics can be evaluated through multiple approaches, including biochemical assays, anatomical and histological assessments, and measurements of body and organ weights.

Example: *Liver*

The liver, being the primary site of drug metabolism, is particularly susceptible to adverse metabolic effects. Biochemical assessment of hepatic function can be performed by measuring serum levels of key liver enzymes, alanine aminotransferase (ALT), aspartate aminotransferase (AST), alkaline phosphatase (ALP), and gamma-glutamyl transferase (GGT), along with total bilirubin.

Anatomical and histological evaluations of the liver provide further insights into toxicity, where alterations such as disruption of lobular architecture, hepatocyte degeneration, or vacuolation indicate hepatocellular damage (Ara et al. 2022; Gonzalez and Chatterjee 2019; Guan et al. 2005; Kue et al. 2015). Additionally, liver weight serves as an important parameter for assessing systemic or organ-specific toxicity.

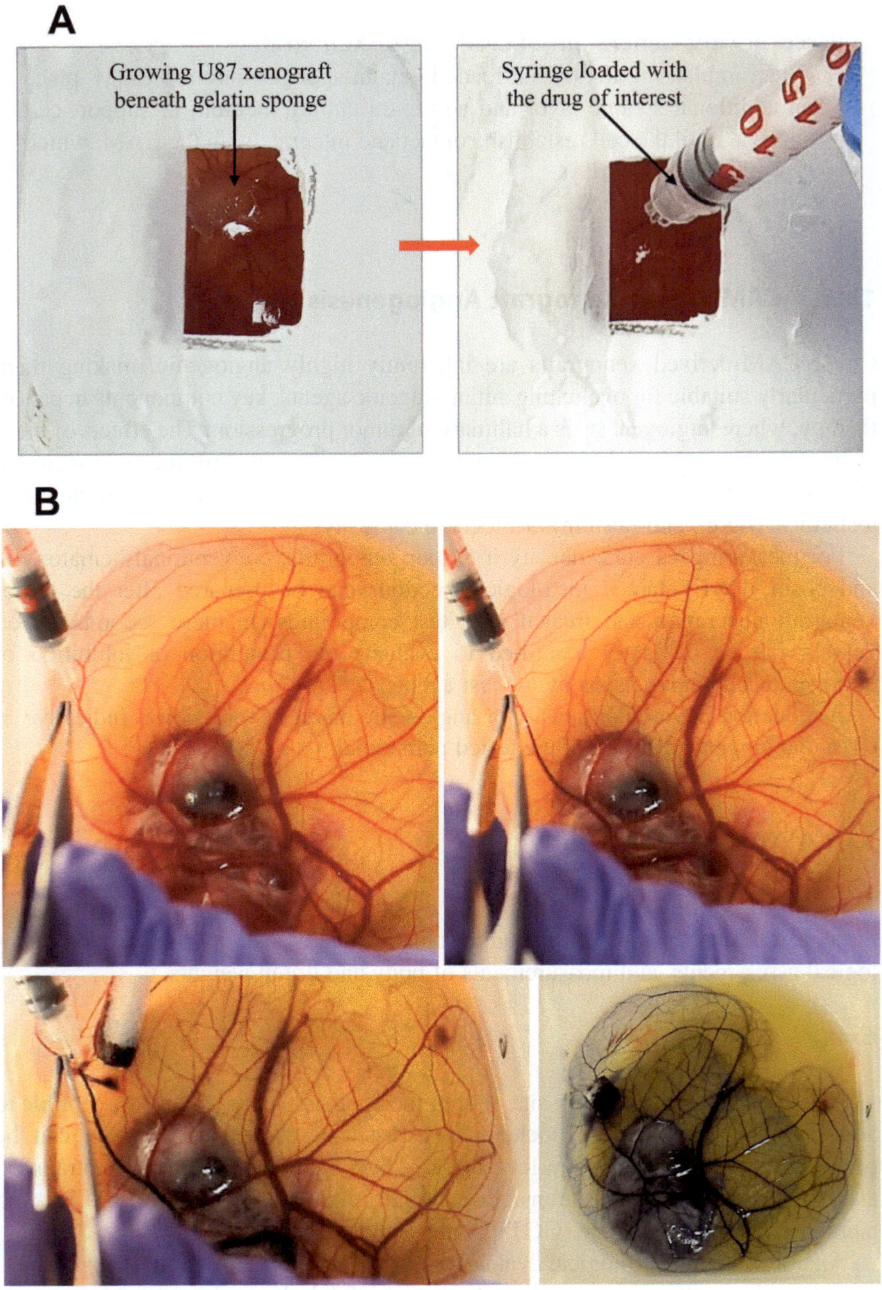

Fig. 13.5 Routes of drug administration. (**a**) Intratumoral administration of the drug. (**b**) Intravenous injection of Evans Blue dye into the CAM vasculature to evaluate drug distribution within the chick embryo following vascular administration

Fig. 13.6 Image showing chick embryos maintained under controlled incubator conditions

13.3.5 Route of Drug Administration

Drugs can be administered topically or intratumorally by direct application to the tumor site (Fig. 13.5a), or systemically via intravenous injection (Fig 13.5b). Sterile Dispovan 1 ml insulin syringes could be used for precise delivery. Additionally, drugs may be co-administered with tumor cells at the time of implantation, as demonstrated in a previous study (Senrung et al. 2023).

The fertilized eggs and the culture of chick embryos are to be maintained in a BOD incubator at an optimum temperature of 37 ± 1 °C and 80 ± 10% relative humidity (Fig. 13.6).

Practical Tips: Drug Administration

- For intratumoral administration of test compounds, avoid deep needle insertion, as it may rupture the tumor xenograft, leading to artifacts in tissue sections and interfering with analysis.
- For intravenous administration, select medium-sized CAM blood vessels to ensure accurate needle insertion and minimize bleeding upon withdrawal.

13.4 Results/Observations

13.4.1 Chick CAM Angiogenesis Analysis

To evaluate the antiangiogenic activity of a test compound, both qualitative and quantitative assessments can be performed. Effective antiangiogenic activity in the CAM assay is typically indicated by reduced vascular density and/or disorganized, tortuous blood vessels around the test site.

Figure 13.7a presents representative images from an experiment involving Axitinib (an FDA-approved antiangiogenic agent) and Piperine, a potential phytochemical with antiangiogenic properties, tested at 1 and 10 μM concentrations, with 1× PBS serving as the control (Senrung et al. 2023). The CAM regions treated with Axitinib and Piperine exhibited a marked, dose-dependent reduction in blood vessel formation compared to the control. Additionally, the treated areas displayed

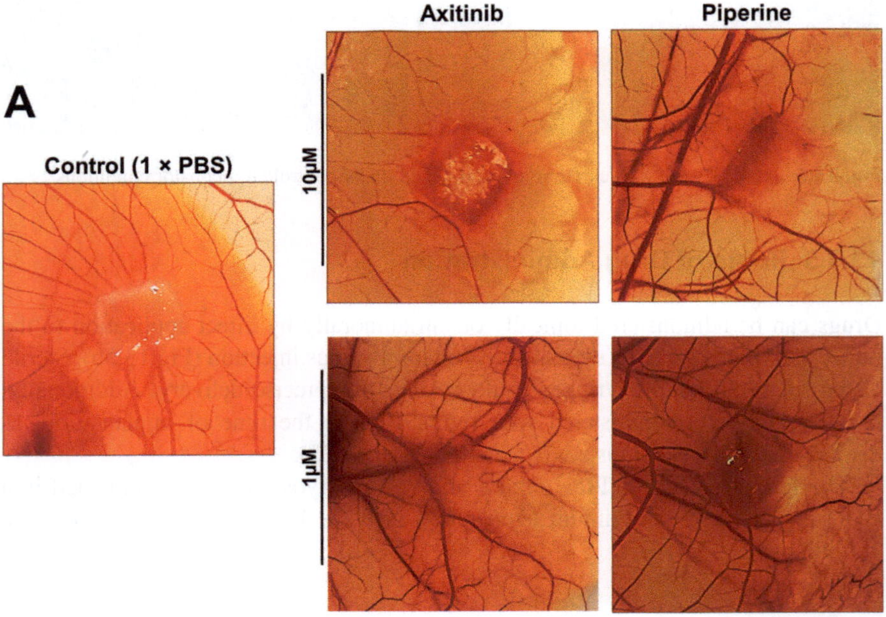

Fig. 13.7 Representative images and quantitative data from an angiogenesis assay. (**a**) Images depict results from an experiment involving Axitinib and Piperine tested at 1 and 10 μM concentrations, with 1× PBS as the control. (**b–d**) Graphical representation of vascular density parameters, including the number of nodes (Nb of nodes), number of meshes (Nb of meshes), and total length (Tot length), respectively. Nodes are pixels with ≥3 neighbors, meshes are interconnected network areas formed between branching structures, and total vessel length is the sum of segments, isolated elements, and branches. At 10 μM Axitinib concentration, no embryos survived for the 48-h study duration. (Images adapted from Senrung et al. (2023). The images are licensed under Creative Commons)

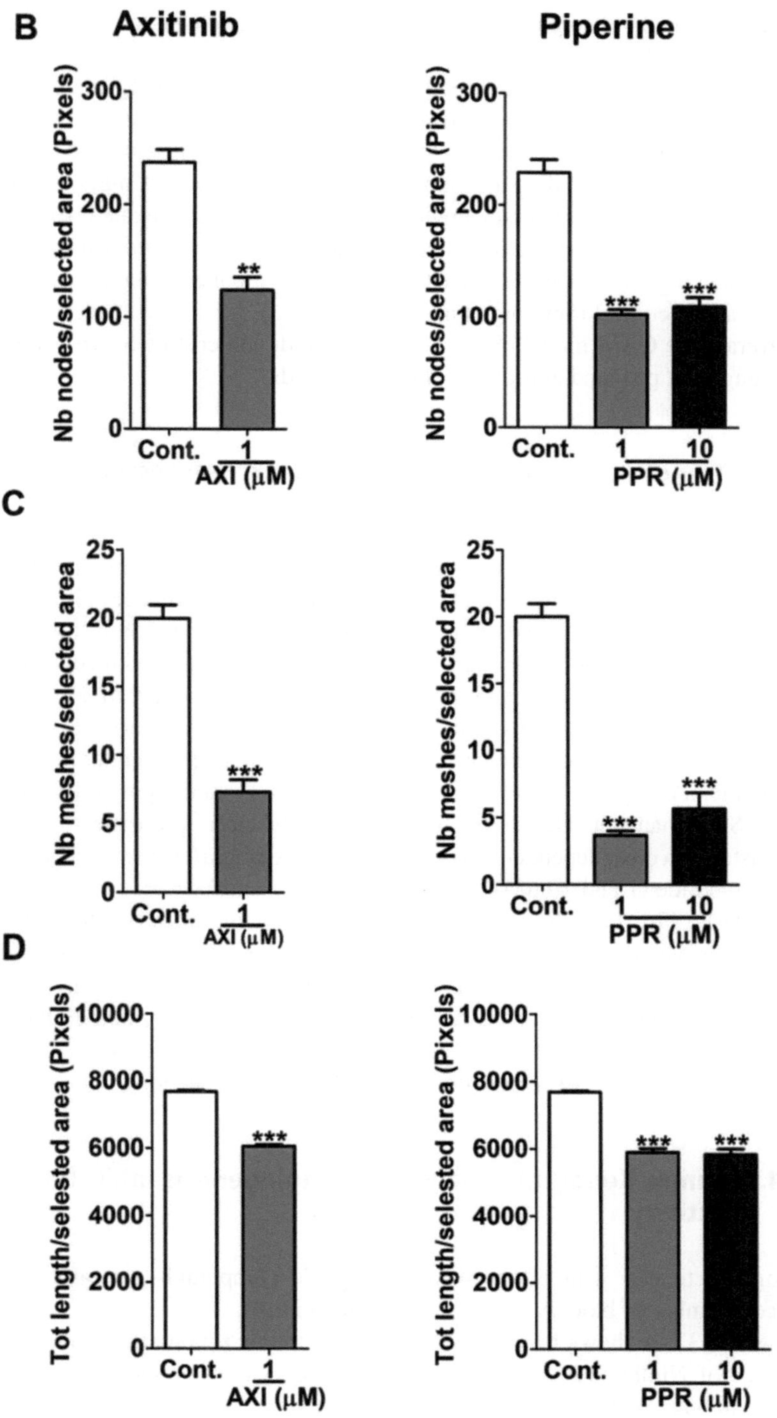

Fig. 13.7 (continued)

noticeable vascular disorganization, unlike the well-ordered vasculature observed in the 1× PBS-treated control.

For quantitative assessment, blood vessel density and branching can be analyzed using ImageJ or similar image analysis software, and the results can be represented graphically. Figure 13.7b–d illustrates the quantitative representation of vascular density parameters, including the number of nodes (Nb of nodes), number of meshes (Nb of meshes), and total vessel length (Tot length), in CAMs treated with Axitinib (an FDA-approved antiangiogenic agent) and Piperine, a potential phytochemical with antiangiogenic properties, tested at 1 and 10 μM concentrations, with 1× PBS serving as the control (Senrung et al. 2023).

Overall, the CAM model offers a reliable, rapid, and cost-effective platform for screening both pro- and anti-angiogenic compounds.

13.4.2 Tumor Xenograft Analysis for Tumor Suppression

The chick CAM provides a highly suitable model for tumor studies. Tumor xenografts can be established with nearly 100% success, making the CAM an excellent platform for screening chemotherapeutic agents. Xenografts treated with anticancer compounds (agents that inhibit or kill cancer cells) typically exhibit a significant reduction in tumor size or growth compared to untreated controls.

Figure 13.8 presents a representative study of tumor xenograft growth in the presence of anticancer agents administered at the time of tumor implantation on CAM, where U87 tumor xenograft growth was evaluated under the influence of Nimbolide, a compound observed (Senrung et al. 2023) and reported (Nagini et al. 2021; Subramani et al. 2016) to possess anticancer properties. The study demonstrated a dose-dependent reduction in tumor xenograft size in embryos treated with Nimbolide (1 and 10 μM) compared with the control.

Similarly, Fig. 13.9 shows the effect of Doxorubicin, an established anticancer drug, administered intratumorally either as a single dose (50 μl of 1 μM) on EDD10 (48 h post tumor implantation) or metronomically at a concentration of 0.3 μM (50 μl) on EDD10 (48 h post tumor implantation), EDD11, and EDD12. The treatment resulted in a noticeable decrease in tumor xenograft size, as evidenced by the reduced tumor tissue area observed in histological sections.

13.4.3 Tumor Xenograft Analysis for Angiogenesis Inhibition Activity

Xenografts treated with effective antiangiogenic compounds typically exhibit a reduced number of blood vessels compared to controls.

Figure 13.10a shows histological sections of tumor xenografts generated in the presence of Nimbolide, a compound reported to possess antiangiogenic activity. Treatment with Nimbolide at 1 and 10 μM resulted in a dose-dependent reduction in vascular density or tumor-induced angiogenesis compared to the control Fig. 13.10b.

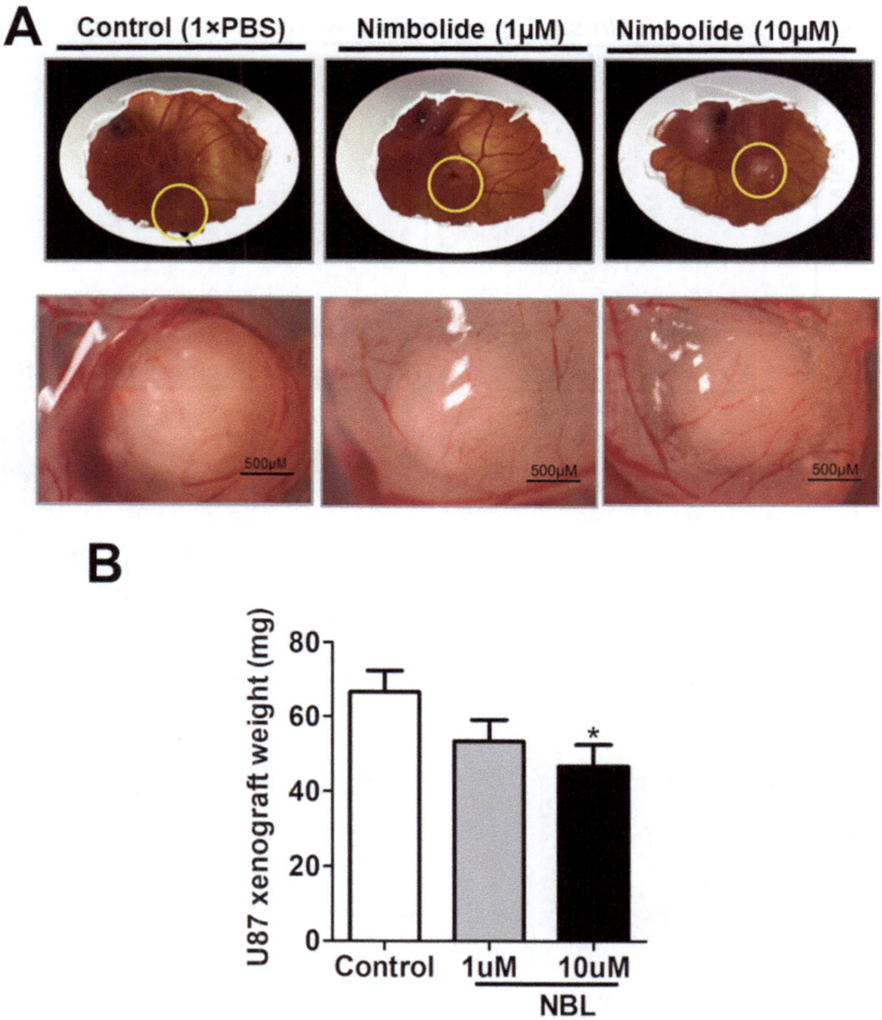

Fig. 13.8 Representative images of tumor (U87) xenograft generated on chick CAM model by the windowing method. (**a**) Xenograft (in situ) growing below the gelatin sponge (upper panel) and the excised xenograft viewed (lower panel). The yellow circles indicate gelatin sponge with a growing U87 xenograft underneath. (**b**) Graphical representation of the tumor xenograft weight. Scale bar = 500 μm. (Images and the graph are adapted from Senrung et al. (2023). The images are licensed under Creative Commons)

13.4.4 Drug Toxicity or Biosafety Evaluation

Chick embryos provide a suitable model system for evaluating drug toxicity or biosafety using multiple parameters, including anatomical, morphological, or histological deviations from normal development. Additional parameters such as organ

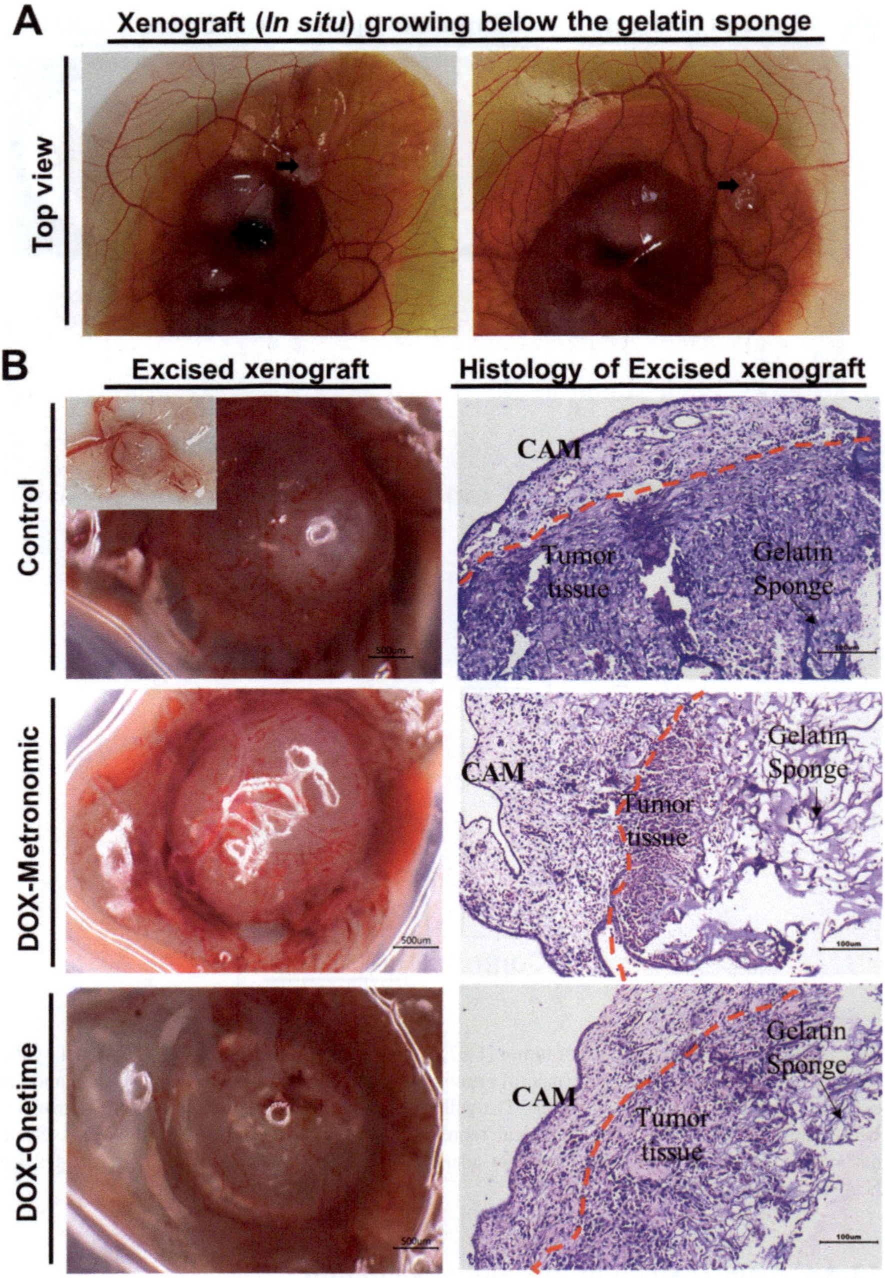

Fig. 13.9 Representative images of tumor (U87) xenograft generated on chick CAM model by the polystyrene boat method. (**a**) Xenograft (in situ) growing below the gelatin sponge (GS). The black arrow indicates GS with a growing U87 xenograft underneath. (**b**) The excised xenograft (left panel) and corresponding histological sections (right panel). The red dashed line indicates the interface between chick CAM and tumor (U87) tissue. Imaging by Radical Stereo Zoom Microscope fitted with a Camera and Nikon Eclipse Ts2R inverted microscope. Scale bar = 100 μm. CAM-Chorioallantoic membrane

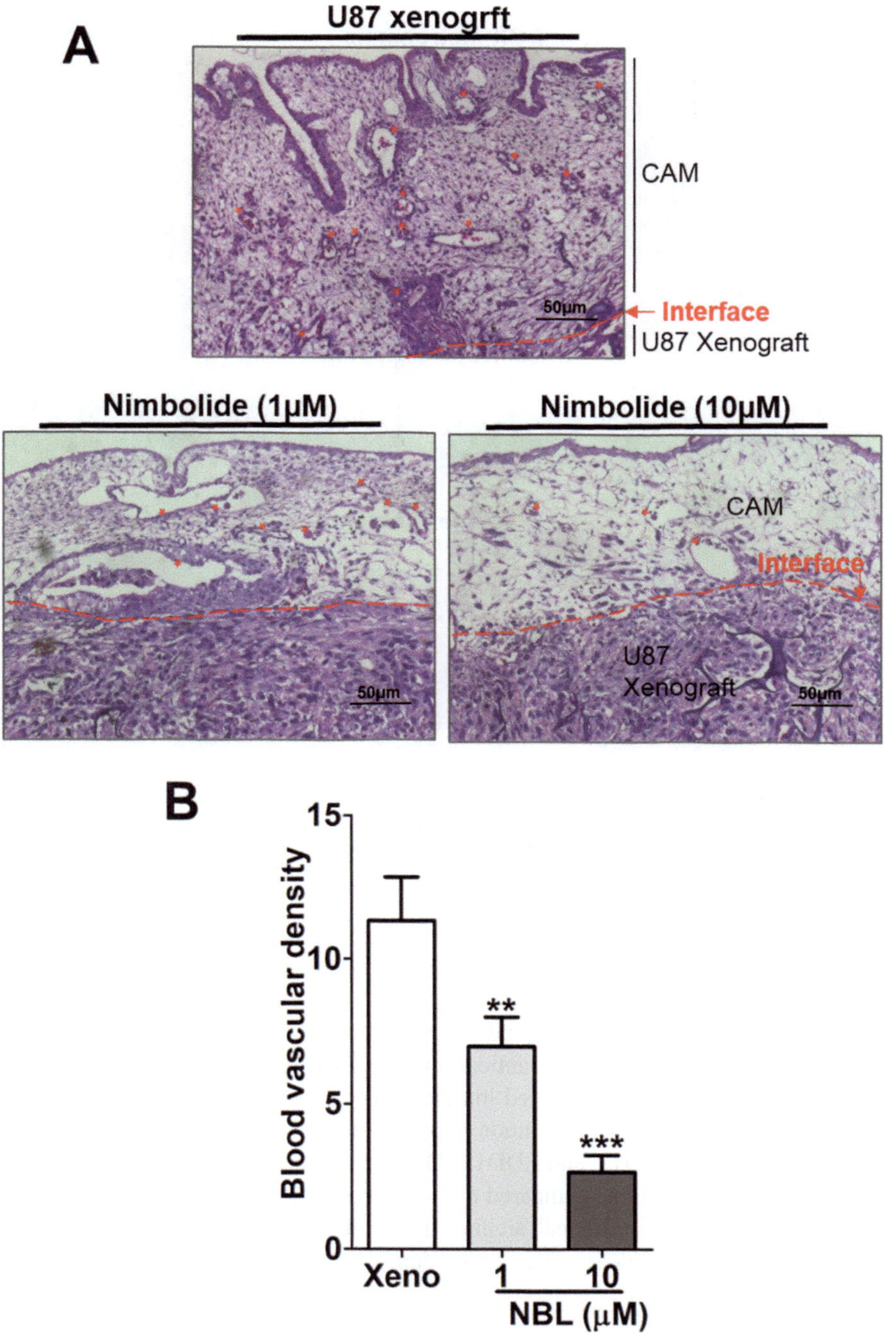

Fig. 13.10 Evaluation of tumor xenograft angiogenesis. (**a**) Representative images of histological sections of tumor (U87) xenograft generated on chick CAM model for angiogenesis study. The red dashed line indicates the interface between Chick CAM and tumor (U87) tissue. (**b**) Graphical representation of blood vascular density. Scale bar = 100 µm. (Images adapted from Senrung et al. (2023). The images are licensed under Creative Commons)

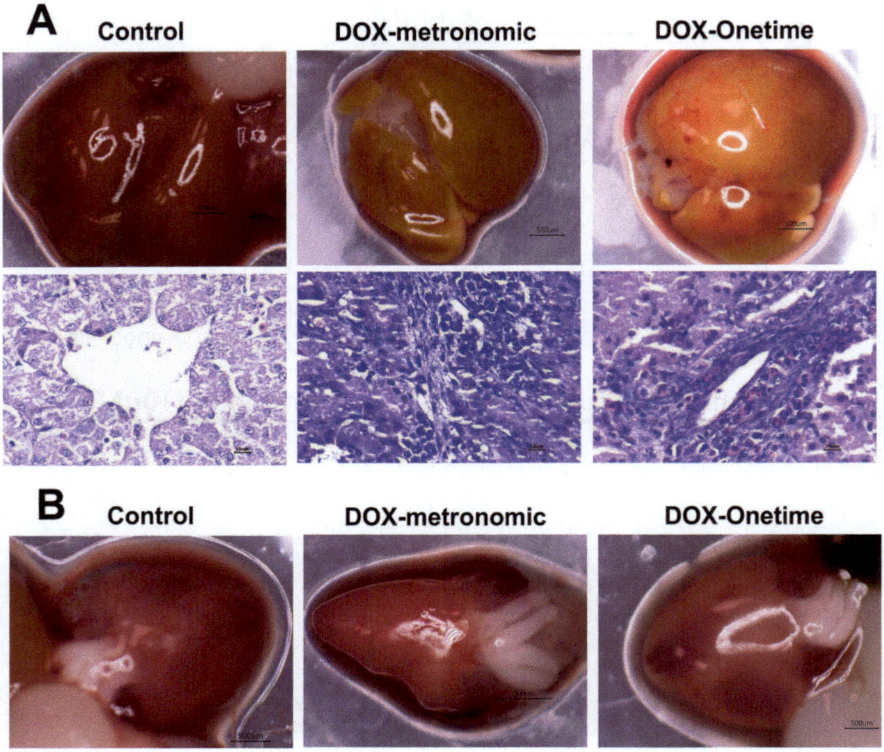

Fig. 13.11 Chick embryo organs at embryonic day 14 (EDD14) at 120 h following Doxorubicin treatment. (**a**) Liver: Whole image of liver (upper panel), Histology of the corresponding specific liver section vertically (upper panel). (**b**) Heart. EDD Embryonic development day

weight and overall embryo weight are also commonly assessed. Drugs that are toxic or non-biocompatible typically induce structural deformities or alterations in these parameters.

Figure 13.11 shows the liver and heart of chick embryos obtained from CAM tumor xenograft treated with the anticancer drug Doxorubicin, known to cause side effects. The drug was administered intratumorally either as a single dose (50 μl of 1 μM) on EDD10 (48 h post-tumor implantation) or metronomically at 0.3 μM (50 μl) on EDD10, EDD11, and EDD12. These treatments resulted in morphological and histological alterations compared to the control. In the liver, a distinct deviation in morphology and histological architecture is observed (Fig. 13.11a). Similarly, notable morphological deformities are evident in the heart (Fig. 13.11b).

13.5 Significance/Research Applications

As vertebrates, chickens and humans share numerous conserved genes governing fundamental biological functions, highlighting the relevance of the chick embryo as a translational research model. The chick embryo model bridges the gap between in vitro studies and mammalian systems, offering a physiologically relevant environment for evaluating drug behavior and safety. Moreover, it provides a robust platform for investigating interactions between potential therapeutic agents and biological tissues, enabling assessment of biocompatibility and early identification of potential adverse effects. Table 13.1 summarizes the applications of the chick embryo across diverse biomedical studies.

Table 13.1 Representative studies exhibiting the suitability and usage of the chick embryo as a biomedical research model

S. no.	Type of study/research	References
1.	Evaluation of the angiogenesis-mediated wound-healing efficacy and toxicological profile (including morphological, histological, and liver enzyme alterations) of chitosan-derived hydrogels using a chick CAM assay	Ara et al. (2022)
2.	Evaluation of the photobiomodulation (PBM) potential in promoting angiogenesis-mediated wound healing using chick CAM assays with fresh human skin xenografts	Winter et al. (2018)
3.	Evaluation of the potential of the chick CAM model as a time- and cost-efficient, ethically compliant (3Rs principle-based) alternative to conventional rodent xenograft models for malignant pleural mesothelioma drug target assessment	Barnett et al. (2022)
4.	Determination of the chick CAM model to support patient-derived xenografts (PDXs) of uveal melanoma	Tsimpaki et al. (2023)
5.	To assess the potential therapeutic benefits of saroglitazar in diabetic retinopathy, a condition characterized by pathological angiogenesis, retinal vascular damage, and inflammation, by evaluating its inhibitory effect on VEGF-induced angiogenesis using the chick CAM assay	Joharapurkar et al. (2021)
6.	Determination of the chick embryo as a suitable alternative model for acute drug formulation toxicity screening before full toxicological investigations in rodents, through evaluation of FDA-approved anticancer agents (paclitaxel, carmustine, camptothecin, cyclophosphamide, vincristine, cisplatin, aloin, mitomycin C, actinomycin-D, melphalan) based on embryo mortality within 48 h following intravenous administration in a 9-day-old CAM model	Kue et al. (2015)

13.6 Precautions

- The BOD incubator should be well sterilized before putting the eggs for incubation.
- The culture should be maintained in a well aerated area. Damp areas often lead to fungal growth on the culture.
- All the apparatus and instruments should be sterilized.
- The culture should be performed inside the hood.
- All setups should be performed in a sterile environment.

13.7 Quick Review Questions

13.7.1 Multiple Choice Questions

1. What is the full form of CAM?
   (a) Chorioallantoic Membrane
   (b) Cell adhesion molecule
   (c) Cancer-associated macrophage
   (d) Central angiogenesis model
   Answer: (a) Chorioallantoic membrane
2. Which of the following routes can be used for drug administration in the CAM tumor model?
   (a) Intratumoral
   (b) Topical
   (c) Intravenous
   (d) All of the above
   Answer: (d) All of the above
3. What is the primary indicator of antiangiogenic activity in CAM tumor xenografts?
   (a) Increased tumor size
   (b) Reduced vascular density within the xenograft
   (c) Elevated liver enzyme levels
   (d) Increased number of blood vessels
   Answer: (b) Reduced vascular density within the xenograft
4. Toxicity refers to:
   (a) A measure of the beneficial effects of a substance
   (b) The degree to which a substance can damage an organism
   (c) The metabolism of drugs in the liver
   (d) The process of wound healing
   Answer: (b) The degree to which a substance can damage an organism
5. Define angiogenesis
   (a) The process of breaking down existing blood vessels in response to injury
   (b) The formation of new blood vessels from pre-existing ones

   (c) The development of new nerve cells in response to chemical signals
   (d) The degeneration of old blood vessels during tissue repair

   Answer: (b) The formation of new blood vessels from pre-existing ones

6. What is the significance of angiogenesis in pathological conditions?
   (a) It helps maintain normal blood circulation during tissue repair
   (b) It plays a key role in the spread and growth of tumors and other diseases
   (c) It prevents the formation of new capillaries in damaged tissues
   (d) It only occurs during embryonic development and has no pathological role

   Answer: (b) It plays a key role in the spread and growth of tumors and other diseases

7. How does treatment with effective anticancer drugs typically affect tumor xenografts on CAM?
   (a) Tumors grow faster than controls
   (b) Tumors fail to grow or are reduced in size than controls
   (c) Tumors become more vascularized
   (d) Tumors show no change

   Answer: (b) Tumors fail to grow or are reduced in size than controls

8. What makes CAM a preferred model for evaluating both chemotherapeutic and antiangiogenic agents?
   (a) Ease of xenograft generation
   (b) High angiogenic potential of the membrane
   (c) Short experimental timeline
   (d) All of the above

   Answer: (d) All of the above

9. Why is toxicity evaluation of test compounds necessary in biomedical research?
   (a) To determine the taste and odor of the compound before testing
   (b) To assess the safety, dosage limits, and potential adverse effects of the compound
   (c) To increase the rate of compound absorption in test models
   (d) To enhance the color contrast during imaging studies

   Answer: (d) To assess the safety, dosage limits, and potential adverse effects of the compound

10. What is the main advantage of using CAM over rodent models in early-stage preclinical cancer research?
    (a) Ethical simplicity and lower cost
    (b) Rapid evaluation of angiogenesis and tumor growth
    (c) High reproducibility of tumor xenografts
    (d) All of the above

    Answer: (d) All of the above

13.7.2 Short Answer Questions

1. What is the full form of CAM in the context of chick embryo studies?
2. Why is the chick CAM considered a suitable model for tumor xenografts?

3. At which embryonic day is the CAM optimum for angiogenesis assay?
4. Name any two routes of drug administration used in CAM tumor models.
5. How can tumor growth be assessed in CAM xenografts?
6. Explain how antiangiogenic activity is measured using CAM xenografts.
7. How can systemic toxicity be assessed in CAM-based experiments?
8. Name two advantages of using the chick embryo model over rodent models in early-stage cancer research.
9. Describe the effect of effective anticancer drugs on CAM tumor xenografts.
10. What is the significance of measuring organ weights in CAM studies after drug administration?

Glossary

Age-related macular degeneration (AMD) A progressive eye disorder that affects the macula, common in older adults after the age of 50.

Alanine aminotransferase (ALT) A liver enzyme catalyzing the conversion of alanine and α-ketoglutarate to pyruvate and glutamate. Elevated ALT levels in blood indicate liver damage or dysfunction.

Alkaline phosphatase (ALP) An enzyme present in the liver, bone, kidney, and intestine. It hydrolyzes phosphate groups from molecules and serves as a key marker of liver and bone health.

Allantois An extraembryonic membrane that develops from the posterior part of the embryonic gut in amniotes. Functions primarily in waste storage, gas exchange, and nutrient transport.

Anti-desmin An antibody that specifically binds to desmin. Anti-desmin staining can help identify and quantify blood vessel-associated pericytes.

Aspartate aminotransferase (AST) An enzyme often used as a biochemical marker for liver damage.

Biochemical assays Laboratory techniques used to measure, quantify, or detect specific chemical substances or enzymatic activities in biological samples: blood, tissues, or cells.

Biocompatibility The ability of a material, drug, or therapeutic agent to interact with biological systems without causing harmful effects.

Biomedical research The systematic and scientific investigation of biological processes and diseases to develop effective diagnostic tools, treatments, and preventive strategies to improve health outcomes.

Chorion The outermost extraembryonic membrane formed from the trophoblast and mesoderm, enclosing the embryo. It functions in protection, gas exchange, and, in mammals, contributes to placenta formation.

Diabetic retinopathy A diabetes-related eye disorder caused by damage to retinal blood vessels. Persistent high blood sugar leads to leakage and subsequent abnormal vessel growth, resulting in progressive vision loss or blindness if untreated.

Doxorubicin A potent anthracycline chemotherapeutic agent that intercalates into DNA, inhibits topoisomerase II, and generates free radicals, thereby disrupting

DNA replication and transcription and inducing apoptosis in rapidly dividing cancer cells.

Embryonic development day 4 (EDD 4) Refers to the fourth day of growth and differentiation in a developing embryo.

Ethical In research, it means adhering to moral values, honesty, integrity, and respect for human and animal welfare, ensuring that actions are responsible and justifiable.

Extraembryonic Structures or tissues that develop from the fertilized egg but lie outside the main body of the embryo.

Gamma-glutamyl transferase (GGT) A key enzyme in glutathione metabolism and amino acid transport. Predominantly found in the liver, its elevated blood levels indicate liver dysfunction, bile obstruction, or alcohol abuse.

Gelatin sponge (GS) A gelatin sponge is a porous, absorbable biomaterial used in biomedical applications to control bleeding, aid healing, and deliver therapeutic agents locally.

Hammock techniques Here, it refers to a chick embryo culture method in which a cling wrap, shaped like an avian egg, is suspended in a "hammock-like" manner from a tripod or a plastic cup to support embryo growth, culture, and observation for testing or implantation studies.

Hematoxylin and eosin (H&E) A widely used staining technique in histology and pathology to visualize tissue architecture, where nuclei appear blue/purple (hematoxylin-A basic dye) and cytoplasm and extracellular matrix appear pink (eosin-An acidic dye).

Histology Reproductive behavior in which the male clasps tightly around the back or behind the forelimbs of the female with forelimbs, to facilitate external fertilization.

Immunodeficient A condition in which the immune system's ability to fight infections or respond to foreign substances is weakened or absent.

Immunohistochemistry (IHC) A laboratory technique used to detect specific proteins or antigens in tissue sections by using antibodies that bind selectively to those target molecules.

Intratumoral Occurring within or administered directly into a tumor.

Intravenous Administration or infusion of substances directly into a vein.

Metronomically Administration of therapeutic agents, particularly anticancer drugs, at regular, low, and frequent doses with minimal or no prolonged breaks.

Patient-derived xenografts (PDX) Experimental cancer models created by implanting tumor tissues obtained directly from a patient into a model organism.

Photobiomodulation (PBM) A therapeutic technique that uses low-level light, typically from lasers or light-emitting diodes (LEDs), to stimulate cellular activity.

Piperine Naturally occurring alkaloid found in black pepper (Piper nigrum) and long pepper (Piper longum).

Preclinical The stage of research conducted before clinical trials in humans.

Primary allantoic vein Major blood vessel that develops within the allantois of the chick embryo.

Topically Application of a substance directly to a specific surface, such as the skin, to produce a localized effect.

Total bilirubin The sum of direct (conjugated) and indirect (unconjugated) bilirubin present in the blood.

Toxicology The scientific study of the adverse effects of chemical, physical, or biological agents on living organisms and the environment.

Trophoblast Outer cell layer of the blastocyst.

Twigs In ImageJ, twigs refer to small, thin branch-like structures in a skeletonized image, typically representing minor extensions or endpoints of a network, such as blood vessels or neural pathways.

Vitelline membrane A thin, transparent layer that surrounds the yolk of an egg.

Warm-blooded physiology The ability of certain animals, such as birds and mammals, to maintain a constant internal body temperature independent of external environmental conditions.

Whatman filter paper A cellulose-based paper used for filtration in laboratories.

Windowing A technique used in chick embryo studies where a small opening or "window" is carefully made in the eggshell to allow direct observation and manipulation of the developing embryo without disrupting its normal growth.

Wound healing The biological process by which the body repairs and regenerates damaged tissue after injury.

Xenograft A tissue or organ transplant derived from one species and implanted into a different species.

Yolk sac An extraembryonic membrane that surrounds the yolk of an egg and provides essential nutrients to the developing embryo during early growth.

References

Ara C, Jabeen S, Afshan G, Farooq A, Akram MS, Asmatullah, Islam A, Ziafat S, Nawaz B, Khan RU (2022) Angiogenic potential and wound healing efficacy of chitosan derived hydrogels at varied concentrations of APTES in chick and mouse models. Int J Biol Macromol 202:177–190. https://doi.org/10.1016/j.ijbiomac.2022.01.047

Ayers GD, McKinley ET, Zhao P, Fritz JM, Metry RE, Deal BC, Adlerz KM, Coffey RJ, Manning HC (2010) Volume of preclinical xenograft tumors is more accurately assessed by ultrasound imaging than manual caliper measurements. J Ultrasound Med 29(6):891–901. https://pmc.ncbi.nlm.nih.gov/articles/PMC2925269/

Barnett SE, Herrmann A, Shaw L, Gash EN, Poptani H, Sacco JJ, Coulson JM (2022) The Chick embryo xenograft model for malignant pleural mesothelioma: a cost and time efficient 3Rs model for drug target evaluation. Cancers (Basel) 14(23). https://doi.org/10.3390/cancers14235836

Carpentier G, Berndt S, Ferratge S, Rasband W, Cuendet M, Uzan G, Albanese P (2020) Angiogenesis analyzer for ImageJ - a comparative morphometric analysis of "endothelial tube formation assay" and "fibrin bead assay". Sci Rep 10(1):11568. https://doi.org/10.1038/s41598-020-67289-8

Carpentiera G, Martinellia M, Courtya J, Casconea I (2012) Angiogenesis analyzer for ImageJ. In: 4th ImageJ user and developer conference proceedings

Chen W, Wong C, Vosburgh E, Levine AJ, Foran DJ, Xu EY (2014) High-throughput image analysis of tumor spheroids: a user-friendly software application to measure the size of spheroids automatically and accurately. J Vis Exp (89). https://doi.org/10.3791/51639

Gonzalez I, Chatterjee D (2019) Histopathological features of drug-induced liver injury secondary to Osimertinib. ACG Case Rep J 6(2):e00011. https://doi.org/10.14309/crj.0000000000000011

Guan LP, Nan JX, Jin XJ, Jin QH, Kwak KC, Chai KY, Quan ZS (2005) Protective effects of chalcone derivatives for acute liver injury in mice. Arch Pharm Res 28(1):81–86. https://doi.org/10.1007/BF02975140

Joharapurkar A, Patel V, Kshirsagar S, Patel MS, Savsani H, Jain M (2021) Effect of dual PPAR-alpha/gamma agonist saroglitazar on diabetic retinopathy and oxygen-induced retinopathy. Eur J Pharmacol 899:174032. https://doi.org/10.1016/j.ejphar.2021.174032

Kue CS, Tan KY, Lam ML, Lee HB (2015) Chick embryo chorioallantoic membrane (CAM): an alternative predictive model in acute toxicological studies for anti-cancer drugs. Exp Anim 64(2):129–138. https://doi.org/10.1538/expanim.14-0059

Nagini S, Nivetha R, Palrasu M, Mishra R (2021) Nimbolide, a neem limonoid, is a promising candidate for the anticancer drug arsenal. J Med Chem 64(7):3560–3577

Senrung A, Tripathi T, Aggarwal N, Janjua D, Chhokar A, Yadav J, Chaudhary A, Thakur K, Singh T, Bharti AC (2024a) Anti-angiogenic potential of trans-chalcone in an in vivo chick chorioallantoic membrane model: an ATP antagonist to VEGFR with predicted blood-brain barrier permeability. Cardiovasc Hematol Agents Med Chem 22(2):187–211. https://doi.org/10.2174/0118715257250417231019102501

Senrung A, Tripathi T, Janjua D, Yadav SK, Chhokar A, Aggarwal N, Yadav J, Chaudhary A, Joshi U, Sethi P (2024b) Chick chorioallantoic membrane: a valuable 3D in vivo model for screening nanoformulations for tumor antiangiogenic therapeutics. Int J Dev Biol 67(4):115–135

Senrung A, Tripathi T, Yadav J, Janjua D, Chaudhary A, Chhokar A, Aggarwal N, Joshi U, Goswami N, Bharti AC (2023) In vivo antiangiogenic effect of nimbolide, trans-chalcone and piperine for use against glioblastoma. BMC Cancer 23(1):1173. https://doi.org/10.1186/s12885-023-11625-4

Subramani R, Gonzalez E, Arumugam A, Nandy S, Gonzalez V, Medel J, Camacho F, Ortega A, Bonkoungou S, Narayan M (2016) Nimbolide inhibits pancreatic cancer growth and metastasis through ROS-mediated apoptosis and inhibition of epithelial-to-mesenchymal transition. Sci Rep 6(1):19819

Tsimpaki T, Bechrakis NE, Seitz B, Kraemer MM, Liu H, Dalbah S, Sokolenko E, Berchner-Pfannschmidt U, Fiorentzis M (2023) Chick chorioallantoic membrane as a patient-derived xenograft model for uveal melanoma: imaging modalities for growth and vascular evaluation. Cancers (Basel) 15(5). https://doi.org/10.3390/cancers15051436

Winter R, Dungel P, Reischies FMJ, Rohringer S, Slezak P, Smolle C, Spendel S, Kamolz LP, Ghaffari-Tabrizi-Wizsy N, Schicho K (2018) Photobiomodulation (PBM) promotes angiogenesis in-vitro and in chick embryo chorioallantoic membrane model. Sci Rep 8(1):17080. https://doi.org/10.1038/s41598-018-35474-5

Pharmacological Regulation of Gametogenesis and Fertilization: To Explore the Effects of Drugs and Environmental Toxins on Gamete Formation and Fertilization in Mammals

14

Tanya Tripathi, Anna Senrung, Alok Chandra Bharti, and Gautam Sethi

Abstract

Reproductive health is strongly influenced by both therapeutic agents and endocrine-disrupting chemicals (EDCs). Chemotherapeutics often impair germ-cell development, while EDCs interfere with hormone-regulated reproductive processes. To assess their impact on gametogenesis, experimental models are essential. In vitro assays such as MTT, comet, and oxidative stress measurements enable precise evaluation of cellular damage and viability in the germ cells, whereas ex vivo testicular and ovarian cultures provide insights into tissue-level effects. On the contrary, in vivo models capture whole-organism outcomes, including fertility endpoints. This chapter highlights the application of these models and assays, the measurable parameters they offer, and their utility in advancing reproductive toxicology and translational research.

Keywords

Spermatogenesis · Oogenesis · Chemotherapeutics · Endocrine-disrupting chemicals · In vitro models · In vivo models

T. Tripathi · A. C. Bharti
Molecular Oncology Lab., Department of Zoology, University of Delhi, Delhi, India

A. Senrung
Department of Zoology, Daulat Ram College, University of Delhi, Delhi, India

G. Sethi (✉)
Department of Pharmacology, National University of Singapore, Singapore, Singapore

A. C. Bharti et al. (eds.), *Textbook of Developmental Biology*,
https://doi.org/10.1007/978-981-95-7269-4_14

Aim This chapter aims to help students understand how chemotherapeutic agents and endocrine-disrupting chemicals impact mammalian gametogenesis and reproductive health. It also equips them with fundamental in vitro, ex vivo, and in vivo approaches, including key assays and culture models, to assess germ-cell toxicity and link cellular effects to overall fertility outcomes.

Learning Objectives

Upon completing this chapter, students will be able to:

- Understand how chemotherapeutics and endocrine-disrupting chemicals (EDCs) affect gametogenesis in mammals.
- Differentiate between in vitro, ex vivo, and in vivo approaches for evaluating reproductive toxicity and their respective applications.
- Utilize key in vitro experimental techniques such as MTT assay, comet assay, and oxidative stress measurement to assess cell viability, DNA damage, and cellular stress in germ cells.
- Utilize ex vivo testicular and ovarian culture methods to study toxicant-induced alterations in tissue structure and function.
- Integrate findings from in vitro, ex vivo, and in vivo studies to better understand impact of chemotherapeutics/EDCs on reproductive outcomes spanning from gamete-level changes to fertility potential.

14.1 Introduction

Gametogenesis is the fundamental biological process through which specialized germ cells, sperm in males and oocytes in females, are produced to ensure sexual reproduction. It involves tightly regulated events of cell division, growth, and differentiation, supported by somatic cells within the gonads. In males, spermatogenesis occurs in the seminiferous tubules with the help of Sertoli and Leydig cells, whereas in females, oogenesis is orchestrated within ovarian follicles comprising oocytes, granulosa cells, and theca cells. The detailed processes and stages of both spermatogenesis and oogenesis have been described in detail in Chap. 9. However, the normal course of gametogenesis and fertilization can be disrupted or influenced by pathological conditions, therapeutic agents, dietary supplements, contraceptives, and environmental contaminants, leading to infertility. Globally, infertility affects approximately 17.5% of adult population, i.e., 1 in 6 couples experience difficulty conceiving during their lifetime. This prevalence is similar across high-, middle-, and low-income countries, highlighting infertility as a widespread public health issue (Mascarenhas et al. 2012). Chemotherapeutics (Koyama et al. 1977; Lu and Meistrich 1979) and endocrine-disrupting chemicals (EDCs) (Sharpe et al. 1995; Susiarjo et al. 2007) are known for their adverse effects on gametogenesis, ultimately leading to loss of fertility and impaired reproductive health. Chemotherapeutics account for infertility in 15–40% of cancer survivors (Zhou et al. 2024), and exposure to EDCs may contribute to fewer cases; however, the exact percentage is not reported.

The chemotherapeutics used indispensably in cancer treatment often compromise germ cells and gonadal function due to their mechanism of action. Alkylating agents such as cyclophosphamide and busulfan form DNA cross-links that lead to irreparable double-strand breaks, resulting in germ-cell apoptosis and loss of reproductive potential (Aguilar-Mahecha et al. 2005). In contrast, platinum-based compounds like cisplatin and carboplatin induce oxidative stress and apoptosis in spermatogonial and oogonial stem cells, thereby reducing the long-term regenerative capacity of testis and ovary (Smart et al. 2018). Similarly, EDCs represent a diverse group of natural and synthetic compounds that interfere with hormone-regulated signaling pathways and disrupt reproductive processes. They are encountered in daily life through industrial products such as Bisphenol A (BPA) and phthalates, agricultural pesticides like DDT, atrazine, and vinclozolin, and environmental pollutants, including heavy metals such as cadmium, arsenic, and lead. Their widespread presence in food, water, and consumer goods makes them particularly concerning as chronic low-level exposures may produce cumulative reproductive toxicity (Delbes et al. 2022).

Given these risks, it becomes essential to evaluate the impact of chemotherapeutics and EDCs on gametogenesis before drawing conclusions about their safety or clinical utility. In this context, carefully designed experimental models provide a reliable framework to assess both structural and functional consequences of these gonado-toxins. In vitro systems allow rapid screening and mechanistic insights, while in vivo models capture the complexity of whole-organism physiology, including hormone feedback loops, fertility potential, and germ-cell regeneration. To illustrate the utility of these approaches, this chapter highlights representative studies employing different methodologies, with emphasis on how the models were applied, what endpoints were measured ranging from histological and molecular changes to hormonal and functional readouts, and how such protocols can be replicated or adapted for future work.

14.2 Methods for Evaluation of Effects of Chemotherapeutics and EDCs on Gametogenesis

14.2.1 In Vitro Methods

14.2.1.1 Cell Lines

Immortalized mouse germ cell lines have been widely utilized as standardized in vitro systems to investigate the cytotoxic and genotoxic effects of chemotherapeutics. These cell lines, which represent distinct stages of spermatogenesis, provide a controlled platform for mechanistic studies on drug-induced germ cell damage. A comprehensive list of commonly used germ cell lines and their tissue origin is summarized in Table 14.1.

Table 14.1 Commercially available germ cell lines to study the impact of chemotherapeutics and EDCs

Gametogenesis	Cell lines	Target tissue/ cells	Model organism	Drug used	Molecular targets	Mechanism of action	References
Spermatogenesis	GC-6spg	Spermatogonial stem cells	Rat	Doxorubicin	–	Induction of oxidative stress	Tremblay and Delbes (2018)
	GC-1 spg	Spermatogonial cells	Mouse	Busulfan	Iron content↑	Induction of testis ferroptosis	Hu et al. (2025)
	GC-2 spd	Spermatocytes	Mouse	Palcitaxel	GRP78, p-ERK, eIF2α↑	Trigger of ER stress pathway	Karadeniz Saygili et al. (2025)
	TM4	Sertoli cells	Mouse	Melamine	N-Cadherin, occludin, vimentin, α-tubulin↓	Induction of apoptosis	Chang et al. (2017)
	Ser-W3	Native Sertoli cells	Rat	Doxorubicin	GSH↓	Induction of oxidative stress	Tremblay and Delbes (2018)
	R2C	Leydig cells	Rat	Bisphenol A	Aromatase, COX-2↑	Reduction in testosterone level	Kim et al. (2010)
	TM3	Leydig cells	Mouse	Bisphenol A	–	Reduction in testosterone level	Goncalves et al. (2018)
	MA-10	Leydig cells	Mouse	Mono-(2-ethylhexyl) phthalate	STAR↓, MMP↓	Reduction in progesterone production, generation of ROS	Traore et al. (2021)
Oogenesis	KGN	Granulosa cells	Human	Cisplatin	↑Bax, Bcl-2↓	Induction of autophagy and apoptosis	Wang et al. (2022)

Abbreviations: *Bax* Bcl-2-associated X protein, *Bcl-2* B-cell lymphoma 2 protein, *eIF2α* Eukaryotic initiation factor 2 alpha, *GRP78* Glucose-regulated protein 78, *GSH* Glutathione, *MMP* Mitochondrial membrane potential, *p-ERK* Phosphorylated extracellular signal-regulated kinase, *ROS* Reactive oxygen species, *STAR* Steroidogenic acute regulatory protein

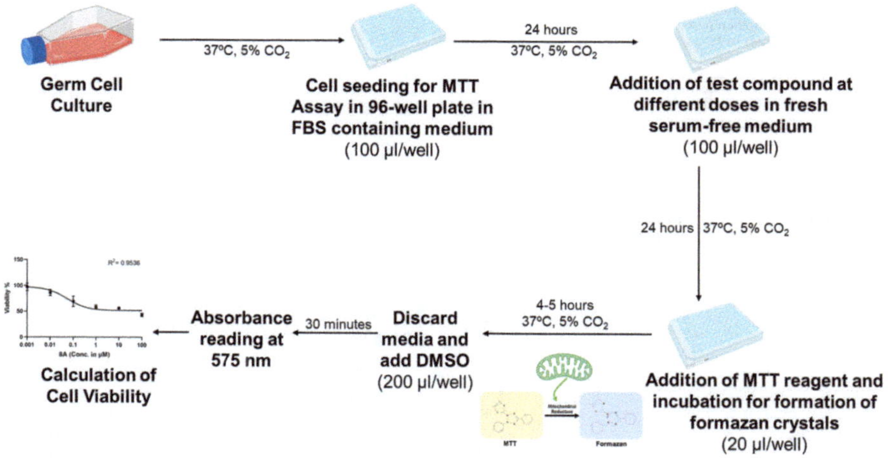

Fig. 14.1 Schematic representation of the MTT assay protocol. The workflow depicts sequential steps including cell seeding, drug treatment, and subsequent incubation with MTT reagent, leading to the formation of intracellular formazan crystals. These crystals are solubilized using DMSO, and the absorbance is measured spectrophotometrically to determine cell viability

Numerous studies have exposed these commercially available germ cell lines to various chemotherapeutics under defined culture conditions. The impact on cell viability is frequently assessed using the MTT assay, a colorimetric method based on the reduction of the yellow tetrazolium salt (MTT) to insoluble purple formazan crystals by mitochondrial dehydrogenases in metabolically active cells. The amount of formazan produced is directly proportional to the number of viable cells, making the *MTT assay* a standard approach for determining cell viability, cytotoxicity, and cell proliferation rates (van Meerloo et al. 2011). The sequential steps of the MTT assay protocol are illustrated in Fig. 14.1.

Experimental Errors- MTT Assay

- Uneven seeding of cells can cause variability in absorbance readings due to inconsistent cell density across wells.
- Contamination during culture or handling may alter cell viability and produce unreliable results.
- Over- or under-incubation with MTT reagent may affect the intensity of color development.
- Incomplete solubilization of formazan crystals can result in inaccurate absorbance measurements.
- Inaccurate pipetting during reagent addition or mixing can introduce significant experimental variation.

MTT is a simple, cost-effective, and works with various cell types providing a quantitative measure of cell metabolic activity, however not always sensitive to early

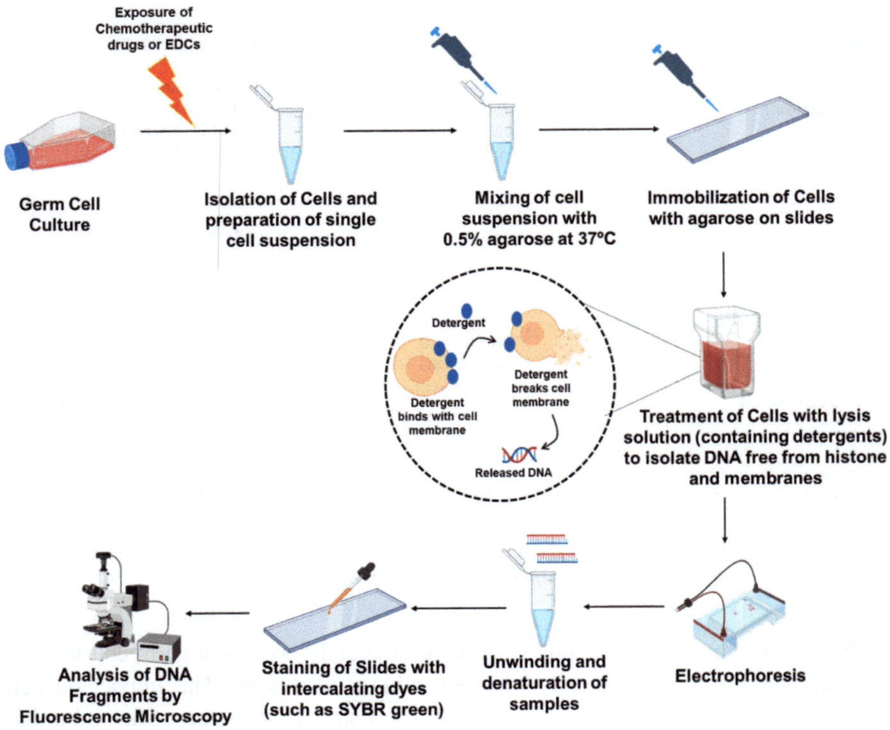

Fig. 14.2 Schematic representation of comet assay protocol. The schematic illustrates the sequential steps of the comet assay, including preparation of single cell suspension, cell embedding in agarose, cell lysis, DNA unwinding under alkaline conditions, and electrophoresis. Following staining, fragmented DNA migrates toward the anode, generating a comet-like structure that is visualized by fluorescence microscopy and quantified to assess DNA damage

apoptosis and can be affected by compounds that alter mitochondrial activity or interfere with formazan formation. In addition to viability, cell cycle arrest is evaluated through flow cytometry, typically using propidium iodide staining to quantify DNA content across different cell cycle phases G0/G1, S, and G2/M. This method allows researchers to detect drug-induced cell cycle disruptions, particularly at the G2/M checkpoint. Evaluation of cell cycle arrest provides a mechanistic insight into cytotoxicity, but may not distinguish between reversible arrest and apoptosis.

Further, to examine DNA damage under the influence of chemotherapeutics or EDCs, the *comet assay* (also known as single-cell gel electrophoresis) is also widely employed. In this assay, cells embedded in agarose on a microscope slide are lysed and subjected to electrophoresis under alkaline conditions. DNA fragments migrate out of the nucleus, forming a "comet tail" when visualized under fluorescence microscopy. The extent of DNA migration reflects the degree of strand breaks and chromatin relaxation, providing a sensitive measure of genotoxicity (Olive and Banath 2006). The sequential steps of the comet assay protocol are illustrated in

Fig. 14.2. It is a sensitive method for detecting DNA strand breaks at the single-cell level, but accuracy can be affected by false positives or negatives due to cytotoxicity, overlapping comets, or incomplete chromatin decondensation. The assay requires careful handling of fragile gels, and procedural variability can introduce inconsistencies.

Experimental Errors: Comet Assay

- Wrong concentration or incomplete melting of agarose may cause poor embedding of cells and weak comet formation.
- Cell damage during sample preparation due to harsh pipetting or centrifugation can lead to artificially increased DNA breaks.
- Excessive lysis can degrade DNA and distort comet morphology.
- Variation in electrophoresis conditions, such as buffer strength, voltage, duration, and temperature, may cause inconsistent DNA migration.
- Overstaining or uneven staining with DNA dyes can obscure comet visualization.
- Drying of gels before staining or imaging may cause cracks and artifacts.
- Photobleaching or overexposure during imaging can reduce clarity of comet tails.

Global cellular redox status in cells resulting from drug exposure is typically quantified using *ROS detection assays*, such as DCFH-DA (2′,7′-dichlorofluorescin diacetate), which fluoresces upon oxidation by intracellular reactive oxygen species. Elevated ROS levels are indicative of mitochondrial dysfunction and oxidative damage, both of which contribute to impaired germ cell function when treated with mono-(2-ethylhexyl) phthalate (MEHP) (Traore et al. 2021) (Fig. 14.3). These assays offer high sensitivity and are relatively easy to perform, enabling real-time monitoring of oxidative stress within cells, which is valuable for evaluating the cytotoxic effects of drugs or chemicals. However, they lack complete specificity since DCFH-DA can react with a variety of reactive oxygen species, and the fluorescence signal may be influenced by factors such as cell density, esterase activity, and probe uptake. Moreover, very high ROS levels can saturate the probe, and the assay does not differentiate between ROS generated in different cellular compartments (Kalyanaraman et al. 2012).

Experimental Errors: Cellular ROS Measurement

- Incorrect dye/probe concentration or incubation time can cause under- or over-estimation of ROS levels.
- Light exposure during staining may lead to auto-oxidation of DCFH-DA, producing false-positive signals.

(continued)

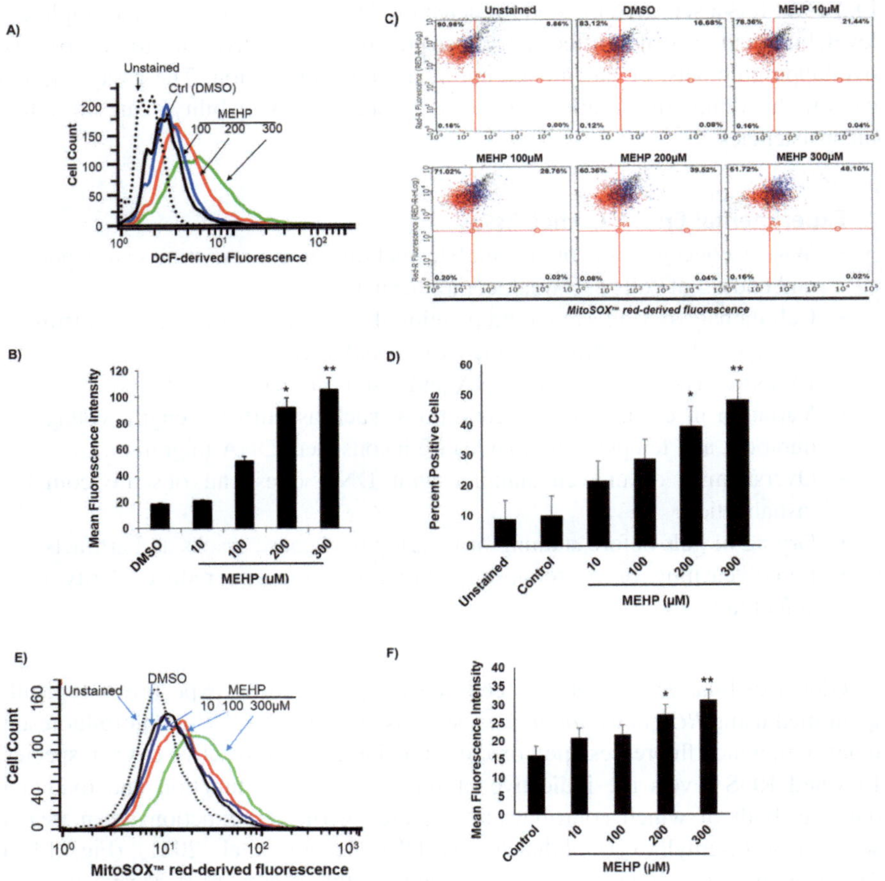

Fig. 14.3 Effect of MEHP on intracellular and mitochondrial-derived ROS. (**a**) Intracellular ROS (hydrogen peroxide) production (DCF-positive cells) in relationship to increasing MEHP dose. (**b**) ROS production per cell in response to increasing MEHP concentration (Mean Fluorescence Intensity). (**c**) Representative dot plots of superoxide production by cells treated with MEHP. Cells in the upper-left quadrant (low fluorescence) migrate toward the upper right quadrant (high fluorescence) in response to MEHP treatment. (**d**) Percent positive superoxide-producing cells. (**e**) Histogram representation of intracellular superoxide production induced by MEHP. (**f**) Bar graph representations of average level of superoxide production per cell (Mean Fluorescence Intensity). (The image/figure is adapted from Traore et al. (2021) with permission)

- Uneven dye/probe loading across cells results in variability in fluorescence intensity.
- Incomplete removal of excess dye/probe after incubation may increase background fluorescence.

(continued)

- Cell stress during handling such as harsh pipetting or temperature shifts can artificially elevate ROS levels.
- Photobleaching during fluorescence imaging may reduce signal accuracy.
- Instrument variability (laser intensity, detector settings) can lead to inconsistent results if not calibrated.
- Cell density differences among wells can influence dye/probe uptake and ROS signal.

Furthermore, gene and protein expression analyses are conducted using *quantitative real-time PCR* (qRT-PCR) and Western blotting, respectively. These methods target key markers involved in apoptosis such as Bax, Bcl-2, and Caspase-3; DNA damage response such as p53, ATM, and γ-H2AX; and oxidative stress regulation such as SOD, GPx, and catalase. Alterations in expression profiles help delineate the pathways through which chemotherapeutics exert their cytotoxic effects on germ cells.

The ultra-structural changes in the cell lines post-treatment can also be visualized through transmission electron microscopy (TEM). A study on the effect of melamine (MA) on Sertoli cells showed that MA exposure induced dose-dependent ultrastructural alterations, visualized by TEM, in TM4 cells (Chang et al. 2017). The structural changes included mitochondrial swelling, endoplasmic reticulum dilation, and nuclear chromatin disruption, indicating cellular stress and potential cytotoxic effects (Fig. 14.4).

These in vitro studies collectively highlight the vulnerability of spermatogonial, oogonial, and premeiotic germ cells to chemotherapeutics as well as EDC-induced damage. Since many of these cell lines recapitulate early and mid-stages of spermatogenesis and oogenesis, they serve as valuable tools for dissecting the molecular events underlying impaired proliferation, meiotic initiation, and germ cell apoptosis following chemotherapeutics or EDC exposure. However, in vitro systems lack the complex tissue architecture and cell-cell interactions present in the native gonad, which can limit the physiological relevance of the findings. To address this, ex vivo models such as cultures of testicular or ovarian tissue are preferred.

14.2.2 Ex Vivo Methods

14.2.2.1 Testicular Culture

Ex vivo testicular culture has emerged as a valuable model system to investigate spermatogenesis. Unlike in vitro models, it preserves three-dimensional structure and somatic cell interactions, enabling a more accurate assessment of germ cell responses to chemotherapeutics or EDCs while maintaining controlled experimental conditions. Among the available methods, the agarose gas-liquid interface technique has been most widely adopted due to its ability to support long-term spermatogenesis in neonatal testicular tissue (Allen et al. 2020). Polydimethylsiloxane (PDMS) disks

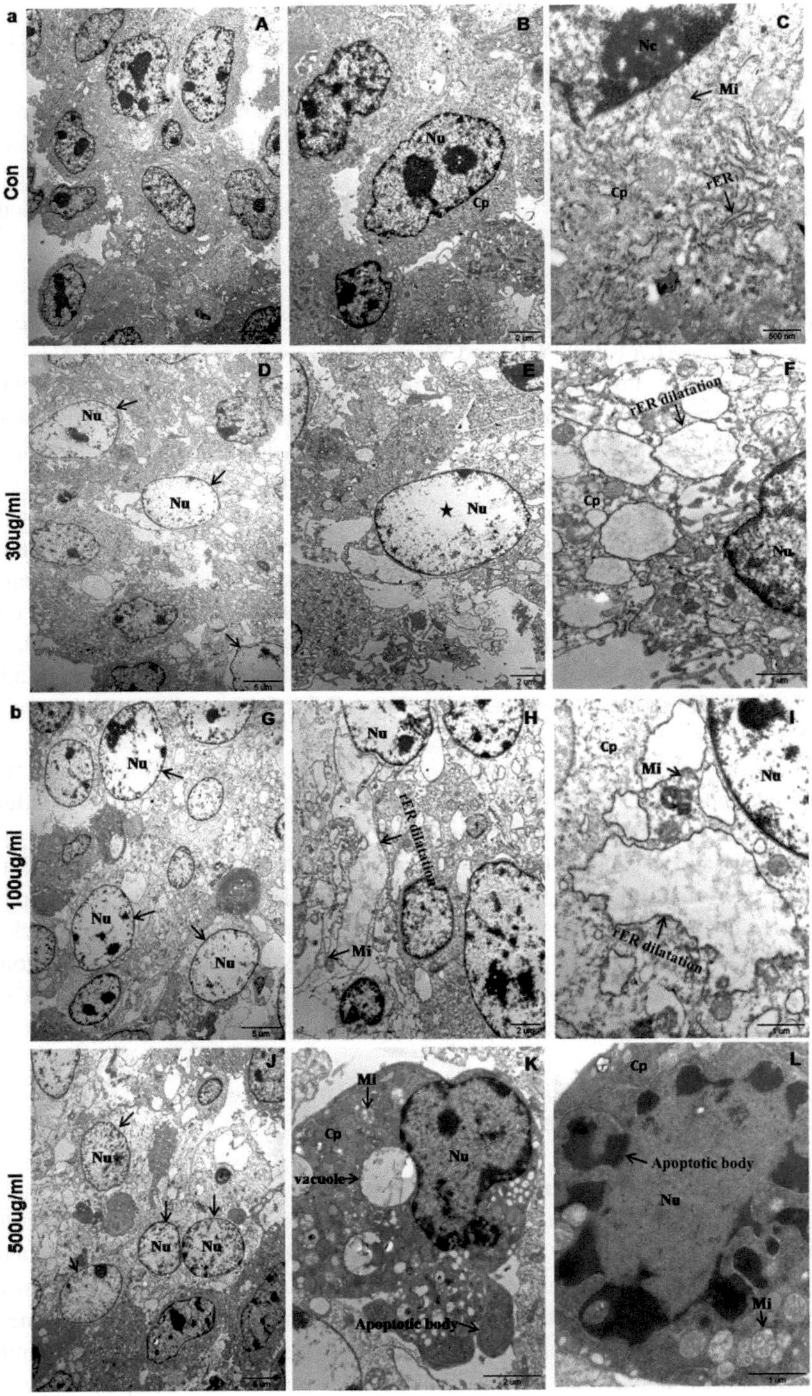

Fig. 14.4 Effects of MA on ultrastructure of TM4 cells. TM4 cells were treated with 0 (control), 30, 100, or 500 μg/mL MA for 24 h. At the end of the incubation period, cell ultrastructural changes were observed by TEM. (**a**) The ultrastructural changes in TM4 cells treated with 0 (control) and

have also been developed to optimize culture conditions and enhance tissue viability (Hashimoto et al. 2023).

A widely cited protocol, described by Nakamura et al. 2017 and later refined by Allen et al. 2020, involves the isolation and culture of neonatal mouse testes to study the progression of spermatogenesis in the presence of chemotherapeutics/EDCs (Fig. 14.5). In this method, testes are harvested from 3–7-day-old mice, and the surrounding tunica albuginea is carefully removed under sterile conditions. The exposed parenchyma is then dissected into small tissue fragments, typically ranging from 0.1 to 1 mm^3 in size. For culture preparation, 1.5% low-melting-point agarose is dissolved and allowed to solidify before being sectioned into 5–7 mm blocks. These agarose blocks are placed in 6-well plates and overlaid with approximately 600–800 μL of basal culture medium, such as α-Minimum Essential Medium (α-MEM) or DMEM/F12, supplemented with serum replacements (e.g., Knock-Out™ Serum Replacement) or other defined factors. The testicular tissue fragments are positioned on top of the agarose pads, creating a gas-liquid interface that facilitates oxygen exchange and supports tissue integrity. Culture is maintained at 34–35 °C in a humidified 5% CO_2 incubator, as this sub-physiological temperature has been shown to better support in vitro spermatogenesis compared to standard 37 °C conditions. Chemotherapeutics or other EDCs are introduced into the culture medium, typically for a 24-hour exposure period, after which the tissues are washed and returned to fresh, drug-free medium. Media changes are conducted every 48–72 hours to maintain nutrient availability and remove waste products. The total culture duration generally spans 28 to 36 days, allowing the evaluation of germ cell development from spermatogonia to more differentiated stages.

Practical Tips: Ex Vivo Testicular and Ovarian Culture and Media Composition

- Tissue damage during dissection may reduce viability and compromise spermatogenesis.

(continued)

Fig. 14.4 (continued) 30 μg/mL MA for 24 h. The control cells showed normal ultrastructural appearance (A–C). However, the MA-treated cells (D–F) showed obvious swelling and degeneration. Note the nuclear enlargement (D, arrow) and dissolution of the chromatin (E, star). In the cytoplasm, dilated rough endoplasmic reticulum was easily seen (F, arrow). (**b**) The ultrastructural changes in TM4 cells treated with 100 and 500 μg/mL MA for 24 h. The degree of the lesions in TM4 cells treated with higher concentrations of MA was much more server than the controls. The nuclear space increased and the nucleus was fragmented and dissolved (G and J, arrow). Extreme dilation of the rough endoplasmic reticulum could be observed (H and I, arrow). In the TM4 cells treated with 500 μg/mL MA, cell apoptosis was easily observed. Note the formation of apoptotic body (K and L, arrow). Nu, nucleus; Cp, cytoplasm; Mi, mitochondria; Nc, nucleoli; rER, rough endoplasmic reticulum. (The image/figure is adapted from Chang et al. (2017) with permission)

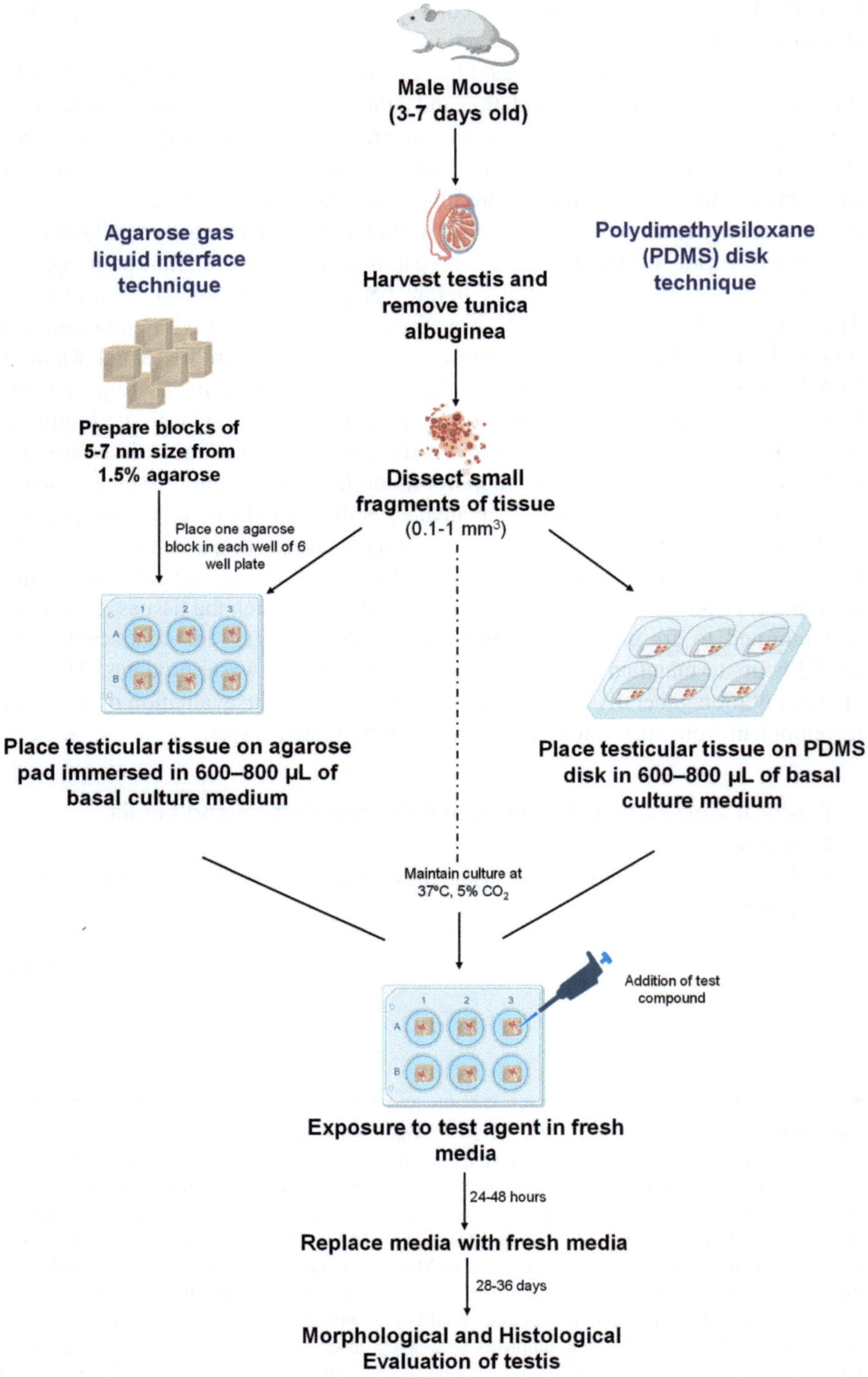

Fig. 14.5 Schematic representation of ex vivo culture of neonatal mouse testes. This figure illustrates the sequential steps of ex vivo testicular culture system. Testes are collected from 3 to 7-day-old neonatal mice, the tunica albuginea is removed, and the parenchymal tissue is cut

- Incomplete removal of tunica albuginea from testis and incomplete removal of surrounding adipose or mesonephric tissue from ovary can hinder nutrient diffusion and tissue growth.
- Incorrect tissue fragment size, either too large or too small, may lead to uneven culture or necrosis.
- Inconsistencies in agarose pad, such as thickness, concentration, or solidification, can affect tissue support and oxygen exchange.
- Uneven or insufficient culture medium coverage may cause tissue desiccation or poor nutrient supply.
- Contamination from bacteria, fungi, or mycoplasma can compromise tissue integrity and experimental outcomes.
- Use media supplemented with KnockOut™ Serum Replacement and growth factors to avoid arrested spermatogenesis.
- Maintain stable pH and osmolarity to avoid abnormal differentiation of germ cells.

Observations: Morphological Evaluation

Morphological and histological assessment of ex vivo cultured testicular tissue provides essential insights into the structural and functional integrity of spermatogenesis under toxicant exposure. Key parameters include the evaluation of seminiferous tubule morphology, such as the maintenance of round tubule profiles, intact basement membranes, presence of luminal spaces, and the orderly organization of germ cell layers progressing from spermatogonia to elongating spermatids. Quantitative staging of seminiferous tubules is also routinely performed to determine the relative abundance and distribution of specific germ cell populations. This involves identifying and counting spermatogonia, spermatocytes, and round or elongating spermatids, either manually or through automated image analysis systems. Allen et al. showed that exposing prepubertal testicular tissue to cisplatin or carboplatin caused a clear dose-related decline in the germ cell population by counting the number of proliferating and non-proliferating germ cells. Even at moderate concentrations, more than two-thirds of the proliferating germ cells were lost, and at the highest doses, only around 12–15% of germ cells could be detected compared with untreated controls (Fig. 14.6).

Fig. 14.5 (continued) into small fragments. These fragments are placed on solidified agarose blocks positioned at the gas-liquid interface, which ensures optimal oxygen diffusion and maintains tissue architecture. The culture is maintained at ~34–35 °C in a humidified CO_2 incubator with serum-supplemented basal medium, providing an environment conducive to germ cell survival and differentiation, and supports the progression of spermatogenesis. This setup enables assessment of spermatogenesis and the impact of chemotherapeutic/EDC exposure under controlled conditions.

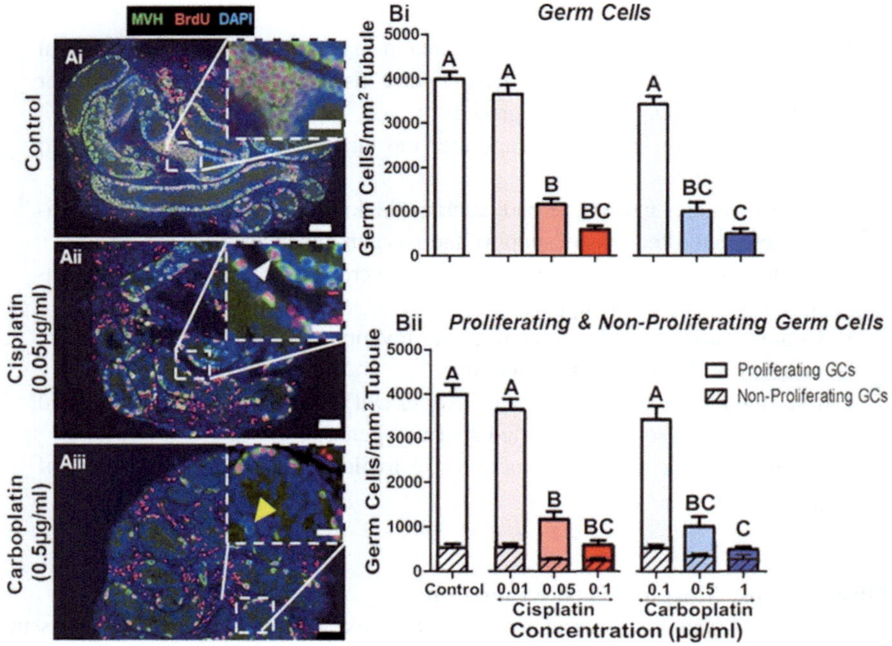

Fig. 14.6 Cisplatin or carboplatin treatment results in a significant loss of the proliferating germ cells. (**a**) Representative images of histological sections from (Ai) control-, (Aii) cisplatin- (0.05 µg/ml), and (Aiii) carboplatin- (0.5 µg/ml)-treated testis, showing a reduction in the density of proliferating germ cells (MVH-positive, green; BrdU-positive, red). Insets show higher magnification of framed areas. White arrowhead shows an example of a proliferating germ cell (MVH-positive/BrdU-positive); yellow arrowhead shows example of a non-proliferating germ cell (MVH-positive/BrdU-negative). Scale bars represent 50 µm, or 25 µm for insets. (**b**) Density of germ cells per seminiferous tubule area (mm2) following cisplatin (red) or carboplatin (blue) treatment. Data show the density of the (Bi) total, (Bii: open bars) proliferating and (Bii: hatched bar) non-proliferating germ cells; $n = 6$ for all experimental conditions. Data are mean + SEM; columns without letters in common are statistically significantly different from each other ($P < 0.05$). For (Bii), letters refer to proliferating germ cells (open bars) only, with no statistically significant difference between groups for the non-proliferating germ cells (hatched bars). (The image/figure is adapted from Allen et al. (2020). All images are licensed under CC BY)

14.2.2.2 Ovarian Culture

Similarly, ex vivo ovarian culture systems offer a valuable platform to investigate the effects of chemotherapeutics and EDCs on early folliculogenesis and ovarian tissue integrity. Similar to testicular culture, these models preserve tissue architecture and enable controlled experimental manipulation, facilitating mechanistic insights into ovarian toxicity. The most commonly employed technique is the floating membrane gas-liquid interface method, in which the ovary is supported at the interface between air and culture medium, enabling optimal oxygen diffusion while maintaining tissue hydration. Variations of this method have also been developed, incorporating extracellular matrix (ECM)-based hydrogels or microfluidic culture platforms to better

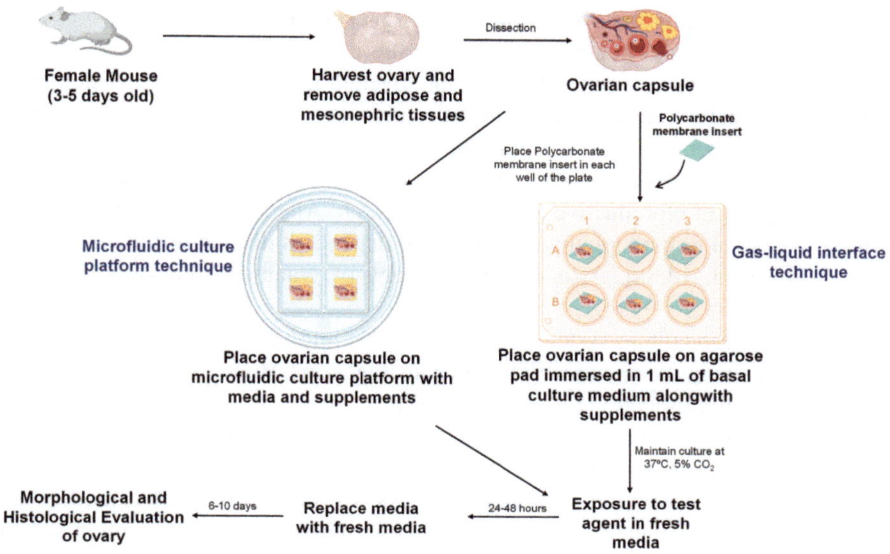

Fig. 14.7 Schematic representation of neonatal mouse ovarian culture. This figure illustrates the sequential steps of ex vivo ovarian culture system. Ovaries are isolated from 3 to 5-day-old mice, when primordial follicles are just forming, and the surrounding adipose or mesonephric tissues are carefully removed to preserve the ovarian capsule. Each ovary is then placed on a polycarbonate membrane insert positioned at the air-liquid interface in a 6-well culture plate, prefilled with basal medium (e.g., α-MEM or DMEM/F12) supplemented with defined components. This arrangement maintains tissue viability and architecture while enabling primordial follicle activation and early follicle growth. The culture is maintained in a humidified CO_2 incubator at 37 °C, with periodic medium changes to sustain nutrient balance. This arrangement preserves ovarian architecture and supports primordial follicle activation, while allowing controlled exposure to chemotherapeutics or EDCs

mimic the in vivo microenvironment and allow real-time monitoring of follicular cell dynamics.

A detailed protocol for neonatal mouse ovarian culture is described by Allen et al., where ovaries are isolated from 3- to 5-day-old mice in which primordial follicles begin to form (Allen et al. 2020). Surrounding adipose and mesonephric tissues are carefully removed under a dissecting microscope to avoid damage to the ovarian capsule. For the culture setup, a polycarbonate membrane insert (Whatman Nuclepore) is placed in each well of a 6-well culture plate, pre-filled with 1 mL of basal medium such as α-MEM or DMEM/F12, typically supplemented with fatty acid-free BSA and other defined components. Each ovary is gently positioned on the membrane at the air-liquid interface, allowing it to be supported without submersion. This setup maintains follicular viability and supports stromal structure. The cultures are maintained at 37 °C in a humidified 5% CO_2 incubator. Chemotherapeutics or EDCs are introduced into the medium for a 24-hour exposure period, after which the tissues are returned to fresh, drug-free medium. The culture is continued for 6–10 days, during which the medium is refreshed every 48–72 hours (Fig. 14.7).

Observations: Follicular Morphology and Ovarian Tissue Response
Following culture and treatment, ovarian tissue is routinely assessed for follicular morphology, granulosa cell integrity, and overall follicle viability. Histological evaluation involves the classification and staging of follicles and the quantification of follicle number across developmental stages. The culture system allows for the evaluation of ovarian responses to toxicants under defined conditions and has been widely applied in reproductive toxicology research. Allen et al. showed that exposure to cisplatin or carboplatin caused a dose-dependent decline in follicle number and health (Fig. 14.8a). Also, the morphology of granulosa cells was damaged (Fig. 14.8b).

These ex vivo culture system provides insights into the mechanisms underlying the effect of chemotherapeutics/EDCs on gametogenesis. However, it is important to acknowledge the limitations of ex vivo systems. These systems lack the systemic physiological interactions present in in vivo conditions, such as endocrine feedback, vascularization, and hepatic metabolism, which can significantly influence drug bioavailability, distribution, and clearance. As a result, though these models are highly informative, their findings should be interpreted with caution when extrapolating to clinical scenarios or whole-animal outcomes. To overcome these limitations, in vivo models are employed, as they capture the full complexity of systemic physiology, including endocrine regulation, blood supply, immune interactions, and metabolism.

14.2.3 In Vivo Methods

In vivo studies utilizing animal models offer a comprehensive and physiologically relevant approach for evaluating the toxic effects of chemotherapeutics and EDCs on gametogenesis. These models allow researchers to assess the integrated effects of drugs or chemicals on germ cell development and reproductive function within the context of a whole organism. The interactions within organism are particularly important when investigating the impact of reproductive toxicants, as they often influence the entire hypothalamic-pituitary-gonadal (HPG) axis, as well as local testicular or ovarian responses.

14.2.3.1 Assessment of Effects on Spermatogenesis

A widely adopted in vivo approach for evaluating the effects of drugs and toxicants on spermatogenesis has been described by Chianese et al. (2018). This protocol utilizes adult male mice and is designed to monitor reproductive outcomes over the course of a complete spermatogenic cycle, which in mice spans approximately 35 days. This duration ensures that all stages of spermatogenesis initiating from spermatogonial proliferation to sperm maturation are represented during the exposure window. In the experimental design, mice are typically randomized into at least two groups: a vehicle control and one or more dose-treated groups, depending on the compound under investigation. The test compound is administered daily via oral gavage, although alternative routes such as intraperitoneal injection may also be

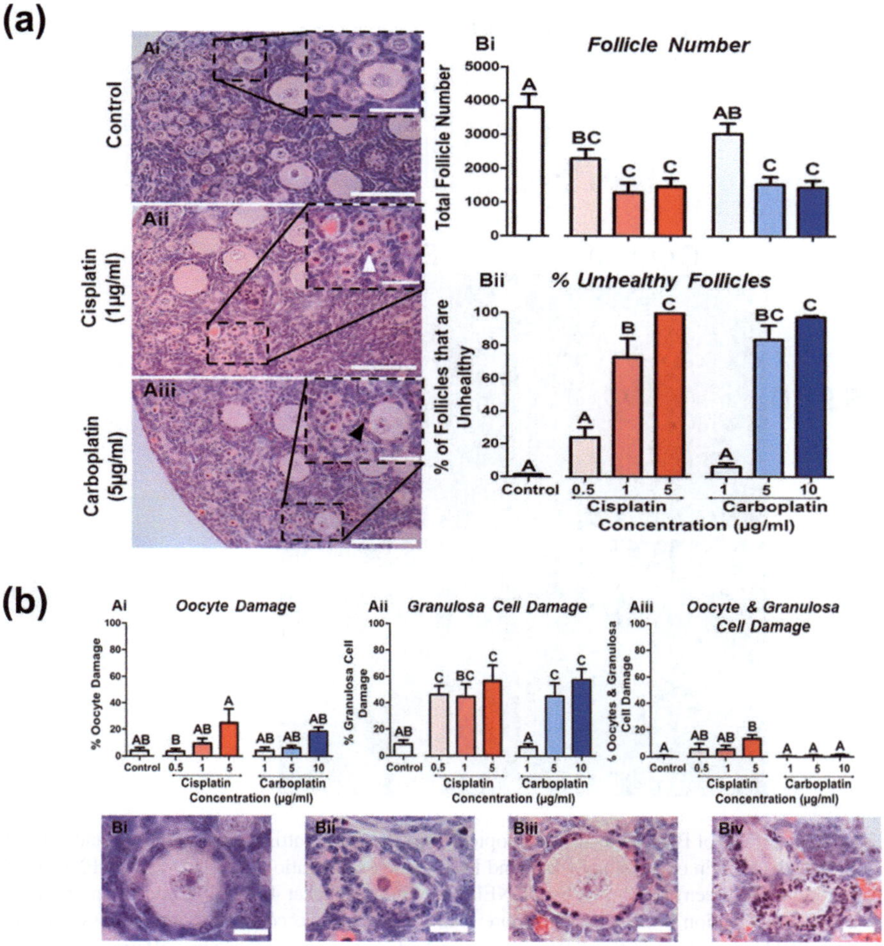

Fig. 14.8 Follicle number and health are reduced following cisplatin or carboplatin treatment. (**a**) Representative histological sections of (Ai) control-, (Aii) cisplatin- (1 μg/ml), and (Aiii) carboplatin (5 μg/ml)-exposed tissue; insets show higher magnification of framed areas. White arrowhead shows example of a damaged oocyte; black arrowhead shows example of a damaged granulosa cell; scale bars represent 50 μm, or 20 μm for insets. Histograms show (Bi) total follicle number and (Bii) the percentage of unhealthy follicles following exposure to cisplatin (red) or carboplatin (blue); $n = 6$ for all experimental conditions. Data are mean + SEM; columns without letters in common are statistically significantly different from each other ($P < 0.05$). (**b**) Granulosa cells of the primary follicles are damaged by cisplatin or carboplatin. Percentage of primary follicles classed as unhealthy due to (Ai) oocyte, (Aii) granulosa cell, or (Aiii) combined (oocyte and granulosa cell) damage following cisplatin (red) or carboplatin (blue) treatment, $n = 6$ for all experimental conditions. Data are mean + SEM; columns without letters in common are statistically significantly different from each other ($P < 0.05$). Representative images show (Bi) healthy primary follicle, (Bii) primary follicle with damaged oocyte, (Biii) primary follicle with damaged granulosa cells, and (Biv) primary follicles with combined (oocyte and granulosa cell) damage. Scale bar = 10 μm. (The image/figure is adapted from Allen et al. (2020). All images are licensed under CC BY)

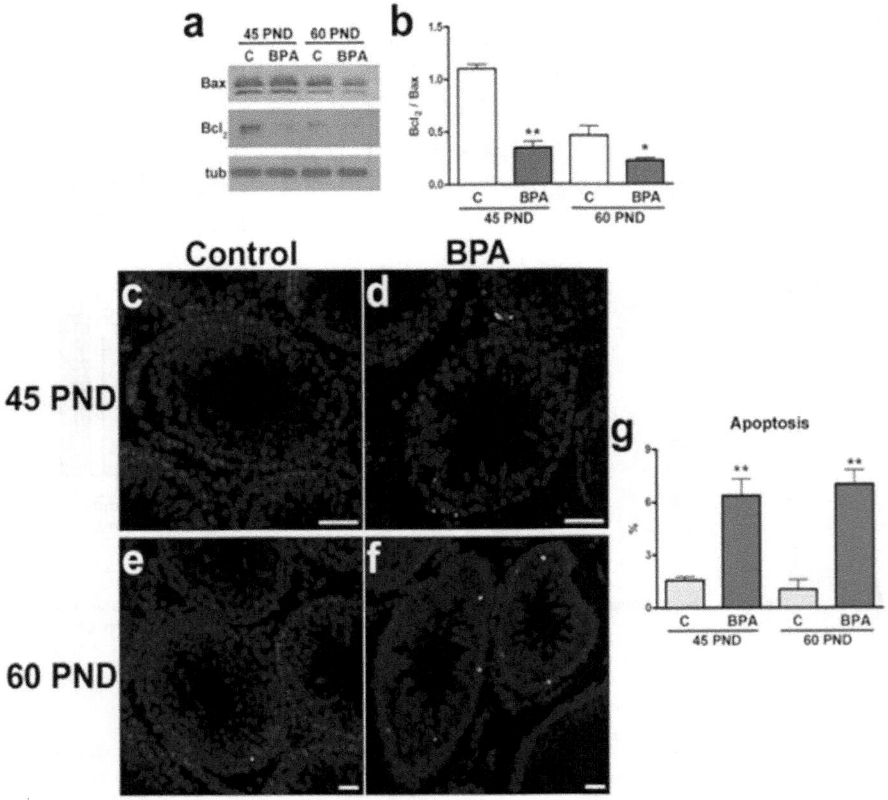

Fig. 14.9 Effects of BPA exposure on apoptosis. (**a**) Western blotting representative bands of Bax and Bcl2. (**b**) Protein expression of Bax and Bcl2 expressed as ratio at 45 PND and 60 PND. (**c**–**f**) Apoptotic cells (green) are shown by TUNEL assay in rat testis at 45 PND (**c** and **d**) and 60 PND (**e** and **f**). (**g**) Fraction of tubules with more than three apoptotic cells. Data are expressed as the mean ± SEM. $*P < 0.05$, $**P < 0.01$ vs. age-matched control group. Scale bar, 50 μm. tub tubulin, TUNEL terminal deoxynucleotidyl transferase-mediated dUTP nick end labelling, C control group, BPA bisphenol A-exposed group, PND postnatal day. (The image/figure is adapted from Chianese et al. (2018). All images are licensed under CC BY)

used depending on pharmacokinetic considerations. At the end of the exposure period (≥35 days), several endpoints are evaluated to assess both local testicular effects and systemic reproductive impacts:

Testis Collection and Processing One testis is fixed in Bouin's solution or 4% paraformaldehyde (PFA) for histological analysis. The other testis is snap-frozen in liquid nitrogen for downstream RNA and protein analysis to evaluate alterations in the expression of key genes and proteins involved in apoptosis, such as Bax, Bcl-2, and Caspase-3. The mice treated with BPA showed reduced levels of Bax and Bcl-2 proteins in testicular tissue analyzed by western blotting and fraction of tubules with more apoptotic cells analyzed by TUNEL assay (Fig. 14.9) (Chianese et al. 2018).

Histological Evaluation Microscopic evaluation of testicular tissue following in vivo exposure to chemotherapeutics or EDCs has consistently revealed structural alterations within the seminiferous tubules. These include germ cell depletion, vacuolization of the seminiferous epithelium, and loss of orderly staging of germ cells. The perfluorooctane sulfonate (PFOS) treatment disrupted the testicular morphology that resulted in increased Sertoli cells vacuolization and dislocation of immature germ cells visualized by light and electron microscopy (Qiu et al. 2013). Similarly, cadmium exposure led to pronounced testicular damage, with seminiferous tubules showing reduced numbers of spermatogonia, spermatocytes, spermatids, and spermatozoa (Ali et al. 2023). The tubular epithelium became disorganized, accompanied by extensive vacuolation and loss of germ cell integrity (Fig. 14.10).

Practical Tips: Histological Evaluation

- Delayed or inadequate fixation can lead to tissue autolysis, poor preservation, and loss of cellular detail. Fix tissues immediately in appropriate fixative such as 4% PFA to avoid autolysis and poor morphology.
- Improper tissue processing or embedding can cause uneven infiltration of paraffin or resin, leading to sectioning difficulties.
- Sectioning errors such as tearing, folding, or chatter lines can distort tissue morphology.
- Uneven or excessive staining may obscure cellular details or cause inconsistent contrast.
- Contamination during processing such as dust, bacteria, fungi can interfere with tissue integrity or imaging.

Hormonal Profiling To estimate the alteration in hormonal level under the influence of chemotherapeutics and EDCs, blood is collected by cardiac puncture or retro-orbital bleeding, followed by serum isolation. Hormonal assays such as ELISA are performed to measure circulating levels of testosterone, luteinizing hormone (LH), and follicle-stimulating hormone (FSH). These data provide insights into the HPG axis regulation and whether endocrine feedback mechanisms are disrupted.

Sperm Collection and Analysis To examine the effect on sperm motility, the cauda epididymis is dissected and minced in pre-warmed medium to release mature spermatozoa. Sperm are evaluated for concentration, motility, and morphological abnormalities, using light microscopy or computer-assisted sperm analysis (CASA). These functional sperm parameters serve as key indicators of fertility potential and testicular output.

14.2.3.2 Assessment of Effects on Oogenesis

To investigate the effects of chemotherapeutics and EDCs on oogenesis and female fertility, several in vivo studies have adopted well-established mouse models. One such protocol has been described by Chang et al. (2015), in which adult female mice

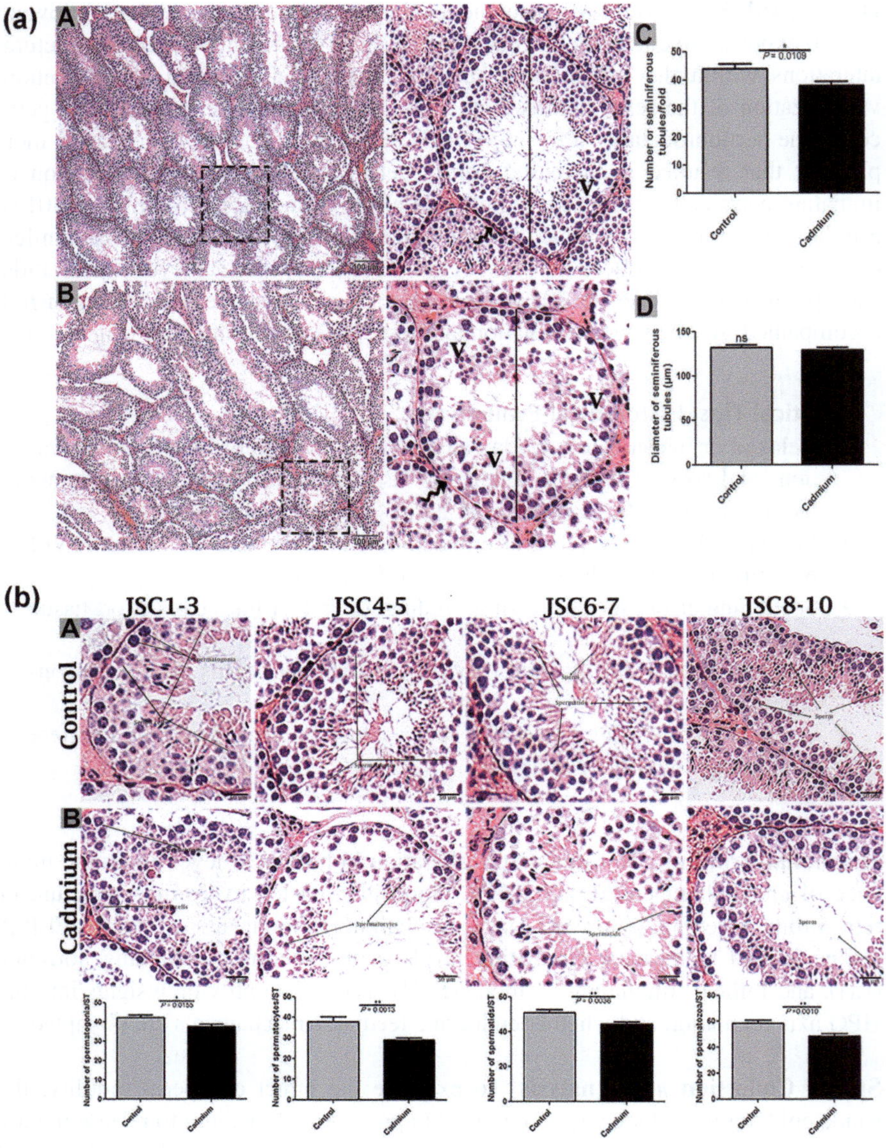

Fig. 14.10 Effects of cadmium on seminiferous tubules. (**a**) Light microscopy of seminiferous tubules of testis of control (A) and cadmium-treated group (B). A large magnification is illustrated in the rectangular area. (C) Number of seminiferous tubules. (D) Quantification of the diameter of seminiferous tubules. V: vacuole; (curved arrow) basement membrane. Scale bars = 100 μm. (**b**) Light microscopy of seminiferous tubules of testis of control (A) and cadmium-treated group (B). JSCs 1–2 lack germ cells, while JSC 3 contains spermatogonia but no spermatocytes. JSCs 4–5 contain a small number of spermatocytes but no spermatids. JSCs 6–7 contain few or many spermatids but no sperm. JSCs 8–10 contain a small or large number of sperms in a seminiferous tubule. Scale bars = 50 μm. (Data presented as Mean ± SEM. The image/figure is adapted from Ali et al. (2023). All images are licensed under CC BY)

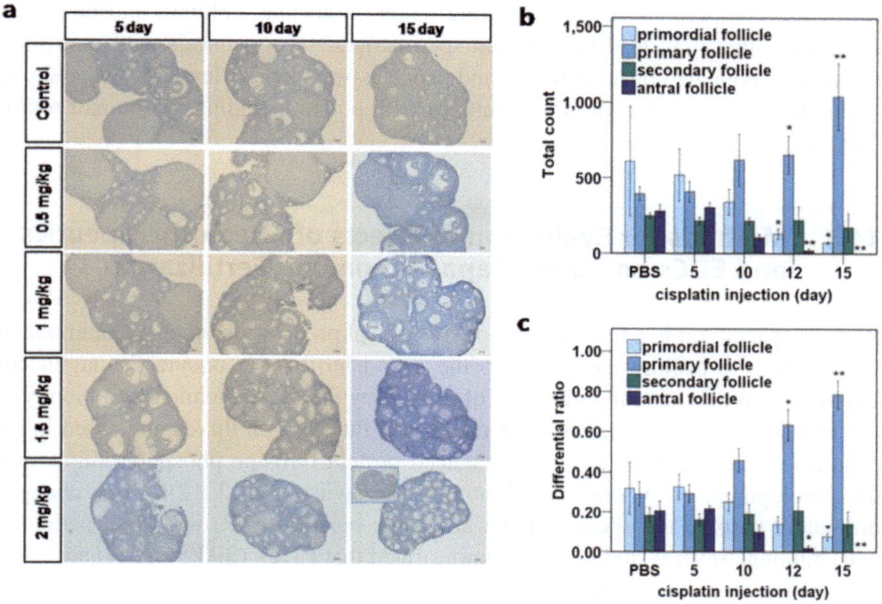

Fig. 14.11 In vivo assessment of impact of cisplatin on oogenesis. Representative histological sections of mouse ovaries at different time points after daily injection of various doses of cisplatin or PBS control. (The image/figure is adapted from Chang et al. (2015). All images are licensed under CC BY)

were used to evaluate both acute and long-term ovarian responses following exposure to toxicants. In this approach, mice were randomly assigned to vehicle control and treatment groups, and the compound of interest was administered via intraperitoneal injection (IP). The timing of animal sacrifice varied based on the endpoint under investigation. Acute effects, such as DNA damage and early apoptosis, were assessed after 24 hours post-treatment, and long-term effects on the ovarian reserve, follicle dynamics, and fertility potential were evaluated over a course of several weeks following exposure. At the time of necropsy, ovaries were harvested, fixed in 4% paraformaldehyde (PFA), embedded in paraffin, and serially sectioned through the entire organ.

Observations In this in vivo model, exposure to chemotherapeutics such as cisplatin resulted in a significant reduction in the number of primordial and growing follicles, indicative of premature ovarian reserve depletion (Chang et al. 2015). Histological analysis of ovarian sections revealed widespread follicular atresia, along with disruption of the stromal architecture, suggesting compromised ovarian integrity following drug exposure (Fig. 14.11).

These in vivo methodology provides a robust platform for evaluating how chemotherapeutics or EDCs interfere with male and female reproductive capacity.

It offers a comprehensive understanding of the molecular, cellular, and hormonal disruptions that may lead to impaired spermatogenesis, oogenesis, and infertility. However, ethical considerations and species-specific differences should be accounted for when interpreting results or extrapolating findings to human reproductive biology.

14.3 Methods for Evaluation of Effects of Chemotherapeutics and EDCs on Sperm Capacitation and Fertilization

Sperm capacitation is a fundamental physiological process that enables mammalian spermatozoa to acquire the ability to move and fertilize an oocyte (Yanagimachi 1970). This process involves a series of tightly regulated biochemical and biophysical changes, including alterations in membrane fluidity, ion fluxes, increased intracellular cAMP and PKA activity, enhanced protein tyrosine phosphorylation, and the development of hyperactivated motility (Ickowicz et al. 2012). These changes ultimately prepare the sperm for the acrosome reaction (AR) and zona pellucida (ZP) binding. Several studies have demonstrated that EDCs (Sikka and Wang 2008) and chemotherapeutics (Delessard et al. 2020) can interfere with different stages of capacitation by affecting calcium signaling, cAMP-PKA pathways, or premature induction of AR. The resulting disruption may impair sperm motility, inhibit zona binding, and reduce fertilization success, ultimately compromising male fertility. A range of in vitro and in vivo approaches have been developed to evaluate the effects of chemotherapeutics and EDCs on sperm capacitation and fertilization.

14.3.1 In Vitro Approaches

14.3.1.1 Assessment of Capacitation via Protein Tyrosine Phosphorylation

Protein tyrosine phosphorylation is widely used as a molecular indicator of sperm capacitation (Naz and Rajesh 2004). This method is based on the observation that capacitation is associated with a time-dependent increase in tyrosine phosphorylation of specific sperm proteins, particularly in the flagellum and acrosomal region. Generally, spermatozoa are incubated in capacitation-supporting media containing bovine serum albumin, calcium, and bicarbonate, with or without exposure to test compounds. Following incubation, proteins are extracted and analyzed via Western blotting or immunofluorescence using anti-phosphotyrosine antibodies. A study showed that human sperm, exposed to lead, had reduced cAMP levels, impaired Ca^{2+} signaling, and suppressed tyrosine phosphorylation (He et al. 2016). In contrast, MEHP (phthalate metabolite) exposure enhanced Ca^{2+} influx and tyrosine phosphorylation in human sperm disrupting normal capacitation dynamics (Sun et al. 2020). Though the method is suitable for small molecular changes, it requires high-quality phospho-specific antibodies for detection, otherwise it may not correlate with the actual outcome.

14.3.1.2 Evaluation of AR

AR is an exocytotic event essential for ZP penetration (Roldan et al. 1994). In in vitro setup, the AR can be induced using physiological stimuli such as progesterone and ZP proteins or pharmacological stimuli. Post-induction, the status of the acrosome is assessed using fluorescently labeled lectins such as PNA-FITC or PSA or antibodies against acrosomal components. Several studies have shown that EDCs and chemotherapeutics can either inhibit or prematurely trigger the AR. EDCs act via non-genomic mechanisms to activate CatSper calcium channels, thereby inducing premature acrosomal exocytosis in human sperm (Rehfeld et al. 2018).

14.3.1.3 Assessment of Motility and Hyperactivation

Sperm motility is a functional outcome of successful capacitation. CASA systems are frequently used to quantify motility parameters, including total motility, curvilinear velocity, amplitude of lateral head displacement, and linearity (Larsen et al. 2000). Hyperactivation is characterized by vigorous, asymmetric flagellar beating and high-amplitude motion. Exposure to toxicants has been shown to affect these parameters. UV-filtering EDCs can mimic progesterone and disturb calcium-dependent motility patterns (Schiffer et al. 2014). CASA enables high-throughput, objective evaluation, although the accuracy depends on sperm quality and software calibration.

14.3.1.4 In Vitro Fertilization (IVF) and ZP Binding Assays

The fertilization potential of sperm following exposure to chemotherapeutics or EDCs can be evaluated using IVF and ZP binding assays. These methods assess functional sperm competence by measuring fertilization rates or the ability of sperm to bind to the oocyte's zona pellucida. Rahman et al. reported that BPA exposure disrupted sperm capacitation and acrosome reaction in mice, resulting in significantly reduced fertilization rates in IVF assays (Rahman et al. 2015). The study also highlighted that even short-term BPA exposure could impair critical fertilization-related processes, including protein tyrosine phosphorylation and calcium signaling.

Although in vitro systems are highly informative for dissecting molecular events in sperm capacitation and fertilization, they cannot fully replicate the intricate physiological environment of the reproductive tract. Therefore, to create the complexity of hormonal regulation, immune interactions, and oviductal milieu, in vivo models are employed to evaluate the impact of chemotherapeutics and EDCs on sperm capacitation and fertilization.

14.3.2 In Vivo Approaches

The different in vivo approaches used to assess the effect of chemotherapeutics and EDCs on sperm capacitation provide an integrative framework within the context of whole-organism physiology. The different in vivo methods are described in the following.

14.3.2.1 Evaluation of Drug and EDCs Impact on Sperm Capacitation In Vivo

In vivo exposure protocols generally involve oral gavage or intraperitoneal injections of the compound of interest over a defined period that spans at least one full spermatogenic cycle in mice (approximately 35–42 days). This ensures that all stages of spermatogenesis, epididymal maturation, and capacitation are represented. At the end of the treatment, spermatozoa are harvested from the cauda epididymis and subjected to both biochemical and functional analyses. Capacitation status was assessed by examining tyrosine phosphorylation patterns and acrosome reaction capacity in response to agonists such as progesterone or zona pellucida proteins.

14.3.2.2 Fertility Outcomes and Functional Sperm Assays

To link molecular alterations to fertility outcomes, mating studies are often performed. Male mice treated with drugs or EDCs are paired with proven fertile females, and endpoints such as mating success, pregnancy rate, litter size, and early embryo development are recorded (Yu et al. 2024). These functional outcomes provide strong evidence for toxicant-induced impairments in sperm capacitation and fertilization competence.

14.3.2.3 Hormonal and Histological Correlates

Sperm capacitation is under hormonal control, particularly regulated by testosterone and FSH (Lopez-Torres and Chirinos 2017). Therefore, blood samples are often collected to measure reproductive hormone levels. These parameters help to interpret whether observed sperm alterations are due to local testicular effects or secondary to HPG axis disruption. In addition, testes and epididymis are harvested for histological analysis, which may reveal degenerative changes in seminiferous tubules, reduced sperm storage in the cauda epididymis, or epithelial vacuolation. The presence of DNA damage, oxidative stress, or apoptosis can be evaluated using TUNEL assay, γH2AX staining, or cleaved caspase-3 immunohistochemistry.

14.4 Translational and Clinical Implications: Fertility Preservation Strategies in Cancer Patients and Future Therapeutic Opportunities

Fertility preservation has become an important consideration for cancer patients, as treatments like chemotherapy and radiotherapy can harm the reproductive system and affect the ability to have children in the future. Protecting fertility gives survivors the opportunity to start a family later, which can greatly improve their emotional well-being. There are several methods available, including gamete cryopreservation, ovarian or testicular tissue preservation, and assisted reproductive techniques. The choice of method depends on factors like age, type of cancer, and treatment schedule. A comparative account of these approaches in males and females are summarized in Table 14.2.

Table 14.2 Comparative summary of male versus female fertility preservation strategies

Aspect	Male strategies	Female strategies
Gamete cryopreservation	Sperm cryopreservation Semen collection by ejaculation and freezing in liquid nitrogen Gold standard for post-pubertal males	*Oocyte Cryopreservation* Ovarian stimulation followed by vitrification of mature oocytes. Standard for unmarried/post-pubertal women.
Embryo cryopreservation	*Not applicable*	*Embryo freezing* after IVF with donor sperm. Highest success rates
Tissue cryopreservation	*Testicular tissue cryopreservation* Surgical biopsy of testicular tissue is frozen Only option for prepubertal boys	*Ovarian tissue cryopreservation* Cortical strips frozen for later reimplantation Only option for prepubertal girls
Surgical methods	Approaches: *PESA* *TESA* *TESE* *Electroejaculation* *Penile vibratory* For azoospermic or ejaculatory dysfunction patients	*Ovarian transposition (Oophoropexy)*—surgical repositioning of ovaries outside radiation field For patients undergoing radiotherapy
Medical protection (GnRH analogs)	Provides limited benefit for fertility preservation	May reduce the risk of premature ovarian failure, though efficacy is variable

Abbreviations: *GnRH* Gonadotropin-releasing hormone, *PESA* Percutaneous epididymal sperm aspiration, *TESE* Testicular sperm extraction, *TESA* Testicular sperm aspiration

In male cancer patients, the most reliable fertility preservation strategy is sperm cryopreservation which involves collecting semen samples prior to the initiation of chemo/radiotherapy and storing them in liquid nitrogen to maintain its viability for future use in assisted reproductive technologies (Borate and Meshram 2022). It has been reported that sperms can maintain their viability for around 40 years in cryopreservation (Szell et al. 2013). Similarly, another study showed that semen frozen before gonadotoxic therapy in testicular cancer resulted in successful pregnancies post-treatment (Zakova et al. 2014). However, achieving optimal outcomes typically requires a minimum of three semen samples, collected at 48-hour intervals, to ensure adequate sperm yield and viability for cryopreservation (Rodriguez-Wallberg and Oktay 2014). In circumstances where ejaculation is not possible due to underlying illness, psychological factors, or any surgical interventions, alternative sperm retrieval methods are employed. This ranges from assisted ejaculation techniques such as electroejaculation and penile vibratory stimulation (PVS) to surgical approaches such as testicular or epididymal sperm extraction. Assisted ejaculation techniques employ gentle electrical or mechanical stimulation to trigger reflex ejaculation. In PVS, mechanical vibration is applied to the penis to trigger the ejaculatory reflex, especially in patients with deformed spinal cord (Ibrahim et al. 2022), while electroejaculation uses a rectal probe to deliver electrical stimulation to the prostate and pelvic nerves to induce ejaculation (Hovav et al. 2001). Overall, while both PVS and electroejaculation methods are effective for sperm retrieval, PVS should be preferred as the initial approach given its non-invasive, simple, and easy-to-perform nature (Schmiegelow et al. 1998). Surgical approaches for sperm retrieval include percutaneous epididymal sperm aspiration (PESA), where a fine needle is used to aspirate sperm from the epididymis, whereas in testicular sperm aspiration (TESA), sperm is aspirated directly from the testicles using a needle. The other surgical methods include testicular sperm extraction (TESE) that requires a small incision to remove seminiferous tubules containing sperms, whereas microdissection TESE (micro-TESE) employs a high-powered microscope to locate and selectively retrieve sperm-rich tubules with greater precision. Although all these methods are effective, they remain invasive and may risk tissue injury. Among these methods, micro-TESE is generally considered the most reliable approach due to its higher retrieval success and reduced damage to testicular tissue. These surgical procedures even allow azoospermic patients with impaired spermatogenesis to father biological children (Esteves et al. 2013). For pre-pubertal boys, who do not yet produce mature spermatozoa, experimental methods like testicular tissue cryopreservation are being explored (Sadri-Ardekani and Atala 2014). Pharmacological strategies, including gonadotropin-releasing hormone (GnRH) analogs, amifostine, and antioxidants, have also been studied to protect spermatogonial stem cells from chemotherapeutic toxicity (Allen et al. 2021). However, most pharmacological agents for male fertility preservation have only been tested in animal models, and those that have reached clinical evaluation remain at very early stages of investigation. Thus, sperm cryopreservation remains the gold standard for adult males.

In female patients, fertility preservation is more challenging due to the finite nature of the ovarian reserve. The commonly employed methods for fertility preservation in females are oocyte and embryo cryopreservation (Kim et al. 2016). In this approach, controlled ovarian stimulation is performed prior to cancer treatment, followed by retrieval of mature oocytes which are either frozen directly or fertilized in vitro to create embryos for cryostorage (von Wolff et al. 2009). A study showed that stimulation using tamoxifen has been successfully applied to minimize estrogen exposure while allowing oocyte retrieval in breast cancer patients (Oktay et al. 2003). Ovarian tissue cryopreservation is another widely employed method where the patient cannot wait for treatment. In this method, cortical ovarian strips containing primordial follicles are surgically removed and frozen, with the possibility of reimplantation after cancer remission. The literature suggests that more than 200 live births have been reported worldwide following ovarian tissue transplantation (Fabbri et al. 2024). Ovarian transposition (oophoropexy) is also used for fertility preservation in females. In this method, the ovaries are surgically repositioned outside the pelvic radiation field to minimize gonadal exposure. While this technique does not prevent chemotherapy-induced toxicity, it effectively reduces the impact of radiotherapy (Tessier et al. 2023). Pharmacological suppression of ovarian function using GnRH agonists during chemotherapy has been tested in hematological malignancies (Clowse et al. 2009) and breast cancer (Bansal et al. 2014). Few studies have suggested partial protection of ovarian reserve and improved post-treatment ovarian function, although its effectiveness remains debated due to insufficient evidence.

Future therapeutic opportunities in protecting and restoring fertility are rapidly evolving, integrating advances in stem cell biology, cryopreservation, tissue engineering, and pharmacological protection. Improved cryopreservation techniques are being optimized for long-term storage of oocytes, embryos, and gonadal tissues without compromising viability. Fertility-protective measures also include the administration of GnRH agonists to shield ovarian reserve during chemotherapy. Furthermore, mesenchymal stem cell-derived exosomes are being investigated for their ability to promote gonadal repair, enhance spermatogenesis, and protect ovarian follicles from chemotherapy-induced damage. Collectively, these emerging strategies, ranging from next-generation cryopreservation to stem cell-based gametogenesis, represent a paradigm shift toward safer, personalized, and regenerative solutions for fertility preservation and restoration in both male and female cancer survivors.

14.5 Case Study: Impact of EDCs on Male Health

Background EDCs such as polychlorinated biphenyls PCBs, dioxins, and organochlorine pesticides are persistent pollutants that accumulate in human tissues. Their estrogenic or anti-androgenic actions raise concerns for male reproductive health.

Study Design A case-control study was conducted in 163 men attending fertility clinics. Semen samples were analyzed for parameters including concentration, motility, and total motile count (TMC). Parallel blood samples were collected and analyzed for levels of multiple EDCs using high-resolution gas chromatography-mass spectrometry (GC-MS).

Findings Men with higher circulating levels of oxychlordane had reduced sperm motility and lower TMC. This established a measurable link between EDC exposure and impaired spermatogenesis in humans.

Clinical and Environmental Relevance The study highlights how toxicological screening methods in the laboratory can directly inform patient outcomes. Men with environmentally relevant EDC exposures showed clinically significant fertility impairments.

14.6 Conclusion

Therefore, the assessment of drug- or EDC-induced reproductive toxicity can be achieved through both in vivo animal models and in vitro culture systems. While in vivo models capture systemic influences on gametogenesis, they are often time-consuming, ethically demanding, and influenced by multiple physiological variables. In contrast, in vitro culture models of testes and ovaries offer a simpler, more controlled, and reproducible platform to directly monitor germ-cell development, follicular growth, and cellular responses to chemotherapeutics/EDCs. Given their relative ease of use and suitability for high-throughput analysis, in vitro methods should be increasingly considered as the primary approach.

For students, understanding these comparative models has practical implications beyond the laboratory exercise. For example, the principles learned here can be applied to predict how chemotherapeutics may impair fertility and to design culture-based approaches for fertility preservation, such as cryopreservation of ovarian follicles or testicular tissue. Similarly, recognizing how in vitro models are used to evaluate the reproductive risks of chemotherapeutics/EDCs provides a foundation for careers in reproductive biology, toxicology, and regulatory science. Thus, this exercise not only trains students in model systems but also prepares them to connect experimental findings with real-world challenges in human health and fertility preservation.

14.7 Precautions

(i) Follow institutional animal ethics guidelines (IAEC/IACUC) for the procurement and use of rodent gonads.
(ii) Dispose of chemical and biological waste (e.g., EDCs, chemotherapeutics, stained gels) in designated hazardous waste containers.
(iii) Always wear lab coats, gloves, and protective eyewear while handling chemicals, dyes, and biological materials.
(iv) Minimize tissue exposure to room temperature to preserve viability during ex vivo cultures.
(v) Label tissues clearly to avoid mix-ups between control and treated groups.
(vi) Ensure all animal procedures are approved by relevant ethics committees and follow 3Rs principles (Replacement, Reduction, Refinement).
(vii) Human sample usage (sperm and oocytes) must comply with informed consent and regulatory guidelines.
(viii) Dyes like Hoechst and PI are mutagenic; minimize exposure and dispose of contaminated materials properly.

14.8 Quick Review Questions

14.8.1 Multiple Choice Questions

1. Which of the following chemotherapeutic agents is most strongly associated with inducing DNA cross-links in germ cells?
   (a) Cisplatin
   (b) Cyclophosphamide
   (c) Doxorubicin
   (d) Melamine
   Answer: (b) Cyclophosphamide
2. In spermatogenesis, which cell type provides structural and nutritional support to developing germ cells?
   (a) Leydig cells
   (b) Sertoli cells
   (c) Spermatocytes
   (d) Granulosa cells
   Answer: (d) Sertoli cells
3. Which assay is best suited for assessing DNA strand breaks in individual cells exposed to toxicants?
   (a) MTT assay
   (b) Comet assay
   (c) Western blotting
   (d) TEM imaging
   Answer: (b) Comet assay
4. Ex vivo testicular culture most commonly uses which support method?
   (a) Microfluidic chip
   (b) Agarose gas-liquid interface
   (c) Polycarbonate membrane floating
   (d) Collagen hydrogel
   Answer: (b) Agarose gas-liquid interface
5. Endocrine-disrupting chemical Bisphenol A (BPA) exerts toxicity in Leydig cells primarily by:
   (a) Inducing ferroptosis
   (b) Decreasing aromatase
   (c) Upregulating aromatase and COX-2
   (d) Blocking CatSper channels
   Answer: (c) Upregulating aromatase and COX-2
6. Which parameter is measured to assess sperm capacitation via CASA?
   (a) Zona pellucida binding
   (b) Beat frequency and hyperactivation
   (c) Tyrosine phosphorylation of proteins
   (d) DNA damage
   Answer: (b) Beat frequency and hyperactivation

7. Which is the most widely accepted fertility preservation method for male cancer patients?
   (a) Micro-TESE
   (b) Sperm cryopreservation
   (c) Ovarian tissue cryopreservation
   (d) GnRH analogs
   Answer: (b) Sperm cryopreservation
8. Cisplatin-induced ovarian damage is most evident as:
   (a) Enhanced primordial follicle reserve
   (b) Follicular atresia and stromal disruption
   (c) Increased granulosa proliferation
   (d) Rise in testosterone production
   Answer: (b) Follicular atresia and stromal disruption
9. Why are in vivo models considered more physiologically relevant compared to in vitro?
   (a) They allow higher throughput
   (b) They capture systemic interactions and hormonal feedback loops
   (c) They eliminate ethical concerns
   (d) They avoid interspecies variation
   Answer: (b) They capture systemic interactions and hormonal feedback loops
10. Which pharmacological agent has been investigated for protecting spermatogonial stem cells during chemotherapy?
   (a) Amifostine
   (b) Cisplatin
   (c) BPA
   (d) Cyclophosphamide
   Answer: (a) Amifostine
11. In ovarian culture, what is the function of the polycarbonate membrane?
   (a) Acts as an electrical insulator
   (b) Maintains tissue at air-liquid interface for oxygenation
   (c) Filters toxins
   (d) Increases follicle count
   Answer: (b) Maintains tissue at air-liquid interface for oxygenation
12. A cancer survivor presents with azoospermia. Which surgical method provides the highest chance of retrieving sperm?
   (a) PESA
   (b) TESA
   (c) TESE
   (d) Micro-TESE
   Answer: (d) Micro-TESE
13. What does a strong "comet tail" in the comet assay indicate?
   (a) Enhanced cell proliferation
   (b) Intact DNA
   (c) High degree of DNA damage
   (d) Oxidative stress only

Answer: (c) High degree of DNA damage

14. Which is a critical limitation of ex vivo ovarian culture models?
    (a) Inability to mimic endocrine regulation and systemic metabolism
    (b) Requirement of high-cost microfluidics
    (c) Incompatibility with mouse ovaries
    (d) Excessive tissue vascularization
    Answer: (a) Inability to mimic endocrine regulation and systemic metabolism
15. Which of the following EDCs can mimic progesterone and prematurely induce acrosome reaction?
    (a) Melamine
    (b) UV-filtering agents (4-MBC, 3-BC)
    (c) Doxorubicin
    (d) Busulfan
    Answer: (b) UV-filtering agents (4-MBC, 3-BC)

14.8.2 Short Answer Questions

1. Define endocrine-disrupting chemicals (EDCs) and give two examples relevant to reproductive health.
2. Compare the utility of in vitro versus in vivo models in assessing reproductive toxicity. Comment on their strength and weakness.
3. What is the principle of the MTT assay, and how does it indicate cell viability? What is the limitation of this assay?
4. Explain how comet assay results are interpreted in the context of DNA damage.
5. List two fertility preservation strategies in female cancer patients and their limitations.
6. Describe the role of protein tyrosine phosphorylation in sperm capacitation.
7. Why is sperm cryopreservation considered the gold standard in male fertility preservation?
8. Discuss one pharmacological strategy being explored for fertility protection during chemotherapy.

Glossary

Acrosome reaction Exocytosis of sperm acrosome enabling zona penetration.

Agarose gas-liquid interface Ex vivo culture system for neonatal testes supporting spermatogenesis.

Alkylating agents Chemotherapeutics like cyclophosphamide that form DNA cross-links causing germ-cell apoptosis.

Amifostine A cytoprotective agent tested for shielding spermatogonial stem cells during chemotherapy.

Apoptosis Programmed cell death pathway activated in germ cells after drug or EDC exposure.

Aromatase Enzyme converting androgens to estrogens, often upregulated by BPA in Leydig cells.

Assisted ejaculation Techniques like PVS or electroejaculation used to retrieve sperm for cryopreservation.

Bax/Bcl-2 Apoptosis regulators; altered expression indicates germ-cell susceptibility to chemotherapeutics.

Bisphenol A (BPA) Common EDC in plastics, reduces testosterone and impairs spermatogenesis.

Capacitation Biochemical sperm maturation enabling fertilization.

Carboplatin Platinum chemotherapeutic causing oxidative stress and ovarian follicle depletion.

CASA Computer-assisted sperm analysis; measures motility and hyperactivation in toxicity studies.

Caspase-3 Executioner enzyme of apoptosis and marker of germ-cell death.

Chemotherapy-induced infertility Loss of reproductive capacity in cancer survivors due to gonadotoxic drugs.

Comet assay Single-cell electrophoresis assay measuring DNA breaks in germ cells.

Cryopreservation Long-term freezing of sperm, oocytes, or tissue to preserve fertility before chemo/radio therapy.

Cyclophosphamide Alkylating chemotherapeutic drug causing severe gonadal toxicity.

DNA cross-links Lesions caused by alkylating agents, block germ-cell division, and survival.

DNA damage Strand breaks and lesions induced by chemotherapy/EDCs, assessed by comet or γH2AX.

EDCs Endocrine-disrupting chemicals like BPA and phthalates that mimic/block hormones.

Ex vivo culture Organ fragment culture preserving tissue architecture for testing gonadotoxicants.

Exosomes Vesicles explored for fertility restoration by repairing chemotherapy-damaged gonads.

Fertility preservation Clinical strategies like cryopreservation or tissue grafting to maintain fertility.

Ferroptosis Iron-dependent germ-cell death pathway.

Folliculogenesis Sequential maturation of ovarian follicles, disrupted by cisplatin and EDCs.

FSH Gonadotropin regulating Sertoli and granulosa function, altered in endocrine disruption.

GnRH agonists Drugs to protect ovarian reserve during chemotherapy by inducing quiescence.

Gonadotoxicity Drug or chemical-induced damage to testicular/ovarian germ cells.

Granulosa cells Ovarian support cells; cisplatin damages them causing follicle atresia.

HPG axis Hypothalamic-pituitary-gonadal endocrine loop regulating reproduction, disrupted in vivo by EDCs.

Hyperactivation Vigorous sperm motility pattern essential for fertilization, affected by toxicants.

In vivo models Whole-animal capturing systemic gonadal effects of drugs and EDCs.

Leydig cells Testicular cells producing testosterone, impaired by BPA and phthalates.

MEHP It is a primary metabolite of DEHP (di-(2-ethylhexyl) phthalate) and is commonly studied for its endocrine-disrupting effects, especially on reproductive cells.

Micro-TESE Microsurgical sperm retrieval technique with high success in azoospermic patients.

Mitochondrial ROS Oxidants generated in germ-cell mitochondria, leading to apoptosis and infertility.

MTT assay Colorimetric cell viability test using mitochondrial activity in germ cells.

Oogenesis Development of oocytes in ovary.

Ovarian tissue cryopreservation Freezing cortical strips with primordial follicles for fertility restoration.

Oxidative stress Imbalance of ROS and antioxidants due to drug/EDC-induced germ-cell injury.

PESA/TESA/TESE Surgical sperm retrieval methods for fertility in azoospermic men.

Platinum compounds Cisplatin and carboplatin causing germ-cell depletion and follicular loss.

Primordial follicle Dormant oocyte reserve; depletion by chemotherapy leads to infertility.

Protein tyrosine phosphorylation Capacitation marker in sperm, disrupted by lead or phthalates.

ROS Reactive oxygen species; indicators of oxidative damage in germ and support cells.

Sertoli cells Nurse cells in testes supporting spermatogenesis.

Sperm cryopreservation Standard male fertility preservation method before chemotherapy.

Sperm retrieval Clinical techniques to obtain sperm surgically for assisted reproduction.

Spermatogenesis Multi-step process producing spermatozoa.

Spermatogonial stem cells Self-renewing progenitors for sperm, extremely sensitive to gonadotoxicants.

TEM Transmission electron microscopy visualizing ultrastructural germ-cell damage.

Testicular tissue cryopreservation Experimental preservation of prepubertal testicular tissue for future fertility.

Testosterone Male steroid hormone from Leydig cells.
Zona pellucida (ZP) Oocyte coat mediating sperm binding and AR.

References

Aguilar-Mahecha A, Hales BF, Robaire B (2005) Effects of acute and chronic cyclophosphamide treatment on meiotic progression and the induction of DNA double-strand breaks in rat spermatocytes. Biol Reprod 72(6):1297–1304. https://doi.org/10.1095/biolreprod.104.038620

Ali W, Bian Y, Ali H, Sun J, Zhu J, Ma Y, Liu Z, Zou H (2023) Cadmium-induced impairment of spermatozoa development by reducing exosomal-MVBs secretion: a novel pathway. Aging (Albany NY) 15(10):4096–4107. https://doi.org/10.18632/aging.204675

Allen CM, Lopes F, Mitchell RT, Spears N (2020) Comparative gonadotoxicity of the chemotherapy drugs cisplatin and carboplatin on prepubertal mouse gonads. Mol Hum Reprod 26(3): 129–140. https://doi.org/10.1093/molehr/gaaa008

Allen CM, Lopes F, Mitchell RT, Spears N (2021) Can antioxidants protect against chemotherapy in a rat spermatogonial stem cell line? Reprod Fertil 2(4):L7–L9. https://doi.org/10.1530/RAF-21-0042

Bansal A, Patel FD, Rai B, Dhanireddy B, Sharma SC (2014) Gonadotrophin releasing hormone analogues for ovarian function preservation in young females undergoing chemotherapy. Asian Pac J Cancer Prev 15(5):2185–2190. https://doi.org/10.7314/apjcp.2014.15.5.2185

Borate GM, Meshram A (2022) Cryopreservation of sperm: a review. *Cureus* 14(11):e31402. https://doi.org/10.7759/cureus.31402

Chang EM, Lim E, Yoon S, Jeong K, Bae S, Lee DR, Yoon TK, Choi Y, Lee WS (2015) Cisplatin induces overactivation of the dormant primordial follicle through PTEN/AKT/FOXO3a pathway which leads to loss of ovarian reserve in mice. PLoS One 10(12):e0144245. https://doi.org/10.1371/journal.pone.0144245

Chang L, Wang J, She R, Ma L, Wu Q (2017) In vitro toxicity evaluation of melamine on mouse TM4 Sertoli cells. Environ Toxicol Pharmacol 50:111–118. https://doi.org/10.1016/j.etap.2017.01.009

Chianese R, Viggiano A, Urbanek K, Cappetta D, Troisi J, Scafuro M, Guida M, Esposito G, Ciuffreda LP, Rossi F (2018) Chronic exposure to low dose of bisphenol A impacts on the first round of spermatogenesis via SIRT1 modulation. Sci Rep 8(1):2961

Clowse ME, Behera MA, Anders CK, Copland S, Coffman CJ, Leppert PC, Bastian LA (2009) Ovarian preservation by GnRH agonists during chemotherapy: a meta-analysis. J Womens Health (Larchmt) 18(3):311–319. https://doi.org/10.1089/jwh.2008.0857

Delbes G, Blazquez M, Fernandino JI, Grigorova P, Hales BF, Metcalfe C, Navarro-Martin L, Parent L, Robaire B, Rwigemera A, Van Der Kraak G, Wade M, Marlatt V (2022) Effects of endocrine disrupting chemicals on gonad development: mechanistic insights from fish and mammals. Environ Res 204(Pt B):112040. https://doi.org/10.1016/j.envres.2021.112040

Delessard M, Saulnier J, Rives A, Dumont L, Rondanino C, Rives N (2020) Exposure to chemotherapy during childhood or adulthood and consequences on spermatogenesis and male fertility. Int J Mol Sci 21(4). https://doi.org/10.3390/ijms21041454

Esteves SC, Miyaoka R, Orosz JE, Agarwal A (2013) An update on sperm retrieval techniques for azoospermic males. Clinics (Sao Paulo) 68 Suppl 1(Suppl 1):99–110. https://doi.org/10.6061/clinics/2013(sup01)11

Fabbri R, Vicenti R, Magnani V, Paradisi R, De Meis L, Raimondo D, Arena A, Venturoli S, Raffone A, Raspollini A, Seracchioli R (2024) Ovarian tissue transplantation: 10 years of experience at the Bologna University. Front Endocrinol (Lausanne) 15:1332673. https://doi.org/10.3389/fendo.2024.1332673

Goncalves GD, Semprebon SC, Biazi BI, Mantovani MS, Fernandes GSA (2018) Bisphenol A reduces testosterone production in TM3 Leydig cells independently of its effects on cell death and mitochondrial membrane potential. Reprod Toxicol 76:26–34. https://doi.org/10.1016/j.reprotox.2017.12.002

Hashimoto K, Arakawa H, Imamura R, Nishimura T, Kitajima S, Sato T, Makiyama K, Ogawa T, Yokota S (2023) Testis organ culture system capable of evaluating testicular toxicity. bioRxiv. 2023.2012. 2001.569566

He Y, Zou Q, Chen H, Weng S, Luo T, Zeng X (2016) Lead inhibits human sperm functions by reducing the levels of intracellular calcium, cAMP, and tyrosine phosphorylation. Tohoku J Exp Med 238(4):295–303. https://doi.org/10.1620/tjem.238.295

Hovav Y, Dan-Goor M, Yaffe H, Almagor M (2001) Electroejaculation before chemotherapy in adolescents and young men with cancer. Fertil Steril 75(4):811–813. https://doi.org/10.1016/s0015-0282(01)01680-6

Hu H, Yuan W, Wang Y, Dong Z, Chen G (2025) Ferroptotic pathway activation in spermatogonia: a novel mechanism of busulfan-induced testicular injury. Biology (Basel) 14(6). https://doi.org/10.3390/biology14060594

Ibrahim E, Brackett NL, Lynne CM (2022) Penile vibratory stimulation for semen retrieval in men with spinal cord injury: patient perspectives. Res Rep Urol 14:149–157. https://doi.org/10.2147/RRU.S278797

Ickowicz D, Finkelstein M, Breitbart H (2012) Mechanism of sperm capacitation and the acrosome reaction: role of protein kinases. Asian J Androl 14(6):816–821. https://doi.org/10.1038/aja.2012.81

Kalyanaraman B, Darley-Usmar V, Davies KJ, Dennery PA, Forman HJ, Grisham MB, Mann GE, Moore K, Roberts LJ 2nd, Ischiropoulos H (2012) Measuring reactive oxygen and nitrogen species with fluorescent probes: challenges and limitations. Free Radic Biol Med 52(1):1–6. https://doi.org/10.1016/j.freeradbiomed.2011.09.030

Karadeniz Saygili S, Sahin MC, Yukcu F, Sanli S (2025) Analysis of endoplasmic reticulum stress proteins in spermatogenic cells after paclitaxel administration. Curr Issues Mol Biol 47(8). https://doi.org/10.3390/cimb47080620

Kim JY, Han EH, Kim HG, Oh KN, Kim SK, Lee KY, Jeong HG (2010) Bisphenol A-induced aromatase activation is mediated by cyclooxygenase-2 up-regulation in rat testicular Leydig cells. Toxicol Lett 193(2):200–208. https://doi.org/10.1016/j.toxlet.2010.01.011

Kim SY, Kim SK, Lee JR, Woodruff TK (2016) Toward precision medicine for preserving fertility in cancer patients: existing and emerging fertility preservation options for women. J Gynecol Oncol 27(2):e22. https://doi.org/10.3802/jgo.2016.27.e22

Koyama H, Wada T, Nishizawa Y, Iwanaga T, Aoki Y (1977) Cyclophosphamide-induced ovarian failure and its therapeutic significance in patients with breast cancer. Cancer 39(4):1403–1409. https://doi.org/10.1002/1097-0142(197704)39:4<1403::aid-cncr2820390408>3.0.co;2-8

Larsen L, Scheike T, Jensen TK, Bonde JP, Ernst E, Hjollund NH, Zhou Y, Skakkebaek NE, Giwercman A (2000) Computer-assisted semen analysis parameters as predictors for fertility of men from the general population. The Danish First Pregnancy Planner Study Team. Hum Reprod 15(7):1562–1567. https://doi.org/10.1093/humrep/15.7.1562

Lopez-Torres AS, Chirinos M (2017) Modulation of human sperm capacitation by progesterone, estradiol, and luteinizing hormone. Reprod Sci 24(2):193–201. https://doi.org/10.1177/1933719116641766

Lu CC, Meistrich ML (1979) Cytotoxic effects of chemotherapeutic drugs on mouse testis cells. Cancer Res 39(9):3575–3582. https://www.ncbi.nlm.nih.gov/pubmed/476683

Mascarenhas MN, Flaxman SR, Boerma T, Vanderpoel S, Stevens GA (2012) National, regional, and global trends in infertility prevalence since 1990: a systematic analysis of 277 health surveys. PLoS Med 9(12):e1001356. https://doi.org/10.1371/journal.pmed.1001356

Naz RK, Rajesh PB (2004) Role of tyrosine phosphorylation in sperm capacitation / acrosome reaction. Reprod Biol Endocrinol 2:75. https://doi.org/10.1186/1477-7827-2-75

Oktay K, Buyuk E, Davis O, Yermakova I, Veeck L, Rosenwaks Z (2003) Fertility preservation in breast cancer patients: IVF and embryo cryopreservation after ovarian stimulation with tamoxifen. Hum Reprod 18(1):90–95. https://doi.org/10.1093/humrep/deg045

Olive PL, Banath JP (2006) The comet assay: a method to measure DNA damage in individual cells. Nat Protoc 1(1):23–29. https://doi.org/10.1038/nprot.2006.5

Qiu L, Zhang X, Zhang X, Zhang Y, Gu J, Chen M, Zhang Z, Wang X, Wang SL (2013) Sertoli cell is a potential target for perfluorooctane sulfonate-induced reproductive dysfunction in male mice. Toxicol Sci 135(1):229–240. https://doi.org/10.1093/toxsci/kft129

Rahman MS, Kwon WS, Lee JS, Yoon SJ, Ryu BY, Pang MG (2015) Bisphenol-A affects male fertility via fertility-related proteins in spermatozoa. Sci Rep 5:9169. https://doi.org/10.1038/srep09169

Rehfeld A, Egeberg DL, Almstrup K, Petersen JH, Dissing S, Skakkebaek NE (2018) EDC IMPACT: chemical UV filters can affect human sperm function in a progesterone-like manner. Endocr Connect 7(1):16–25. https://doi.org/10.1530/EC-17-0156

Rodriguez-Wallberg KA, Oktay K (2014) Fertility preservation during cancer treatment: clinical guidelines. Cancer Manag Res 6:105–117. https://doi.org/10.2147/CMAR.S32380

Roldan ER, Murase T, Shi QX (1994) Exocytosis in spermatozoa in response to progesterone and zona pellucida. Science 266(5190):1578–1581. https://doi.org/10.1126/science.7985030

Sadri-Ardekani H, Atala A (2014) Testicular tissue cryopreservation and spermatogonial stem cell transplantation to restore fertility: from bench to bedside. Stem Cell Res Ther 5(3):68. https://doi.org/10.1186/scrt457

Schiffer C, Muller A, Egeberg DL, Alvarez L, Brenker C, Rehfeld A, Frederiksen H, Waschle B, Kaupp UB, Balbach M, Wachten D, Skakkebaek NE, Almstrup K, Strunker T (2014) Direct action of endocrine disrupting chemicals on human sperm. EMBO Rep 15(7):758–765. https://doi.org/10.15252/embr.201438869

Schmiegelow ML, Sommer P, Carlsen E, Sonksen JO, Schmiegelow K, Muller JR (1998) Penile vibratory stimulation and electroejaculation before anticancer therapy in two pubertal boys. J Pediatr Hematol Oncol 20(5):429–430. https://doi.org/10.1097/00043426-199809000-00004

Sharpe RM, Fisher JS, Millar MM, Jobling S, Sumpter JP (1995) Gestational and lactational exposure of rats to xenoestrogens results in reduced testicular size and sperm production. Environ Health Perspect 103(12):1136–1143. https://doi.org/10.1289/ehp.951031136

Sikka SC, Wang R (2008) Endocrine disruptors and estrogenic effects on male reproductive axis. Asian J Androl 10(1):134–145. https://doi.org/10.1111/j.1745-7262.2008.00370.x

Smart E, Lopes F, Rice S, Nagy B, Anderson RA, Mitchell RT, Spears N (2018) Chemotherapy drugs cyclophosphamide, cisplatin and doxorubicin induce germ cell loss in an in vitro model of the prepubertal testis. Sci Rep 8(1):1773. https://doi.org/10.1038/s41598-018-19761-9

Sun X, Chen W, Weng S, Pan T, Hu X, Wang F, Xia T, Chen H, Luo T (2020) Effects of the environmental endocrine disruptors di-2-ethylhexyl phthalate and mono-2-ethylhexyl phthalate on human sperm function in vitro. Reprod Fertil Dev 32(6):629–636. https://doi.org/10.1071/RD19164

Susiarjo M, Hassold TJ, Freeman E, Hunt PA (2007) Bisphenol A exposure in utero disrupts early oogenesis in the mouse. PLoS Genet 3(1):e5. https://doi.org/10.1371/journal.pgen.0030005

Szell AZ, Bierbaum RC, Hazelrigg WB, Chetkowski RJ (2013) Live births from frozen human semen stored for 40 years. J Assist Reprod Genet 30(6):743–744. https://doi.org/10.1007/s10815-013-9998-9

Tessier L, McKechnie T, Lee Y, Park LJ, Gangam N, Eskicioglu C (2023) Laparoscopic ovarian transposition prior to pelvic radiation in young women with anorectal malignancies: a systematic review and meta-analysis of prevalence. Color Dis 25(7):1336–1348. https://doi.org/10.1111/codi.16571

Traore K, More P, Adla A, Dogbey G, Papadopoulos V, Zirkin B (2021) MEHP induces alteration of mitochondrial function and inhibition of steroid biosynthesis in MA-10 mouse tumor Leydig cells. Toxicology 463:152985. https://doi.org/10.1016/j.tox.2021.152985

Tremblay AR, Delbes G (2018) In vitro study of doxorubicin-induced oxidative stress in spermatogonia and immature Sertoli cells. Toxicol Appl Pharmacol 348:32–42. https://doi.org/10.1016/j.taap.2018.04.014

van Meerloo J, Kaspers GJ, Cloos J (2011) Cell sensitivity assays: the MTT assay. Methods Mol Biol 731:237–245. https://doi.org/10.1007/978-1-61779-080-5_20

von Wolff M, Thaler CJ, Frambach T, Zeeb C, Lawrenz B, Popovici RM, Strowitzki T (2009) Ovarian stimulation to cryopreserve fertilized oocytes in cancer patients can be started in the luteal phase. Fertil Steril 92(4):1360–1365. https://doi.org/10.1016/j.fertnstert.2008.08.011

Wang R, Wang L, Wang L, Cui Z, Cheng F, Wang W, Yang X (2022) FGF2 is protective towards cisplatin-induced KGN cell toxicity by promoting FTO expression and autophagy. Front Endocrinol (Lausanne) 13:890623. https://doi.org/10.3389/fendo.2022.890623

Yanagimachi R (1970) The movement of golden hamster spermatozoa before and after capacitation. J Reprod Fertil 23(1):193–196. https://doi.org/10.1530/jrf.0.0230193

Yu Y, Guo Y, Zhu J, Shen R, Tang J (2024) Chemotherapy drug combinations induced maternal ovarian damage and long-term effect on fetal reproductive system in mice. Eur J Pharm Sci 201: 106860. https://doi.org/10.1016/j.ejps.2024.106860

Zakova J, Lousova E, Ventruba P, Crha I, Pochopova H, Vinklarkova J, Tesarova E, Nussir M (2014) Sperm cryopreservation before testicular cancer treatment and its subsequent utilization for the treatment of infertility. ScientificWorldJournal 2014:575978. https://doi.org/10.1155/2014/575978

Zhou B, Kwan B, Desai MJ, Nalawade V, Henk J, Viravalli N, Murphy JD, Nathan PC, Ruddy KJ, Shliakhtsitsava K, Su HI, Whitcomb BW (2024) Association of platinum-based chemotherapy with live birth and infertility in female survivors of adolescent and young adult cancer. Fertil Steril 121(6):1020–1030. https://doi.org/10.1016/j.fertnstert.2024.01.039

FSC
www.fsc.org
MIX
Papier aus verantwortungsvollen Quellen
Paper from responsible sources
FSC® C105338

If you have any concerns about our products,
you can contact us on
ProductSafety@springernature.com

In case Publisher is established outside the EU,
the EU authorized representative is:
Springer Nature Customer Service Center GmbH
Europaplatz 3, 69115 Heidelberg, Germany

Printed by Libri Plureos GmbH
in Hamburg, Germany